Optoelectronics

An introduction

Second edition

J. WILSON
J. F. B. HAWKES

Department of Physics,
Newcastle upon Tyne Polytechnic

PRENTICE HALL

New York   London   Toronto   Sydney   Tokyo   Singapore

First published 1983
This second edition first published 1989 by
Prentice Hall International (UK) Limited
Campus 400, Maylands Avenue
Hemel Hempstead
Hertfordshire, HP2 7EZ
A division of
Simon & Schuster International Group

Printed and bound in Great Britain at
the University Press, Cambridge

Library of Congress Cataloging-in-Publication Data

Wilson, J. (John), 1939–
    Optoelectronics, an introduction / J. Wilson and
J. F. B. Hawkes. – 2nd ed.
    p.   cm – (Prentice Hall international series in
optoelectronics)
    Includes bibliographies and index.
    ISBN 0-13-638461-7
    I. Optoelectronics. I. Hawkes, J. F. B., 1942–
II. Title. III. Series.
QC673.W54   1988
621.38′0414 – dc 19   88-21914

British Library Cataloguing in Publication Data

A catalogue record for this book is available from the
British Library

ISBN 0-13-638461-7 (hbk)
ISBN 0-13-638495-1 (pbk)

9 10 11 12 13   98 97 96 95 94

Optoelectronics

An introduction

Prentice Hall International Series in Optoelectronics

Consultant editors: John Midwinter, University College London, UK
Allan Snyder, Australian National University
Bernard Weiss, University of Surrey, UK

Fundamentals of Optical Fiber Communications
W. van Etten and J. van der Plaats

Optical Communication Systems (Second Edition)
J. Gowar

Optical Fiber Communications: Principles and Practice (Second Edition)
J. M. Senior

Lasers: Principles and Applications
J. Wilson and J. F. B. Hawkes

Optoelectronics: An Introduction (Second Edition)
J. Wilson and J. F. B. Hawkes

Contents

Preface to the second edition

In writing the second edition, we have taken the opportunity to react to some of the changes which have occurred in the field of optoelectronics in the past five years. Optoelectronics has become even more firmly established as a subject in its own right, and an increasing number of courses in physics and electronic engineering include aspects of optoelectronics.

Our interpretation of what is meant by optoelectronics has remained unchanged. That is, we take it to cover the interaction of light (in the wavelength range from about 0.2 μm to 20 μm) with matter in gaseous, liquid or solid form and the devices which depend on these interactions.

Fiber optic and optical communications provide good examples of systems which incorporate a wide range of devices, including those based on semiconductors and those based on the behavior of light in crystals subject to external force fields. The continued growth of optical communications is largely responsible for the large sales of optoelectronic devices, but the ever increasing applications of lasers and the developments of optical recording and data storage have also contributed.

Although we originally wrote the book with the final year undergraduate in mind, it has also proved a useful introduction to the subject for postgraduate students and to a wide range of engineers and scientists who require a treatment which is neither too advanced nor too popularist. It also serves as an introduction to the many very good, specialist books on optoelectronics, fiber optic communications and lasers that have been published in the last few years.

In an introductory text of this nature we have not attempted either to cover all aspects of optoelectronics or to deal with topics in full theoretical rigor. Rather, we have tried to put special emphasis on the fundamental principles underlying the operation of the devices and systems covered. This should enable the reader to appreciate the operation of devices and systems not covered here, and indeed to appreciate future developments as they arise.

Optoelectronics relies heavily on the disciplines of optics and solid state physics and the reader is expected to have some background knowledge of these. For those who have not had recent experience in these subjects, we have provided a brief review of relevant topics in Chapters 1 and 2, respectively.

In this second edition Chapters 3 to 7 are largely unchanged, although the opportunity has been taken to update the text. Chapter 3 deals with methods of modulating light that rely on the interaction of light with solids. A new appendix has been included which is intended to enhance the reader's appreciation of this chapter. Display devices are covered in Chapter 4, with special emphasis being placed on the light emitting diode (LED) because of its importance in optical communication systems.

Chapters 5 and 6 are devoted, respectively, to a discussion of the fundamentals of laser action and some of the uses to which lasers may be put. The ways in which the laser output may be modified to make it more useful are also described in Chapter 6.

In Chapter 7 we discuss the wide range of optical radiation detectors in common use, though semiconductor detectors are covered more fully in view of their relative importance.

Chapter 8, which covers fiber optical waveguides, has been modified to emphasize the growing importance of monomode fibers in communications. The section on planar dielectric waveguides has been substantially rewritten. Chapter 9 deals with the requirements for complete optical communication links. In this chapter, we draw upon the discussion of many of the subsystems (e.g. emitters and detectors) covered in previous chapters.

Finally. we have included a new chapter, Chapter 10, which deals with applications of fibers other than communications. These include fiber optic sensors and the use of fibers as light guides.

Interspersed throughout the text are worked examples intended both to illustrate the use of relevant equations and to provide typical numerical values for the parameters encountered. At the end of each chapter we have provided a number of problems intended to test the reader's understanding of the text and, in some cases, to extend the range of the text and to derive some of the results. References are given to most of those equations which we have quoted without derivation; the references also include suggestions for further reading. There are also a number of appendices which amplify some topics that are referred to rather briefly in the text. A teachers' manual containing solutions to the problems can be obtained from the publisher. The book has been written in SI units throughout.

We would like to thank our colleagues and students in the Department of Physics at Newcastle upon Tyne Polytechnic for many helpful discussions on the text and for pointing out errors and misprints. In this respect we would like to thank the many people who kindly wrote to us, from many parts of the world, commenting on the first edition. We would also like to thank Mrs Pat Weddell for again typing the manuscript and our families for their further forbearance and encouragement while we worked on the second edition.

Glossary of symbols

Wherever possible we have endeavored to use the commonly accepted symbols for the various physical parameters needed. Inevitably many symbols have duplicate meanings, and if in any doubt the reader should note carefully both the context and the dimensions. The following list of symbols does not include all the variants formed by adding suffices, nor does it include all the (fairly frequent) cases where a symbol is used as a measure of physical dimensions.

$\mathscr{A}$	source strength
A	area, electric field amplitude, spontaneous transition rate (A_{21})
a	Richardson–Dushman constant, fiber radius, periodicity of lattice
B	magnetic flux density, Einstein coefficient (B_{21}, B_{12}), electron–hole recombination parameter, luminance, 'flicker' noise constant
C	capacitance, waveguide coupling factor
D	diffusion coefficient (D_e, D_h), specific detectivity (D^*)
d	mode volume thickness
$\mathscr{E}$	electric field
E	energy, bandgap (E_g), Young's modulus
e	electron charge
F	fractional transmission, lens f number, APD excess noise factor ($F(M)$), Fermi–Dirac distribution function ($F(E)$)
f	modulation frequency, focal length
G	thermal conductance, gain
g	degeneracy, electron-hole generation rate, lineshape function ($g(v)$)
$\mathscr{H}$	magnetic field
H	heat capacity
h	Planck's constant
$\hbar$	$= h/2\pi$
h_{fe}	transistor common emitter current gain
$\mathscr{I}$	radiant or luminous intensity
I	irradiance
i, i	current, $\sqrt{-1}$
$\mathbf{i}$	unit vector (x direction)

$\mathscr{J}$	molecular rotational quantum number
J	current density
$\mathbf{j}$	unit vector (y direction)
$\mathscr{K}$	diffraction factor
K	Kerr constant, electron beam range parameter
$\mathbf{k}, k$	wavevector, wavenumber, small signal gain coefficient, Boltzmann's constant
L	diffusion length (L_e, L_h), radiance
$\mathscr{L}$	inductance
l_c	coherence length
M	mass, avalanche multiplication factor
m	mass, effective mass (m_e^*, m_h^*)
$\mathscr{N}$	number of photons
N	population inversion, donor/acceptor densities (N_d, N_a), effective density of states in conduction/valence band (N_c, N_v)
NA	numerical aperture
NEP	noise equivalent power
n	electron concentration, intrinsic carrier concentration (n_i), refractive index, quantum number, mode number
$\mathscr{P}$	phase factor
P	power, dipole moment, electrical polarization, quadratic electro-optic coefficient
p	hole concentration, momentum, probability
Q	charge, 'quality factor', trap escape factor, profile dispersion parameter, radiant or luminous energy
R	electrical resistance, radius of curvature, reflectance, frequency response ($R(f)$), responsivity, electron range (R_e), Fresnel reflection loss (R_F)
r	linear electro-optic coefficient, ratio of electron to hole ionization probabilities, electron–hole generation/recombination rates (r_g, r_r)
S/N	signal-to-noise ratio
T	transmittance, temperature, period
t	time, active region thickness
$U_0(x, y)$	electric field amplitude
$\mathscr{V}$	fringe visibility
V	voltage, potential energy, Verdet constant, normalized film thickness, eye relative spectral response (V_λ)
υ	velocity, group velocity (υ_g), molecular vibration quantum number
W	power, total depletion layer width, spectral radiant emittance
$x_{n,p}, x$	n, p depletion layer widths, coordinate distance
y	coordinate distance
Z	depth of field, density of states ($Z(E)$)
z	coordinate distance
α	absorption coefficient, temperature coefficient of resistance, transistor common base current gain, angle

β	diode ideality factor, electron–hole generation efficiency factor
γ	loss coefficient, mutual coherence function (γ_{12}), fiber refractive index profile parameter
Δ	fiber refractive index ratio
Δt	coherence time
δ	phase angle, secondary electron emission coefficient
ε	relative permittivity/dielectric constant (ε_r), emissivity
η	efficiency
θ, θ_B	angle, Brewster angle
Λ	acoustic wavelength
λ	light wavelength
μ	electron/hole mobility (μ_e, μ_h), relative permeability (μ_r)
ν	lightwave frequency
ϱ	charge density, radiation density (ϱ_v), resistivity
σ	conductivity, Stefan's constant
τ	time constant, lifetime, time
Φ	phase angle, light flux
ϕ	phase angle, work function
χ	electric susceptibility, electron affinity
Ψ	time-dependent wave function
ψ	time-independent wave function, phase change
Ω	solid angle
ω	angular frequency

Chapter 1

Light

In discussing the various topics which we have brought together in this text under the title *Optoelectronics*, of necessity we rely heavily on the basic physics of light, matter and their interactions. In this and the next chapter we have described rather briefly those concepts of optics and solid state physics which are fundamental to optoelectronics. The reader may be familiar with much of the content of these two chapters though those who have not recently studied optics or solid state physics may find them useful. For a more detailed development of the topics included, the reader is referred to the many excellent texts on these subjects, a selection of which is given in refs. 1.1 and 2.1.

In this chapter we shall describe phenomena such as polarization, diffraction, interference and coherence; we have assumed that the basic ideas of the reflection and refraction of light and geometrical optics are completely familiar to the reader. In this context it is worth noting that the term 'light' is taken to include the ultraviolet and near-infrared regions as well as the visible region of the spectrum.

1.1 Nature of light

During the seventeenth century two emission theories on the nature of light were developed, the wave theory of Hooke and Huygens and the corpuscular theory of Newton. Subsequent observations by Young, Malus, Euler and others lent support to the wave theory Then in 1864 Maxwell combined the equations of electromagnetism in a general form and showed that they suggest the existence of transverse electromagnetic waves. The speed of propagation in free space of these waves was given by

$$c = \sqrt{\frac{1}{\mu_0 \varepsilon_0}} \qquad (1.1)$$

where μ_0 and ε_0 are the permeability and permittivity of free space, respectively. Substitution of the experimentally determined values of μ_0 and ε_0 yielded a value for c in very close agreement with the value of the speed of light *in vacuo* measured independently. Maxwell therefore proposed that light was an electromagnetic wave having a

Table 1.1 Electromagnetic spectrum

Type of radiation	Wavelength	Frequency (Hz)	Quantum energy (eV)
Radio waves	100 km	3×10^3	1.2×10^{-11}
Microwaves	300 mm	10^9	4×10^{-6}
Infrared	0.3 mm	10^{12}	4×10^{-3}
Visible	0.7 μm	4.3×10^{14}	1.8
Ultraviolet	0.4 μm	7.5×10^{14}	3.1
X rays	0.03 μm	10^{16}	40
γ rays	0.1 nm	3×10^{18}	1.2×10^4
	1.0 pm	3×10^{20}	1.2×10^6

Note: The divisions into the various regions are for illustration only; there is no firm dividing line between one region and the next. The numerical values are only approximate; the upper and lower limits are somewhat arbitrary.

speed c of 3×10^8 m s^{-1}, a frequency of some 5×10^{14} Hz and a wavelength of about 500 nm. Maxwell's theory suggested the possibility of producing electromagnetic waves with a wide range of frequencies (or wavelengths). In 1887 Hertz succeeded in generating non-visible electromagnetic waves, with a wavelength of the order of 10 m, by discharging an induction coil across a spark gap thereby setting up oscillating electric and magnetic fields. Visible light and Hertzian waves are part of the *electromagnetic spectrum* which, as we can see from Table 1.1, extends approximately over the wavelength range of 1.0 pm to 100 km. The wave theory thus became the accepted theory of light. However, while the wave theory, as we shall see below, provides an explanation of optical phenomena such as interference and diffraction, it fails completely when applied to situations where energy is exchanged, such as in the emission and absorption of light and the photoelectric effect. The photoelectric effect, which is the emission of electrons from the surfaces of solids when irradiated, was explained by Einstein in 1905. He suggested that the energy of a light beam is not spread evenly but is concentrated in certain regions, which propagate like particles. He called these 'particles' *photons*.

Einstein was led to the concept of photons by the work of Planck on the emission of light from hot bodies. Planck found that the observations indicated that light energy is emitted in multiples of a certain minimum energy unit. The size of the unit, which is called a *quantum*, depends on the wavelength λ of the radiation and is given by

$$E = \frac{hc}{\lambda} \tag{1.2}$$

where h is Planck's constant. Planck's hypothesis did not require that the energy should be emitted in *localized* bundles and it could, with difficulty, be reconciled with the electromagnetic wave theory. When Einstein showed, however, that it seemed necessary to assume the concentration of energy traveling through space as particles, a wave solution was excluded. Thus we have a particle theory also; light apparently has a *dual* nature!

The two theories of light are not in conflict but rather they are complementary. For our purposes it is sufficient to accept that in many experiments, especially those involving the exchange of energy, the particle (photon or quantum) nature of light dominates the wave nature. On the other hand, for experiments involving interference or diffraction, where light interacts with light, the wave nature dominates.

1.2 Wave nature of light

Light as an electromagnetic wave is characterized by a combination of time varying electric ($\vec{\mathscr{E}}$) and magnetic ($\vec{\mathscr{H}}$) fields propagating through space (see, for example, ref. 1.1b, Chapters 19–21). Maxwell showed that both these fields satisfy the same partial differential equation:

$$\nabla^2(\vec{\mathscr{E}}, \vec{\mathscr{H}}) = \frac{1}{c^2}\frac{\partial^2}{\partial t^2}(\vec{\mathscr{E}}, \vec{\mathscr{H}})\tag{1.3}$$

This is called the *wave equation*; it is encountered in many different kinds of physical phenomena such as mechanical vibrations of a string or in a rod. The implication of eq. (1.3) is that *changes* in the fields propagate through space with a speed c, the speed of light. The frequency of oscillation of the fields, ν, and their wavelength in vacuum, λ_0, are related by

$$c = \nu\lambda_0\tag{1.4}$$

In any other medium the speed of propagation is given by

$$\upsilon = \frac{c}{n} = \nu\lambda\tag{1.5}$$

where n is the refractive index of the medium and λ is the wavelength in the medium (later in the text we often drop the subscript from the vacuum wavelength λ_0 to simplify the notation) n is given by

$$n = \sqrt{\mu_r\varepsilon_r}\tag{1.5i}$$

where μ_r and ε_r are the relative permeability and relative permittivity of the medium respectively.

The electric and magnetic fields vibrate perpendicularly to one another and perpendicularly to the direction of propagation as illustrated in Fig. 1.1; that is, light waves are transverse waves. In describing optical phenomena we often omit the magnetic field vector. This simplifies diagrams and mathematical descriptions but we should always remember that there is also a magnetic field component which behaves in a similar way to the electric field component.

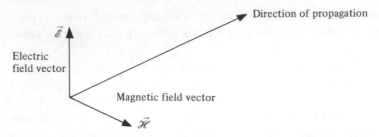

Fig. 1.1 Electromagnetic wave: the electric vector $(\vec{\mathscr{E}})$ and the magnetic vector $(\vec{\mathscr{H}})$ vibrate in orthogonal planes and perpendicularly to the direction of propagation.

The simplest waves are sinusoidal waves, which can be expressed mathematically by the equation:

$$\mathscr{E}(x, t) = \mathscr{E}_0 \cos(\omega t - kx + \phi) \tag{1.6}$$

where $\mathscr{E}$ is the value of the electric field at the point x at time t, $\mathscr{E}_0$ is the amplitude of the wave, ω is the angular frequency ($\omega = 2\pi\nu$), k is the wavenumber ($k = 2\pi/\lambda$) and ϕ is the phase constant. The term ($\omega t - kx + \phi$) is the phase of the wave. Equation (1.6), which describes a perfectly monochromatic plane wave of infinite extent propagating in the positive x direction, is a solution of the wave equation (1.3).

We can represent eq. (1.6) diagrammatically by plotting $\mathscr{E}$ as a function of either x or t as shown in Fig. 1.2(a) and (b), where we have taken $\mathscr{E} = \mathscr{E}_0$ at x and t equal to zero so that $\phi = 0$. Figure 1.2(a) shows the variation of the electric field with distance at a given instant of time. If, as a representative time we take t equal to zero, then the spatial variation of the electric field is given by:

$$\mathscr{E} = \mathscr{E}_0 \cos kx \tag{1.6i}$$

Similarly Fig. 1.2(b) shows the variation of the electric field as a function of time at some specific location in space. If we take x equal to zero then the temporal variation of electric field is given by:

$$\mathscr{E} = \mathscr{E}_0 \cos \omega t \tag{i.6ii}$$

Equations (1.6) can be written in a variety of equivalent forms using the relationships between ν, ω, λ, k and c already given. We note also that the time for one cycle is the period T ($T = 1/\nu$) as shown in Fig. 1.2(b).

If the value of $\mathscr{E}$ at $x = 0$, $t = 0$ is not $\mathscr{E}_0$ then we must include the arbitrary phase constant ϕ. Equations (1.6) can also be expressed using a sine rather than a cosine function, or alternatively using complex exponentials.

In the plane waves described above and in other forms of wave there are surfaces of constant phase, which are referred to as wavesurfaces or wavefronts. As time elapses the wavefronts move through space with a velocity υ given by

$$\upsilon = \frac{\omega}{k} = \nu\lambda \tag{1.7}$$

(a)

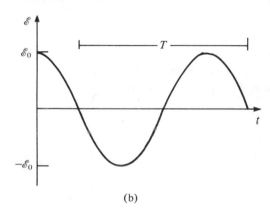

(b)

Fig. 1.2 Electric field ($\mathscr{E}$) of an electromagnetic wave plotted as a function of (a) the spatial coordinate x and (b) the time t.

which is called the *phase* velocity. As it is impossible in practice to produce perfectly monochromatic waves we often have the situation where a group of waves of closely similar wavelength is moving such that their resultant forms a packet. This packet moves with the *group* velocity v_g. A discussion of this phenomenon based on the combination of two waves of slightly different frequencies moving together, which is illustrated in Fig. 1.3, shows that the group velocity is given by (see Problem 1.2)

$$v_g = \frac{\partial \omega}{\partial k} \tag{1.8}$$

Equations (1.6) represent plane waves moving along the x axis; we can generalize our mathematical description to include plane waves moving in arbitrary directions. Such a wave can be characterized by a wavevector k where $|k| = 2\pi/\lambda$ and eq. (1.6) becomes

$$\mathscr{E}(x, y, z, t) = \mathscr{E}_0 \cos(\omega t - k \cdot r + \phi) \tag{1.9}$$

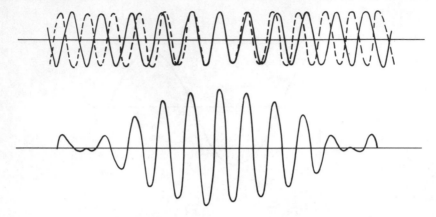

Fig. 1.3 The concept of a wave group or wave packet is illustrated by the combination of two progressive waves of nearly equal frequency. The envelope of the group moves with the group velocity v_g.

where r is a vector from the origin to the point (x, y, z). Thus, for example, if we have a plane wave propagating in a direction θ to the x axis with its wavefronts normal to the (x, y) plane as shown in Fig. 1.4, we can write

$$k = ik_x + jk_y \tag{1.10}$$

and

$$r = ix + jy \tag{1.11}$$

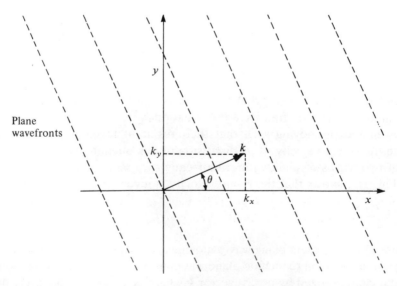

Fig. 1.4 Plane wave with its propagation vector k in the (x, y) plane. The components of the propagation vector are $k_x = |k| \cos \theta$ and $k_y = |k| \sin \theta$.

where i and j are unit vectors in the x and y directions respectively. Combining eqs. (1.10) and (1.11) we have

$$\boldsymbol{k} \cdot \boldsymbol{r} = xk_x + yk_y = xk \cos \theta + yk \sin \theta$$

Hence we can write eq. (1.9) in this case as

$$\mathscr{E}(x, y, t) = \mathscr{E}_0 \cos (\omega t - xk \cos \theta - yk \sin \theta + \phi) \tag{i.9i}$$

An equally important concept is that of spherical waves which, we can imagine, are generated by a point source of light. If such a source is located in an isotropic medium it will radiate uniformly in all directions; the wavefronts are thus a series of concentric spherical shells. We can describe this situation by

$$\mathscr{E} = \frac{\mathscr{A}}{r} \cos (\omega t - \boldsymbol{k} \cdot \boldsymbol{r})$$

where the constant $\mathscr{A}$ is known as the *source strength*. The factor $1/r$ in the amplitude term accounts for the decrease in amplitude of the wave as it propagates further and further from the source. As the irradiance[†] is proportional to the square of the amplitude, there is an inverse-square-law decrease in irradiance. If the medium in which the source is located is anisotropic, then the wave surfaces are no longer spheres; their shapes depend on the speed of propagation in different directions. We shall return to this point in section 3.2.

1.2.1 Polarization

If the electric field vector of an electromagnetic wave propagating in free space vibrates in a specific plane, the wave is said to be plane polarized. Any real beam of light comprises many individual waves and in general the planes of vibration of their electric fields will be randomly oriented. Such a beam of light is unpolarized and the resultant electric field vector changes orientation randomly in time. It is possible, however, to have light beams characterized by highly oriented electric fields and such light is referred to as being *polarized*. The simplest form of polarization is plane polarized light, which is similar to the single wave shown in Fig. 1.1. Other forms of polarization are discussed in section 3.1.

Light can be polarized in a number of different ways and here we consider two, namely polarization by reflection and by absorption. When unpolarized light is incident on a material surface as shown in Fig. 1.5(a) we find that the light with its polarization vector perpendicular to the plane of incidence (denoted by $\rightarrow\bullet\bullet\rightarrow$) is preferentially reflected in comparison to light polarized parallel to the plane of incidence (denoted by $+\!\!\!+\!\!\!\rightarrow$). We can resolve the electric field vector of each wave into components parallel and perpendicular to any convenient direction: here we choose the plane of incidence. Thus by symmetry we may think of unpolarized light as comprising two equal plane polarized components with orthogonal orientations.

[†] In the past the word *intensity* has been used for the flow of energy per unit time per unit area. However, by international agreement that term is being replaced by *irradiance*.

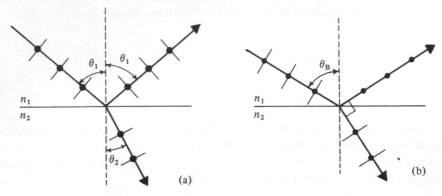

Fig. 1.5 In (a), a light reflected from the interface between two media of refractive index n_1 and n_2 is partially plane polarized (there is less of the parallel component ─+─▸ than the perpendicular component ─•─•▸), while in (b) for incidence at the Brewster angle θ_B, the reflected light is completely plane polarized.

We find further that the reflectance[†] of the surface for the perpendicular and parallel components varies as a function of the angle of incidence as shown in Fig. 1.6. In particular we note that for the parallel component the reflectance is zero at the specific angle of incidence $\theta = \theta_B$. For this angle of incidence, which is called the *Brewster angle*, all of the parallel component is transmitted. It is found also that for incidence at the Brewster angle the reflected and refracted (transmitted) rays are perpendicular to one another as illustrated in Fig. 1.5(b). Thus, using Snell's law for the refraction of light at the interface between two media of refractive indices n_1 and n_2, that is,

$$n_1 \sin \theta_1 = n_2 \sin \theta_2 \tag{1.12}$$

where θ_1 and θ_2 are the angles of incidence and refraction respectively, we have

$$n_1 \sin \theta_B = n_2 \cos \theta_B$$

or

$$\tan \theta_B = \frac{n_2}{n_1} \tag{1.13}$$

Equation (1.13) is known as Brewster's law. One way of polarizing light is simply to pass it through a series of glass plates orientated at the Brewster angle. Then each surface, some of the light polarized perpendicularly to the plane of incidence is reflected, as shown in Fig. 1.7, while all of the parallel component is transmitted. After passing through about six plates, the transmitted light is highly plane polarized. We shall see in Chapter 5 that lasers often include surfaces inclined at the Brewster angle to minimize

---

[†] In the literature on optics one meets the terms reflectivity, reflection coefficient and reflectance. Unfortunately, these terms are often defined in different ways. For example, reflectance is often used for the ratio of the reflected energy per unit area (or flux) to the incident flux and also for the ratio of the reflected to the incident amplitude. We shall adopt the practice suggested by the Symbols Committee of the Royal Society (1971 and 1975) and use reflectance for the ratio of the reflected to the incident flux.

(a)

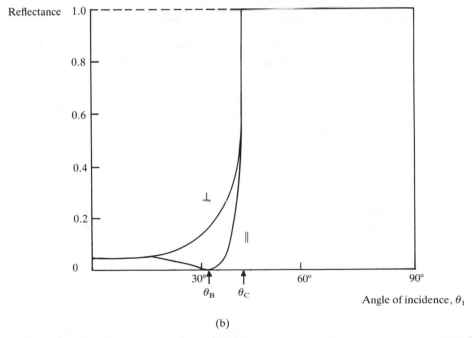

(b)

Fig. 1.6 Reflectance as a function of angle of incidence for light polarized parallel (∥) and perpendicular (⊥) to the plane of incidence incident on the interface from (a) the less optically dense side and (b) the more optically dense side.

Fig. 1.7 Schematic illustration of the production of plane polarized light by reflection using a 'pile of plates'. The light is incident on the plates at the Brewster angle so that with about six plates the emergent light is plane polarized parallel to the plane of incidence.

optical losses for one particular orientation of polarization of the light passing through the surfaces. Such lasers thus emit plane polarized light.

Example 1.1 Brewster angle

We may calculate the Brewster angle for glass given that its refractive index in air is 1.5.

From eq. (1.13) and assuming that the glass has an interface with air ($n = 1$) we have $\theta_B = \tan^{-1} 1.5 = 56.3°$ (see Fig. 1.6).

In Fig. 1.5, and in eqs. (1.12) and (1.13), we have tacitly assumed that $n_2 > n_1$, but for incidence at a boundary separating a more optically dense medium from a less optically dense medium the reverse is obviously the case. Hence the Brewster angle 'in air' is different from the Brewster angle 'in the material'; for air ($n = 1$) to glass ($n = 1.5$) the two angles are $\tan^{-1} 1.5$ (or $56.3°$) and $\tan^{-1} 0.67$ (or $33.7°$). Figure 1.7 should, of course, really be modified to take into account the refraction which occurs at the two surfaces of each plate.

The reflectance of light polarized parallel or perpendicular to the plane of incidence varies with the angle of incidence in the more dense medium (that is $n_1 > n_2$) as shown in Fig. 1.6(b). We see here that for angles greater than a certain angle θ_c, the critical angle, there is no transmitted light, i.e. *total internal reflection* has taken place.

At $\theta = \theta_c$ the angle of refraction (in the less dense medium) is $90°$, so from eq. (1.12) we can write

$$n_1 \sin \theta_c = n_2 \sin 90°$$

or

$$\theta_c = \sin^{-1} n_2/n_1 \tag{1.14}$$

Total internal reflection is discussed in more detail in relation to the propagation of light along optical waveguides in Chapter 8.

Light can also be polarized by selective absorption in various materials such as tourmaline, which occurs naturally, and 'polaroid'. Polaroid consists of a plastic sheet of polyvinyl alcohol impregnated with iodine. Molecules of iodine polyvinyl alcohol are orientated into long chains by stretching the sheet. This material then transmits about 80% of the light polarized perpendicular to the chains of molecules but less than 1% of the light polarized parallel to these chains. Evidently the chains of molecules interact with this component and effectively absorb it. Polaroid is widely used for producing plane polarized light in optical systems, but various polarizing prisms such as the Nicol prism and the Glan–Thompson prism are more efficient in terms of the degree of polarization (which is defined as the ratio of the irradiance of the polarized component to the total irradiance).

If the light incident on such a 'linear' polarizing device is already plane polarized, then the amount of light transmitted depends on the angle θ between the plane of polarization of the incident light and the plane of polarization of the light transmitted by the polarizer (often called the polarizing axis). We can easily show (see Problem 1.4) that if $\mathscr{E}_0$ is the amplitude of the incident light then the transmitted amplitude is $\mathscr{E}_0 \cos \theta$ and the irradiance of the transmitted light is given by[†]

$$I = \mathscr{E}_0^2 \cos^2 \theta = I_0 \cos^2 \theta \tag{1.15}$$

This relationship is known as *Malus's law*.

1.2.2 Principle of superposition

In discussing optical phenomena we are often confronted with the problem of finding the resultant produced when two or more waves act together at a point in space. It is found that the principle of superposition applies. This states that 'the resultant electric field at a given place and time due to the simultaneous action of two or more sinusoidal waves is the algebraic sum of the electric fields of the individual waves'. That is,

$$\mathscr{E} = \mathscr{E}_1 + \mathscr{E}_2 + \mathscr{E}_3 + \ldots$$

where $\mathscr{E}_1, \mathscr{E}_2, \mathscr{E}_3$ are the electric fields of the individual waves at the specified time and place. The summation can be carried out in several different ways, the choice of method for a particular problem being a matter of mathematical convenience (ref. 1.2).

Let us consider the simple case of the superposition of two waves of the same frequency propagating in the same direction given by:

$$\left. \begin{array}{l} \mathscr{E}_1 = \mathscr{E}_{01} \sin (\omega t - kx + \phi_1) \\[2mm] \text{and} \\[2mm] \mathscr{E}_2 = \mathscr{E}_{02} \sin (\omega t - kx + \phi_2) \end{array} \right\} \tag{1.16}$$

[†] Here, and elsewhere in the literature, for simplicity the irradiance is taken to be the square of the amplitude of the electric vector. In reality, the irradiance is given by the Poynting vector $\mathbf{S} = \mathscr{E} \times \mathscr{H}$. For example, the time average of the Poynting vector for a plane wave is the quantity to which our eyes or a detector would be sensitive and is given by $\langle S \rangle = \dfrac{\varepsilon_0 c}{2} |\mathscr{E}_0|^2$, where $\mathscr{E}_0$ is the maximum amplitude of the electric field oscillations. Thus $\langle S \rangle = 1.33 \times 10^{-3} |\mathscr{E}_0|^2$ Wm^{-2} in free space (see, for example, ref. 1.1e).

The resultant is

$$\mathscr{E} = \mathscr{E}_1 + \mathscr{E}_2$$

$$\mathscr{E} = \mathscr{E}_{01} \sin (\omega t - kx + \phi_1) + \mathscr{E}_{02} \sin (\omega t - kx + \phi_2)$$

$$\mathscr{E} = (\mathscr{E}_{01} \cos \phi_1 + \mathscr{E}_{02} \cos \phi_2) \sin (\omega t - kx)$$
$$+ (\mathscr{E}_{01} \sin \phi_1 + \mathscr{E}_{02} \sin \phi_2) \cos (\omega t - kx)$$

This is identical with

$$\mathscr{E} = \mathscr{E}_0 \sin (\omega t - kx + \phi) \tag{1.17}$$

provided that

$$\mathscr{E}_0^2 = (\mathscr{E}_{01} \cos \phi_1 + \mathscr{E}_{02} \cos \phi_2)^2 + (\mathscr{E}_{01} \sin \phi_1 + \mathscr{E}_{02} \sin \phi_2)^2$$

or

$$\mathscr{E}_0^2 = \mathscr{E}_{01}^2 + \mathscr{E}_{02}^2 + 2 \mathscr{E}_{01} \mathscr{E}_{02} \cos (\phi_2 - \phi_1) \tag{1.17i}$$

and

$$\tan \phi = \frac{(\mathscr{E}_{01} \sin \phi_1 + \mathscr{E}_{02} \sin \phi_2)}{(\mathscr{E}_{01} \cos \phi_1 + \mathscr{E}_{02} \cos \phi_2)} \tag{1.17ii}$$

The form of eq. (1.17) implies that the resultant of adding two sinusoidal waves of the same frequency is itself a sinusoidal wave with the same frequency as the original waves given by eqs. (1.16). The resultant is, of course, a solution of the wave equation (1.3). By repeated application of this process we can show that the resultant of any number of sinusoidal waves of the same frequency is a sinusoidal wave of that frequency. In this case the resultant is given by eq. (1.17), but now we have

$$\mathscr{E}_0^2 = \left(\sum_i \mathscr{E}_{0i} \cos \phi_i \right)^2 + \left(\sum_i \mathscr{E}_{0i} \sin \phi_i \right)^2$$

or

$$\mathscr{E}_0^2 = \sum_i \mathscr{E}_{0i}^2 + \sum_{\substack{i \ j \\ i \neq j}} \mathscr{E}_{0i} \mathscr{E}_{0j} \cos (\phi_j - \phi_i) \tag{1.18}$$

and

$$\tan \phi = \frac{\sum_i \mathscr{E}_{0i} \sin \phi_i}{\sum_i \mathscr{E}_{0i} \cos \phi_i}$$

Now if the original light waves are from completely independent sources, including separated regions of an extended source, the phase difference $(\phi_j - \phi_i)$ in eq. (1.18) will vary in a random way such that the average value of $\cos (\phi_j - \phi_i)$ is zero. That is, for every possible positive value of phase difference there is a corresponding negative value. Therefore, the resultant value is given by:

$$\mathscr{E}_0^2 = \sum_i \mathscr{E}_{0i}^2 \tag{1.19}$$

The resultant irradiance is seen to be the sum of the irradiances which would be produced by the individual sources acting separately. In this case the sources are said to be *incoherent* and an area on which the light from such sources falls will be uniformly illuminated. If, on the other hand, the waves from the different sources maintain constant phase relationships they are said to be *coherent*. When an area is illuminated simultaneously by two or more coherent sources, the irradiance usually varies from point to point giving rise to *interference fringes*.

If we consider a surface illuminated by two coherent sources, we can see that as we move across the surface the relative distances to the sources changes and the phase difference also changes. Hence the term $\cos(\phi_2 - \phi_1)$ varies periodically between plus one and minus one and the irradiance varies from a maximum value I_{max}, where

$$\begin{aligned} I_{max} = \mathscr{E}_0^2 &= \mathscr{E}_{01}^2 + \mathscr{E}_{02}^2 + 2\,\mathscr{E}_{01}\,\mathscr{E}_{02} \\ &= (\mathscr{E}_{01} + \mathscr{E}_{02})^2 \end{aligned} \tag{1.20}$$

when the waves are in phase and $\cos(\phi_2 - \phi_1) = +1$, to a mimimum value I_{min}, where

$$\begin{aligned} I_{min} &= \mathscr{E}_{01}^2 + \mathscr{E}_{02}^2 - 2\,\mathscr{E}_{01}\,\mathscr{E}_{02} \\ &= (\mathscr{E}_{01} - \mathscr{E}_{02})^2 \end{aligned}$$

when the waves are exactly out of phase and $\cos(\phi_2 - \phi_1) = -1$. Thus a regular system of alternate bright and dark interference fringes is formed over the surface. Extending the argument to a large number of coherent sources we can see from eq. (1.20) that the resulting irradiance at any point where all the waves are in phase is

$$I = \left(\sum_i \mathscr{E}_{0i} \right)^2 \tag{1.21}$$

At a nearby point the resultant irradiance may be very small, or indeed zero, depending on the relative magnitudes of the terms $\mathscr{E}_{0i}$. The interaction of waves from coherent sources thus alters the distribution of energy without altering the total amount. The topics of interference and coherence will be pursued further in the next section and section 6.5.1.3 respectively. The superposition of polarized waves is considered in section 3.2.

Example 1.2 Maximum irradiance of several coherent and incoherent sources

Here we compare the maximum irradiance resulting from the superposition of equal contributions I from four (a) coherent sources and (b) incoherent sources.

(a) From eq. (1.21) we have $I_{max} = n^2 I = 16I$, while (b) from eq. (1.19) we have $I_{max} = nI = 4I$.

Standing waves: let us now consider the superposition of two waves of the same frequency traveling in opposite directions. The waves may be represented by the equations

$$\mathscr{E}_1 = \mathscr{E}_0 \sin (\omega t - kx)$$

and

$$\mathscr{E}_2 = \mathscr{E}_0 \sin (\omega t + kx),$$

where, for convenience, we have taken $\mathscr{E}_{01} = \mathscr{E}_{02} = \mathscr{E}_0$ and ϕ_1 and $\phi_2 = 0$. Then $\mathscr{E} = \mathscr{E}_1 + \mathscr{E}_2$, and hence

$$\mathscr{E} = 2 \mathscr{E}_0 \cos kx \sin \omega t \qquad (1.22)$$

Equation (1.22) represents a standing wave and is illustrated in Fig. 1.8. In such a wave there is no net transfer of energy in either direction. The energy density in the medium, being proportional to the square of the amplitude of the motion (that is, $4 \mathscr{E}_0^2 \cos^2 kx$), varies from a maximum at the antinodes to zero at the nodes. Standing waves of this type may be set up in optical cavities (or resonators), which consist of a pair of parallel, highly reflecting mirrors. Such a cavity forms an integral part of any laser.

1.2.3 Interference

The basic mathematical description of 'two beam' interference is given by eq. (1.17i); that is,

$$I = \mathscr{E}_0^2 = \mathscr{E}_{01}^2 + \mathscr{E}_{02}^2 + 2 \mathscr{E}_{01} \mathscr{E}_{02} \cos (\phi_2 - \phi_1)$$

If $\mathscr{E}_{01} = \mathscr{E}_{02}$ then

$$I = 2 \mathscr{E}_{01}^2 [1 + \cos (\phi_2 - \phi_1)]$$

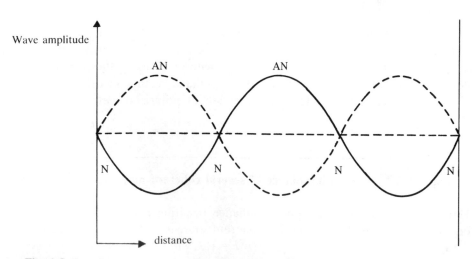

Fig. 1.8 Amplitude profile in a standing wave as a function of distance (N are nodes and AN antinodes).

or

$$I = 4 \mathscr{E}_{01}^2 \cos^2 \left(\frac{\phi_2 - \phi_1}{2} \right) \qquad (1.23)$$

Equation (1.23) shows that the irradiance distribution of the fringes is given by a (cosine)^2 function. If the contributions from the coherent sources are equal the irradiance of the fringes varies from $4I_1^2$ to zero as $(\phi_2 - \phi_1)$ varies between 0 and π. If $\mathscr{E}_{01} \neq \mathscr{E}_{02}$, the resultant irradiance varies between $(I_1 + I_2)^2$ and $(I_1 - I_2)^2$.

To obtain the coherent wave trains required for the observation of interference before the advent of lasers one had to ensure that:

(i) The sets of wave trains were derived from the same small source of light and then brought together by different paths.

(ii) The differences in path were short enough to ensure at least partial coherence of the wave trains (that is, the differences in path were less than the coherence length of the source – see section 6.5.1.3).

The basic ways of satisfying these requirements and demonstrating interference can be classified into two groups, namely 'division of wavefront' and 'division of amplitude'. The classic experiment of Young, in 1802, falls into the former group. In this experiment, which is illustrated in Fig. 1.9, monochromatic light is passed through a pinhole S so as to illuminate a screen containing two further identical pinholes or narrow slits placed close together. The presence of the single pinhole S provides the necessary mutual coherence between the light beams emerging from the slits S_1 and S_2. The wavefronts from S intersect S_1 and S_2 simultaneously so that the light contributions emerging from S_1 and S_2 are derived from the same original wavefront and are therefore coherent. These contributions spread out from S_1 and S_2 (if these are long, thin

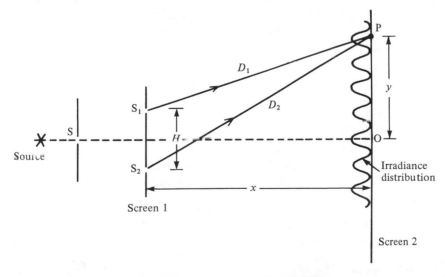

Fig. 1.9 Schematic layout and geometry for a Young's double slit interference experiment.

slits) as 'cylindrical' wavefronts and interfere in the region beyond the screen. If a second screen is placed as shown then an interference pattern consisting of straight line fringes parallel to the slits is observed on it.

To find the irradiance at a given point P it is necessary to find the phase difference ϕ between the two sets of waves arriving at P from S_1 and S_2. This in turn depends on the path difference $(D_2 - D_1)$ as in general

$$\text{Phase difference} = \frac{2\pi}{\lambda} \text{ (optical path difference)}$$

$$\therefore \phi = \phi_2 - \phi_1 = \frac{2\pi}{\lambda} (D_2 - D_1)$$

where D_1 and D_2 are the distances from S_1 and S_2 to P respectively.

Bright fringes occur when the phase difference is zero or $\pm 2p\pi$, where p is an integer; that is, when

$$\frac{2\pi}{\lambda} (D_2 - D_1) = \pm 2p\pi$$

which is equivalent to $|D_2 - D_1| = p\lambda$. Therefore bright fringes occur if the path difference is an integral number of wavelengths. Similarly, dark fringes occur when $\phi = \pm(2p + 1)\pi$, or the path difference is an odd integral number of half-wavelengths. It is left as an exercise for the reader to show, with the parameters given in Fig. 1.9, that bright fringes occur at points P a distance y from O such that

$$y = \pm \frac{p\lambda x}{H} \tag{1.24}$$

provided both y and H are small compared to x, Here H is the slit separation and x is the distance from the screen containing the slits to the observing screen.

Equation (1.23) suggests that in such interference fringe patterns the irradiance of the fringe maxima will be equal. This is not the case, however, due to diffraction effects which are discussed in section 1.2.4.

Interference effects involving amplitude division can be observed in thin films or plates as illustrated in Fig. 1.10. In this case, interference occurs between the light reflected at A on the front surface of the plate and at B on the rear surface. If the plate has parallel faces then the two sets of waves from A and C are parallel and a lens must be used to bring them together to interfere. Using elementary geometry and Snell's law (eq. 1.12) the reader may show that the optical path difference $(AB + BC)n - AD$ is equal to $2nL \cos \theta_2$ where θ_2 is the angle of refraction and L is the plate thickness (see Problem 1.7). The phase difference is then $(2\pi/\lambda) (2nL \cos \theta_2)$ and therefore bright fringes occur when

$$\frac{4\pi nL \cos \theta_2}{\lambda} = 2p\pi$$

that is

$$p\lambda = 2nL \cos \theta_2 \tag{1.25}$$

Fig. 1.10 Schematic diagram of interference effects in a thin film or plate of refractive index n.

Likewise, dark fringes occur when

$$(2p + 1)\lambda/2 = 2nL \cos \theta_2 \qquad (1.25i)$$

If the plate is optically denser than the surrounding medium, there is a phase change of π on reflection at the upper surface, thereby causing the above conditions to be interchanged.

For a given fringe p, λ, L and n are constant and therefore θ_2 must be constant; the fringes are known as 'fringes of equal inclination'. If the angle of incidence is not too large and an extended monochromatic source is used then the fringes are seen as a set of concentric circular rings in the focal plane of the observing lens.

When the optical thickness of the plate is not constant and the optical system is such that θ_2 is almost constant, the fringes are contours of equal optical thickness nL. The situation may be illustrated by considering a small angled wedge. If the wedge is uniform the fringes are approximately straight lines parallel to the apex of the wedge. Again it is left to the reader to show that the apex angle α is given by

$$\alpha \simeq \tan \alpha = \frac{\lambda}{2S} \qquad (1.26)$$

where S is the fringe spacing. The fringes are often close together and are conveniently viewed with a low power microscope.

Multiple beam interference: if the reflectances of the surfaces of the plate shown in Fig. 1.10 were increased there would be many reflected beams to contribute to the interference pattern rather than just the two shown. In practice the resultant interference pattern is seen more clearly if the transmitted beams rather than the reflected ones are used, as shown in Fig. 1.11. If the plate has parallel sides then the multiple beams are parallel and are brought together to interfere in the focal plane of the lens. The resultant of superposing these beams can be calculated quite easily as the phase difference from one beam to the next is constant. The phase difference is due to two additional traversals of the plate plus any phase changes which may occur on reflection at the surfaces of the plate. If the latter are ignored then the condition for the formation of fringe maxima is identical with the two beam case given in eq. (1.25). The irradiance distribution is no longer the $(\text{cosine})^2$ distribution of eq. (1.23), however, but rather it is given by (ref. 1.3)

$$I = \frac{I_0 T^2}{1 - R^2} \left(\frac{1}{1 + \dfrac{4R}{(1 - R)^2} \sin^2 \dfrac{\delta}{2}} \right) \tag{1.27}$$

where I_0 is the irradiance of the incident beam, R and T are the reflectance and transmittance respectively of the plate surfaces and δ is the total phase change between successive beams.

If R is large, greater than about 0.8 say, then the fringe maxima are very sharp as shown in Fig. 1.12. Fringes of this type are formed in the Fabry–Perot interferometer.

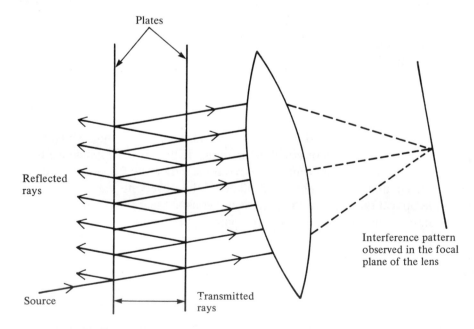

Fig. 1.11 Paths of light rays resulting from multiple reflections from *two parallel plates* (for simplicity the rear surfaces of the plates and refraction at the surfaces are not shown).

This instrument includes a pair of highly reflecting surfaces set accurately parallel to one another forming an optical (or Fabry–Perot) resonator as mentioned above and which is discussed further in section 5.5. The resolving power of the Fabry–Perot interferometer can be very high and it is widely used in the accurate measurement of the hyperfine structure of spectral lines.

Multiple beam interference effects can also be produced by division of the wavefront. This is achieved by increasing the number of slits from two, the ultimate being reached in the diffraction grating where there is a very large number of equal slits. As before, the condition for interference maxima is that the contributions to the total irradiance emerging from successive slits should be in phase. It can be shown using elementary theory that, for normal incidence, maxima in the interference pattern produced by a diffraction grating occur when

$$p\lambda = (a + b) \sin \theta = d \sin \theta \tag{1.28}$$

where $d \ (= a + b)$ is the grating constant, with a and b the width and separation of the slits respectively and θ the angle of diffraction as shown in Fig. 1.13. For a more complete discussion of the theory of the diffraction grating the reader is referred to the texts given in ref. 1.1.

1.2.4 Diffraction

If an opaque object is placed between a source of light and a screen it is found that the shadow cast on the screen is not perfectly sharp. Some light is present in the dark zone of the geometrical shadow. Similarly light which emerges from a small aperture or narrow slit is observed to spread out. This failure of light to travel in straight lines is called *diffraction*; it is a natural consequence of the wave nature of light.

The essential features of diffraction can be explained by Huygens' principle, which states that: 'The propagation of a light wave can be predicted by assuming that each

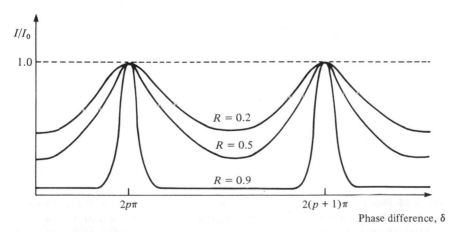

Fig. 1.12 Schematic diagram showing the variation in the irradiance distribution of multiple beam interference fringes as a function of mirror reflectance R.

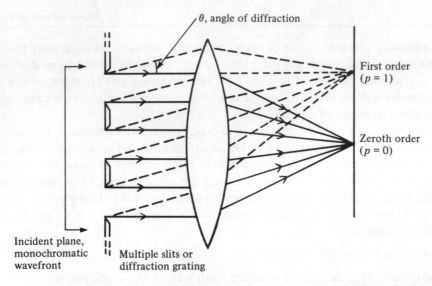

θ, angle of diffraction

First order
($p = 1$)

Zeroth order
($p = 0$)

Incident plane,
monochromatic
wavefront

Multiple slits or
diffraction grating

Fig. 1.13 Schematic illustration of the formation of interference maxima by a diffraction grating. If the incident light is not monochromatic then a 1st, 2nd, ... order interference fringe forms for each wavelength present at different values of θ so that each can therefore be distinguished.

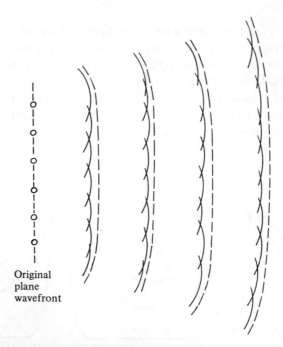

Original
plane
wavefront

Fig. 1.14 Illustration of the application of Huygens' principle to the propagation of a plane wavefront (dotted lines). The small circles are typical 'point sources' on the wavefront which emit secondary wavelets. The wavefront at a later time is the resultant of the secondary wavelets. The diagram shows (in an exaggerated way) that an initially plane wavefront will diverge as it propagates.

point on the wavefront acts as a source of secondary wavelets which spread out in all directions. The envelope of these secondary wavelets after a small interval of time is the new wavefront.' In effect the spherical wavelets emitted from the point sources are summed using the principle of superposition; the result of the summation is equivalent to the wavefront at a later time. The propagation of a plane wave according to Huygens' principle is illustrated in Fig. 1.14. We see that the wavefronts develop some curvature at the edges due to the radiation from the end points being directed away from the axis. Succeeding wavefronts become more and more curved so that the beam diverges.

A quantitative description of diffraction can be obtained by setting Huygens' principle in a mathematical form known as the *Fresnel–Kirchhoff formula* (ref. 1.4).

In the detailed treatment of diffraction it is customary to distinguish between two general cases known as *Fraunhofer* and *Fresnel* diffraction. Qualitatively, Fraunhofer diffraction occurs when the incident and diffracted waves are effectively plane, while in Fresnel diffraction the curvature of the wavefront is significant. Clearly there is not a sharp distinction between the two cases.

Of particular importance is the Fraunhofer diffraction produced by a narrow slit; the arrangement for observing this is shown in Fig. 1.15(a). Over 80% of the light passing through the aperture falls within the central maximum of the pattern, on either side of which there is a series of low irradiance secondary maxima. These maxima are separated by minima which occur at angles of diffraction θ given by

$$\sin \theta = \frac{p\lambda}{D} \tag{1.29}$$

where D is the width of the slit. A rather more important case is that of the circular aperture which produces a pattern consisting of a central bright area surrounded by concentric dark and bright rings. About 84% of the light is concentrated within the central spot, which is called the *Airy disk*. A measure of the amount of diffraction is given by the angle θ at which the first dark ring occurs. It may be shown, using the Fresnel–Kirchhoff formula, that θ is given by

$$\sin \theta = 1.22 \frac{\lambda}{D} \approx \frac{\lambda}{D} \tag{1.30}$$

where D is the diameter of the aperture. In many applications the output from a laser or other source is focused, expanded, or otherwise passed through lenses, prisms, stops and the like. In each case the edges of the optical component may serve as an aperture limiting the extent of the beam and thereby introducing additional divergence due to diffraction.

In passing, it is noteworthy that diffraction sets a theoretical limit to the resolving power of optical instruments and it is also responsible for the occurrence of 'missing orders' in the interference patterns produced by double and multiple split arrangements. The irradiance distribution of such patterns has an envelope which is the single slit diffraction pattern shown in Fig. 1.15(b). Thus at certain angles θ where an interference maximum is expected none appears because a diffraction minimum occurs at the same angle.

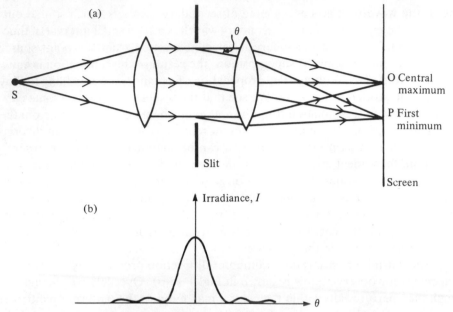

Fig. 1.15 Schematic arrangement for observing the Fraunhofer diffraction produced by a single slit or aperture (a) and the resultant irradiance distribution as a function of the angle of diffraction θ (b).

Example 1.3 Resolving power: the Rayleigh criterion

We may estimate (a) the minimum separation of two point sources that can just be resolved by a telescope with an objective lens of 0.1 m diameter which is 500 m from the sources and (b) the minimum wavelength difference which may be resolved by a diffraction grating which is 40 mm wide and has 600 lines mm^{-1}, in the first order. (Assume $\lambda = 550$ nm in both cases.)

We define resolving power in terms of the Rayleigh criterion, which states that two objects (or wavelengths) are just resolved if their diffraction patterns when viewed through the optical system are such that the principal maximum of one falls in the first minimum of the other.

In (a) this criterion therefore requires that the sources have an angular separation of $\theta \simeq \lambda/D$ (see eq. (1.29)), i.e. $\theta \simeq \lambda/D$. Now θ is also given by $\theta = S_{min}/500$, where S_{min} is the minimum separation of the sources. Hence $\lambda/D = S_{min}/500$, whence

$$S_{min} = \frac{500 \times 550 \times 10^{-9}}{0.1} = 2.75 \text{ mm}$$

For (b) the chromatic resolving power $\lambda/\delta\lambda$ of a grating is given by $\lambda/\delta\lambda = pN$, where N is the number of lines used. Therefore, in the first order $p = 1$, $\lambda/\delta\lambda = 1 \times 40 \times 600$; that is, $\delta\lambda = 0.023$ nm.

1.3 Light sources – blackbody radiation

The sources discussed in this section are the so-called classical or *thermal sources*. These are so-named because they radiate electromagnetic energy in direct relation to their temperature. Thermal sources can be divided into two classes, namely blackbody radiators and line sources. The former are opaque bodies or hot, dense gases which radiate at virtually all wavelengths. Line sources, on the other hand, radiate at discrete wavelengths.

Blackbody sources. The radiation from opaque objects and dense gases was widely studied in the late nineteenth century resulting in the formulation of the following empirical laws. Firstly it was found that the rate at which energy is emitted is proportional to the fourth power of the absolute temperature, that is

$$W = \sigma T^4 \tag{1.31}$$

where W is the total radiated power per unit area and σ is Stefan's constant. Equation (1.31) is known as the *Stefan–Boltzmann law*. It is only strictly true for an ideal blackbody, which is the most efficient emitter of thermal radiation. An approach to an ideal blackbody can be made by piercing a small hole in an otherwise closed cavity, then, if the cavity is maintained at a uniform, constant temperature, the radiation leaving the hole is essentially that of a blackbody. Most hot surfaces only approach the ideal, and eq. (1.31) is in general modified to

$$W = \varepsilon \sigma T^4 \tag{1.31i}$$

where ε is the emissivity of the surface; ε has values between zero and unity (the value for a blackbody).

Example 1.4 Radiated power

We may estimate the total power radiated from a source of area 10^{-5} m^2 at a temperature of 2000 K, given that the emissivity of the surface is 0.7.

From eq. (1.31i) we have

$$W = 0.7 \times 5.67 \times 10^{-8} \times 10^{-5} \times (2000)^4 = 6.35 \text{ W}$$

Secondly it was noted that the spectral distribution of the energy emitted at a given temperature has a definite maximum and that this maximum shifts to shorter wavelengths as the temperature increases, as illustrated in Fig. 1.16. This shift is given by *Wien's displacement law*, which can be expressed as

$$\lambda_m T = \text{constant} \tag{1.32}$$

where λ_m is the wavelength at which the radiated power is a maximum for a given temperature T.

The curves shown in Fig. 1.16 and the above laws are accurately described by Planck's formula which can be written as:

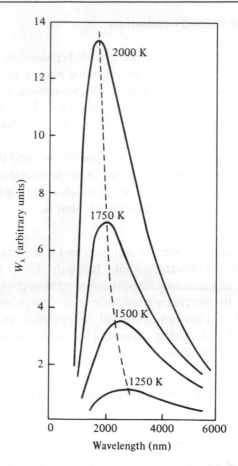

Fig. 1.16 Distribution of energy in the spectrum of a blackbody radiator at various temperatures.

$$W_\lambda = \frac{2\pi hc^2}{\lambda^5}\left[\frac{1}{\exp(hc/\lambda kT) - 1}\right]$$ (1.33)

where W_λ is the spectral radiant emittance (see Table 1.2). To derive this equation, Planck considered the possible standing wave patterns or modes which can exist within a cavity and assumed that the energy associated with each mode was quantized; that is, the energy could only exist in integral multiples of some lowest amount or quantum. Thus, according to Planck, matter could only emit discrete quantities of radiation which were called photons. The success of this assumption, as we mentioned in the introduction to this chapter, provided the foundation for the development of modern quantum theory.

Line sources. In the case of excited gases in which there is little interaction between the individual atoms, ions or molecules, the electromagnetic radiation is emitted at well-defined wavelengths. This can be understood quite easily on the basis of the simple Bohr model of the atom in which it is considered that the atom consists of a positive nucleus of charge Ze (Z being the *atomic number* of the atom) with electrons of mass m and charge e in certain 'allowed' bound orbits around it. Each of these orbits corresponds to a well-define energy level. The energy is given by

$$E_n = \frac{-mZ^2e^4}{8n^2h^2\varepsilon_0^2} \tag{1.34}$$

where n is an integer known as the *principal quantum number* (see, for example, ref. 2.1b). The outermost electron may be excited from its normal or ground state orbit to higher energy orbits which are normally unoccupied. When an electron undergoes a transition from one of these excited orbits (or energy levels) to a lower orbit it emits a quantum of radiation. The energy of the quantum is just the difference ΔE between the energies of the initial and final orbits. Thus the quantum energy is

$$h\nu = \frac{hc}{\lambda} = \Delta E \tag{1.35}$$

from which, using eq. (1.34), we see that

$$\nu = \frac{\Delta E}{h} = \frac{me^4Z^2}{8h^3\varepsilon_0^2}\left(\frac{1}{n_f^2} - \frac{1}{n_i^2}\right)$$

n_f and n_i being the values of the principal quantum number corresponding to the final and initial orbits (or energy levels) involved in the transition.

The spectral lines emitted in this way can have a very narrow frequency spread; that is, they are almost monochromatic. In practice, however, there is a number of causes of spectral broadening which increase the spread of the wavelength (or frequency) associated with the emitted photons (see section 5.7).

As long as the atoms are in thermal equilibrium with their surroundings, the energy radiated by an intense line radiator can never exceed that of a blackbody at the same temperature as the line source. This is true despite the very different wavelength distributions of the emitted energy of the two sources, even when comparing the energies they emit per unit wavelength range (or their spectral radiant emittances). This rule is broken in the case of lasers, where, as we shall see in Chapter 5, the atoms are *not* in thermal equilibrium.

Example 1.5 Ionization energy of the hydrogen atom

Here we calculate the ionization energy of the hydrogen atom ($Z = 1$) given the physical constants in Appendix 6. The ionization energy is the energy required to excite the electron from the ground state ($n = 1$) to infinity ($n = \infty$).

Thus from eq. (1.34) we have

$$E_{ion} = \frac{9.1 \times 10^{-31} \times (1.6 \times 10^{-19})^4}{8 \times (6.6 \times 10^{-34} \times 8.85 \times 10^{-12})^2}$$

$$= 2.176 \times 10^{-18} \text{ J or } 13.6 \text{ eV}$$

In concluding this section we return briefly to the photoelectric effect, which dramatically illustrates the photon or particle nature of light. Einstein explained the emission of electrons from metal surfaces exposed to light as being due to a transfer of energy from a single photon to a single electron. It was found that (a) the energy of the emitted electrons depends not on the irradiance of the incident light but rather on its frequency, (b) for light of a given frequency the photoelectrons have a maximum energy E_{max} and (c) for a given metal there is a minimum frequency for the light which will cause electrons to be emitted. These observations are summarized in Einstein's photoelectric equation:

$$E_{max} = h\nu - e\phi \tag{1.36}$$

where e is the electron charge and ϕ is a constant for a particular metal known as the *work function*. The quantity $e\phi$ represents the energy required to free an electron from the surface (section 2.6). The difference between the incident photon energy $h\nu$ and $e\phi$ then appears as the kinetic energy of the emitted electron.

Example 1.6   Work function of metals

Given that the lowest frequency of light which will eject electrons from a tungsten surface is 1.1×10^{15} Hz, we may calculate its work function.

From eq. (1.36) we have $h\nu_0 = e\phi$, where ν_0 is the *threshold frequency* and we assume that the ejected electrons have zero kinetic energy.

Hence $\phi = 4.5$ eV.

1.4 Units of light

The measurement of the energy of electromagnetic radiation when all wavelengths are treated equally is known as *radiometry*. The measurement of those aspects of radiation which effect vision is referred to as *photometry*. The link between these is the standard luminosity curve shown in Fig. 1.17, which shows the spectral response V_λ of the average eye to light of different wavelengths. The value of V_λ (often called the relative luminous efficiency) is taken as unity at $\lambda = 555$ nm where the eye has its maximum sensitivity. The value of V_λ falls to near-zero at the extremes of the visible spectrum; that is, at about 400 nm and 700 nm. For normal photopic vision (when the eye is adapted for high levels of stimulus) at the peak sensitivity of the eye (555 nm) one watt of radiant energy equals 680 lumens by definition. The watt is a radiometric unit whereas the lumen is a photometric unit. At any other wavelength this conversion is scaled by the value of the relative luminous efficiency at that wavelength.

To obtain a measure of the relative brightness of various sources, we must use photometric units. In laser technology and safety, however, radiometric units are widely

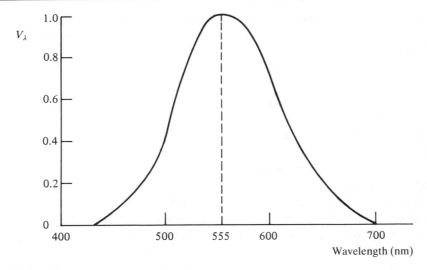

Fig. 1.17 Relative luminous efficiency curve for normal photopic vision (that is, when the eye is adapted for high levels of stimulus).

used. We shall therefore briefly describe the most important radiometric units and give their photometric equivalents in Table 1.2.

Of course, light is a form of energy, and radiant energy and power are measured in joules and watts respectively. Radiant power is sometimes referred to as *flux* Φ; the flux per unit area delivered to a surface is the *irradiance I*. It should be noted at this point that many texts on optics and lasers call the power per unit area the *intensity*. As we mentioned in the footnote on page 7, however, this practice should not be

Table 1.2 Radiometric and photometric units

Symbol (SI units)	Radiometric term and units	Photometric term and units	Definition
Q	Radiant energy (J)	Luminous energy (talbot)	
ϕ	Radiant power or flux (W)	Luminous power or flux (lm)	
$I(E)$	Irradiance (W m^{-2})	Illuminance (lm m^{-2})	Total power falling on unit area
$\mathscr{I}(I)$	Radiant intensity (W sr^{-1})	Luminous intensity (lm sr^{-1})	Power radiated by a point source into unit solid angle
L	Radiance (W m^{-2} sr^{-1})	Luminance or brightness (lm m^{-2} sr^{-1})	Power radiated from unit area into unit solid angle
W	Radiant emittance (W m^{-2})		Total power radiated in all directions from unit area

Note: Any of these quantities can be expressed per unit frequency or wavelength. In these cases the word 'spectral' is added as a prefix to that term and a ν or λ is added as a subscript to the symbol. Thus, for example, the spectral radiance L_λ is the radiance divided by the bandwidth in wavelength units (μm, nm, etc.). The spectral radiance may then have units of W m^{-2} sr^{-1} nm^{-1} or, if the bandwidth is given in frequency units, the spectral radiance would have units W m^{-2} sr^{-1} Hz^{-1}.

encouraged and accordingly in this text we shall use irradiance. The recommended symbol for irradiance is E (or E_e) but to avoid confusion with E used for energy we shall use the symbol I.

The energy emitted from a point source is described in terms of the *radiant intensity* $\mathcal{J}_e$,[†] which is the radiant flux emitted by a *point* source within a unit solid angle in a given direction. $\mathcal{J}_e$ is therefore measured in watts per steradian (W sr^{-1}).

For small *plane* sources, the equivalent term to radiant intensity is the *radiance* L_e, which is the power radiated into unit solid angle per unit area of source; that is, L_e has units W m^{-2} sr^{-1}. Several laser texts refer to this quantity as the brightness, which is possibly the most frequently used and also the most confusing term in this subject. The photometric quantity corresponding to irradiance is *illuminance*, defined as the flux delivered to a surface per unit area. The visual effect of this energy is characterized by the *luminance* or *brightness*. Brightness is also often used to describe psychological perception. The safest procedure would appear to be to take careful note of the *units* of the terms used in a given discussion.

The more commonly used radiometric quantities and their photometric (or luminous) equivalents are shown in Table 1.2. To reduce the profusion of nomenclature which existed in the recent past, the same symbols are used for corresponding radiometric and photometric quantities. When it is necessary to distinguish between radiant and luminous quantities the subscripts e (for energy) and v (for visual) are used.

The photometric quantity luminous intensity $\mathcal{J}_v$ is one of the seven quantities chosen as dimensionally independent *base quantities* in the SI system of units. The units of luminous intensity, namely lumens per steradian (lm sr^{-1}), are often referred to as candela (cd). The luminance, therefore, has units of cd m^{-2} or, more commonly in practice, cd cm^{-2} despite the centimeter not being a recognized SI unit. Quite often British units are used and the situation is further complicated depending on whether the emitting or reflecting surface is a Lambertian surface or a uniform diffuser – see section 4.8 and ref. 1.5.

Problems

1.1   Express equation (1.6) in alternative forms such as $\mathcal{E} = \mathcal{E}_0 \cos 2\pi(t/T - x/\lambda) = \mathcal{E}_0 \cos k(vt - x)$. Write down the equation of the wave which is (a) 90° out of phase with these equations and (b) traveling in the negative x direction.

1.2   By considering the superposition of two waves which differ in angular frequency and wave number by the *small* amounts $\delta\omega$ and δk, derive eq. (1.8).

1.3   Given that the average solar constant is of the order of 700 W m^{-2} in the United Kingdom calculate the corresponding electric field amplitude.

1.4   Derive Malus's law, eq. (1.15).

[†] The recommended symbol for radiant intensity is I, but we intend to use that symbol for irradiance; the radiant intensity is not frequently used in the text.

1.5　Derive the results of eq. (1.17) for the superposition of two waves by (a) treating the two waves as vectors of magnitudes proportional to the wave amplitude and direction given by the phase of the wave and (b) by expressing the waves as complex exponential functions. Show that the energy of the resultant of adding five waves of equal amplitudes $\mathcal{E}_0$ and phase constants 0, $\pi/4$, $\pi/2$, $3\pi/4$, π is $\mathcal{E}_0^2(3 + 2\sqrt{2})$. Show also that the resultant energy of superposing an infinite series of waves of amplitudes $\mathcal{E}_0$, $\mathcal{E}_0/2$, $\mathcal{E}_0/4$, $\mathcal{E}_0/8$, ... and phase constants 0, $\pi/2$, π, $3\pi/2$, ... is $4\mathcal{E}_0^2/5$.

1.6　Show that the fringe spacing in a Young's slits experiment is given by $\delta y = \lambda x/H$. If the aperture to screen distance is 1.5 m and the wavelength is 632.8 nm, what slit separation is required to give a fringe spacing of 1.2 mm? If a glass plate ($n = 1.5$) of 0.05 mm thickness is placed over one slit, what is the lateral displacement of the fringe system?

1.7　Verify eq. (1.25) for interference effects in thin films; explain the colored appearance of thin films of oil on wet surfaces or of soap bubbles.

　　In an experiment to observe fringes of equal inclination, the source emits wavelengths of 500 nm and 600 nm. It is observed that a bright fringe is obtained for both wavelengths when the light passes through the film at $\cos^{-1} 0.7$ to the normal, and that the next such coincidence occurs at an angle $\cos^{-1} 0.8$. Deduce the optical thickness of the film and the orders of the fringes for which the two coincidences occur.

1.8　Explain the phenomenon of 'blooming'. Calculate the thickness of a 'blooming' layer of refractive index 2.0 deposited onto a glass substrate of refractive index 1.5 which will give (a) maximum and (b) minimum reflected light. Assume normal incidence and that $\lambda = 500$ nm.

1.9　What is the maximum number of orders that can be observed using a plane grating of 300 lines mm^{-1} for normally incident light of wavelength 546 nm? If light of all wavelengths from 400 to 700 nm were used, what wavelengths would be superposed on the 546 nm wavelength in the highest of these orders?

1.10　A collimated beam of light from a He–Ne laser ($\lambda = 632.8$ nm) falls normally onto a circular aperture of 0.5 mm diameter. A lens of 0.5 m focal length placed just behind the aperture focuses the diffracted light onto a screen. Calculate the distance of the first dark ring from the center of the diffraction pattern. Using the same arrangement, what is the minimum separation of another source which could just be resolved if the two sources are 10 m from the aperture?

1.11　Calculate the difference in energy between the Bohr orbits for hydrogen ($Z = 1$) for which $n = 4$ and $n = 2$. What is the wavelength associated with the photon emitted by an electron which undergoes a transition between these levels?

1.12　Calculate the minimum frequency of light which will cause the photoemission of electrons from a metal of 2.4 eV work function. What is the maximum kinetic energy of the photoelectrons emitted by light of 300 nm wavelength?

References

1.1　(a) R. W. Ditchburn, *Light* (2nd edn), Blackie, Glasgow, 1962; (3rd edn), Academic Press, New York, 1976.

　　(b) R. S. Longhurst, *Geometrical and Physical Optics* (3rd edn), Longman, 1973.

(c) G. R. Fowles, *Introduction to Modern Optics* (2nd edn), Holt Rinehart & Winston, 1975.

(d) E. Hecht and A. Zajac, *Optics*, Addison-Wesley, Reading, Mass., 1974.

(e) M. V. Klein and T. C. Furtak, *Optics* (2nd edn), John Wiley, 1986.

1.2   R. W. Ditchburn, *op. cit.*, Chapter 3.

1.3   *Ibid*, pp. 137–40.

1.4   G. R. Fowles, *op. cit.*, pp. 108–19.

1.5   M. Young, *Optics and Lasers – An Engineering Physics Approach*, Springer-Verlag, Berlin, 1977, Chapter 2.

Chapter 2

Elements of solid state physics

To understand the operation of many of the optoelectronic devices discussed in later chapters, we need at least an appreciation of the solid state physics of homogeneous materials and of the junctions between different materials. Accordingly, we now examine the mechanisms by which current flows in a solid, why some materials are good conductors of electricity and others are poor conductors, and why the conductivity of a semiconductor varies with temperature, with the number of impurities it contains, and with exposure to light.

One of the physical models which has been quite successful in explaining these and related phenomena is the *energy band* model of solids. Before describing this, however, we shall review some relevant concepts from quantum physics. It is assumed that the reader is familiar with much of the basic physics of isolated atoms including the description of atomic energy states in terms of four quantum numbers and the application of the Pauli exclusion principle leading to the electron configuration of atoms (ref. 2.1).

Because of space limitations, several of the equations presented in later sections will not be derived in detail. The reader will be guided through their derivation, however, via the problems and the references presented at the end of the chapter.

2.1 Review of some quantum mechanical concepts

In Chapter 1 we saw that light displays a dual nature, that of particle and wave. Now the energy of a light particle, the photon, can be written as $E = h\nu$, where ν is the frequency of the wave associated with the photon. As the rest mass of the photon is zero, its momentum p can be written as:

$$p = \frac{E}{c}$$

Therefore

$$p = \frac{h\nu}{c} = \frac{h}{\lambda} \tag{2.1}$$

where λ is its wavelength. This equation, known as the *de Broglie relation*, can be applied quite generally to particles such as electrons and atoms, which also exhibit a dual nature as demonstrated, for example, by electron diffraction.

The question as to what is meant by the wave to be associated with a particle immediately arises. Born (1926) showed that the wave amplitude is related to the probability of locating the particle in a given region of space. More specifically, in quantum mechanical problems, as we shall see below, we attempt to find a quantity Ψ called the *wave function*. While Ψ itself has no direct physical meaning, it is defined in such a way that the probability of finding the particle in the region of space between x and $x + \mathrm{d}x$, y and $y + \mathrm{d}y$, z and $z + \mathrm{d}z$ is given by $\Psi^*\Psi \, \mathrm{d}x \, \mathrm{d}y \, \mathrm{d}z$, where Ψ^* is the complex conjugate of Ψ (Ψ may be complex, that is, involve the square root of -1). Clearly if the particle exists then

$$\int_{-\infty}^{\infty} \Psi^*\Psi \, \mathrm{d}x \, \mathrm{d}y \, \mathrm{d}z = 1 \tag{2.2}$$

$\Psi^*\Psi$ or $|\Psi|^2$ is called the probability density function.

A real particle, which is localized, cannot be described by a wave equation of the form of eq. (1.6) say, which is of infinite extent. Rather, the wave description of a particle is given in terms of a wave packet (see section 1.2), which is the sum of individual waves with varying amplitudes and frequencies. These waves interfere destructively except for a certain region in space where the probability of finding the particle is high (Fig. 2.1). The speed of the particle υ is the same as the group velocity of the wave packet, that is, from eq. (1.8)

$$\upsilon = \upsilon_g = \frac{\partial \omega}{\partial k}$$

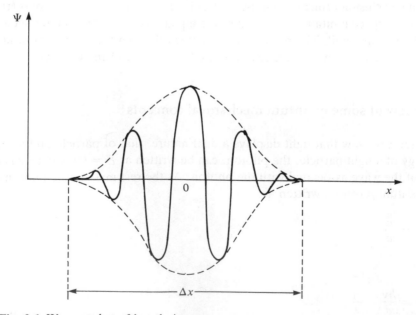

Fig. 2.1 Wave packet of length Δx.

The probability of finding the particle is greatest at the center of the wave packet, $x = 0$ in Fig. 2.1; however, there is a smaller but finite probability of finding it anywhere in the region Δx. Now it may be shown from the Fourier integral (ref. 2.2) that the narrower is Δx, that is, the more accurately the position of the particle is know, the wider must be the range of wavelengths $\Delta \lambda$ comprising the wave packet. From eq. (2.1) this implies a greater uncertainty, Δp, in the momentum of the particle. Heisenberg showed that in any simultaneous measurement of the position and momentum of a particle, the uncertainties in the two measured quantities are related by an equation of the form

$$\Delta x \, \Delta p \geqslant \hbar/2 \tag{2.3}$$

Equation (2.3) is called Heisenberg's *uncertainty principle*. This implies that it is impossible to describe with absolute precision events involving individual particles: we can only talk of the *probability* of finding a particle at a certain position at a given instant of time.

2.1.1 Schrödinger equation

In general, the Schrödinger equation includes both space and time dependencies and is of the form.

$$\frac{\hbar^2}{2m}\left(\frac{\partial^2 \Psi}{\partial x^2} + \frac{\partial^2 \Psi}{\partial y^2} + \frac{\partial^2 \Psi}{\partial z^2}\right) - V\Psi - i\hbar \frac{\partial \Psi}{\partial t} \tag{2.4a}$$

where $i = \sqrt{-1}$. V is the potential energy of the particle and Ψ the wavefunction, both of which may depend on position and time. That is, we have $\Psi = \Psi(x, y, z, t)$ and $V = V(x, y, z, t)$. Equation (2.4a) is often referred to as the time-dependent Schrödinger equation. Unfortunately, eq. (2.4a) is quite complicated and can only be solved in a limited number of cases. For our purposes, it is sufficient to consider situations in which the potential energy of the particle does not depend explicitly upon time. The forces that act upon it, and hence V, then vary only with the position of the particle. In this case, eq. (2.4a) reduces to the time-independent form, which, in one dimension, may be written as:

$$\frac{d^2 \psi(x)}{dx^2} + \frac{2m}{\hbar^2}(E - V(x))\psi(x) = 0 \tag{2.4b}$$

where ψ is the time-independent wave function and E is the total energy of the particle.

2.1.1.1 POTENTIAL WELL

It is quite difficult to find solutions of the Schrödinger equation for most realistic potential distributions. However, there are several physical situations, for example a free electron trapped in a metal or charge carriers trapped by the potential barriers of a double heterojunction (section 5.10.2.3), which can be approximated by an electron in an infinitely deep, one-dimensional potential well. The situation is illustrated in Fig. 2.2, where for simplicity we have taken $V(x) = 0$ except at the boundaries where $V(x)$ is infinitely large. That is, the boundary conditions are

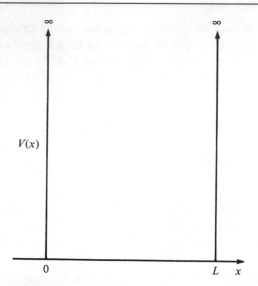

Fig. 2.2 One-dimensional potential well of infinite depth.

$$V(x) = 0, \quad 0 < x < L \tag{2.5a}$$

and

$$V(x) = \infty, \quad x \leqslant 0, x \geqslant L \tag{2.5b}$$

Inside the potential well, eq. (2.4b) becomes

$$\frac{d^2\psi(x)}{dx^2} + \frac{2m}{\hbar^2} E \psi(x) = 0 \tag{2.6}$$

This is the wave equation for a free particle in a region where $V(x) = 0$.

A possible solution to eq. (2.6) is

$$\psi(x) = A \sin kx + B \cos kx \tag{2.7}$$

where A and B are constants and

$$k^2 = \frac{2mE}{\hbar^2} \tag{2.8}$$

Applying the boundary conditions given by eqs. (2.5) we see that $\psi(x)$ must be zero at the boundaries of the well. Otherwise, there would be a nonzero value of $|\psi|^2$ outside the well, which is impossible because a particle cannot penetrate an infinitely high potential barrier. Thus as $\psi(x) = 0$ at $x = 0$, B must be zero, and as $\psi(x) = 0$ at $x = L$, k must be defined so that $\sin kx$ is zero at $x = L$. That is, kL must be an integral multiple of π. We can therefore write

$$\psi(x) = A \sin kx \tag{2.9}$$

where

$$k = \frac{n\pi}{L}, \quad n = 1, 2, 3 \ldots \tag{2.10}$$

Substituting for k from eq. (2.8) we have

$$\left(\frac{2mE_n}{\hbar^2}\right)^{\frac{1}{2}} = \frac{n\pi}{L}$$

or

$$E_n = \frac{n^2h^2}{8mL^2} \tag{2.11}$$

Thus for each value of n the particle energy is described by eq. (2.11). We note that the total energy is *quantized*; n is a *quantum number*.

The value of A in eq. (2.9) can be obtained by using the *normalization condition* expressed by eq. (2.2). We find that $A = \sqrt{(2/L)}$ and

$$\psi_n = \left(\frac{2}{L}\right)^{\frac{1}{2}} \sin \frac{n\pi x}{L} \tag{2.12}$$

The wave functions ψ_n and the corresponding energies E_n, which are often called *eigenfunctions* and *eigenvalues* respectively, describe the quantum state of the particle. The forms of ψ_n and $|\psi_n|^2$ are shown in Figs. 2.3(a) and (b) respectively for the first three quantum states ($n = 1, 2, 3$).

This solution for a one-dimensional potential well can be extended quite easily to the more realistic case of three dimensions, where, assuming that the sides of the potential

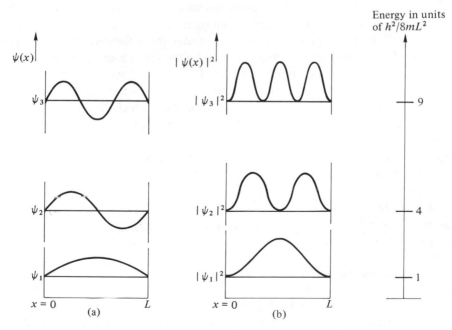

Fig. 2.3 Ground state and first two excited states of an electron in a potention well: (a) the electron wave functions and (b) the corresponding probability density functions. The energies of these three states are shown on the right.

well are all the same length, the eigenfunctions are given by

$$\psi_n = \left(\frac{8}{L}\right)^{\frac{1}{2}} \sin k_1 x \sin k_2 y \sin k_3 z$$

where

$$k_1 = \frac{n_1 \pi}{L}, \quad k_2 = \frac{n_2 \pi}{L}; \quad k_3 = \frac{n_3 \pi}{L}$$

The eigenvalues are still given by eq. (2.11), that is

$$E_n = \frac{n^2 h^2}{8mL^2}$$

but now we have

$$n^2 = n_1^2 + n_2^2 + n_3^2 \tag{2.13}$$

We see that each energy state is defined by a set of three quantum numbers, each of which yields a different eigenfunction ψ. From other evidence, it is found that each electron eigenfunction has one of two states of opposite spin sign associated with it. These two states are described in terms of the quantum number m_s, which can take either of the two values $+\frac{1}{2}$ or $-\frac{1}{2}$. Thus four quantum numbers (that is, n_1, n_2, n_3, m_s) are needed to define a quantum state completely. In fact the need for four quantum numbers is general in any three-dimensional potential, although each potential will give rise to its own unique set of quantum numbers, and these may well describe the quantization of quite different physical parameters. For example, in a hydrogen atom an electron may be considered to be moving under the influence of a single positive point charge. Its quantum state is characterized by the set of four numbers n, l, m_l, and m_s. Here, n is the *principal* quantum number, which determines the electron energy, l is the *orbital momentum* quantum number, which determines the angular momentum of the electron, while m_l is the *magnetic* quantum number, which determines the orientation of the angular momentum vector. Only the spin quantum number is the same in these two cases.

Whilst solutions of the Schrödinger equation tell us what states are available to a single electron, we still need to know which of these states are occupied, especially if we have a situation where a number of electrons are involved. At first sight we might expect that all the electrons would occupy the lowest energy states possible. The implications of this are that in a multielectron atom, for example, the electrons would all occupy the same energy level. Since the chemical properties of an atom are determined by the types of levels the electrons occupy, this assumption would give rise to the atoms of the various elements all having very similar chemical properties. Manifestly this is not the case! In fact electrons are found to obey the *Pauli exclusion principle*. This forbids any two electrons having the same set of quantum numbers. Thus if we have several electrons (assumed to be noninteracting for simplicity) within a three-dimensional potential well, not more than two of them can have the same values for n_1, n_2, and n_3, and in that case they must then have different values of m_s.

2.2 Energy bands in solids

As isolated atoms are brought together to form a solid various interactions occur between neighboring atoms. The forces of attraction and repulsion between atoms find a balance at the proper interatomic spacing for the crystal. In the process important changes occur in the electron energy levels, which result in the varied electrical properties of different classes of solids.

Qualitatively, we can see that as the atoms are brought closer together the application of the Pauli principle becomes important. When atoms are isolated, as in a gas, there is no interaction of the electron wave functions, each atom can have its electrons in identical energy levels. As the interatomic spacing decreases, however, the electron wave functions begin to overlap and, to avoid violating the Pauli principle, there is a splitting of the discrete energy levels of the isolated atoms into new levels belonging to the collection of atoms as a whole. In a solid many atoms are brought together so that the split energy levels form a set of *bands* of very closely spaced levels with forbidden energy gaps between them, as illustrated in Fig. 2.4. The lower energy bands are occupied by electrons first; those energy bands which are completely occupied (that is, full) are not important, in general, in determining the electrical properties of the solid. On the other hand, the electrons in the higher energy bands of the solid are important in determining many of the physical properties of the solid. In particular the two highest energy bands, called the *valence* and *conduction* bands, are of crucial importance in this respect, as is the forbidden energy region between them which is referred to as the *energy gap*, E_g. In different solids the valence band might be completely filled, nearly filled or only half filled with electrons, while the conduction band is never more than slightly filled. The extent to which these bands are occupied and the size of the energy gap determines the nature of a given solid.

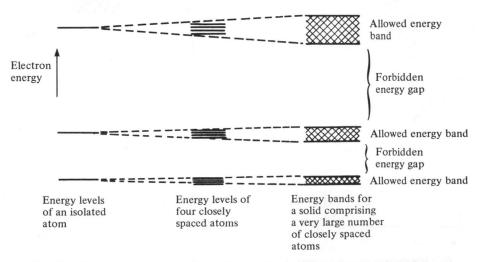

Fig. 2.4 Schematic representation of how the energy levels of interacting atoms form energy bands in solids.

We may further reinforce our model by considering that in an ideal crystalline solid the atoms are arranged in a perfectly periodic array. The potential experienced by an electron in the solid is correspondingly spatially periodic, so that, after a distance in the crystal equal to the lattice spacing, the potential V repeats itself, that is

$$V(x) = V(x + a) = V(x + 2a) = \ldots$$

where a is the periodicity of the lattice.

We could now attempt to apply the Schrödinger equation to the problem of electron motion within this system. This is difficult unless we choose a very simple array of atoms. One such approach, the Kronig–Penney model (ref. 2.3), considers a single electron impinging on a one-dimensional array of atoms represented by a periodic series of rectangular potential wells. Application of the Schrödinger equation in this case shows that not every value of electron energy is allowed. In fact we find that there are ranges of allowed energies separated by ranges of disallowed energies, that is, the electrons in a solid can occupy certain bands of energy levels which are separated by forbidden energy gaps.

The discontinuities between allowed and forbidden energy values occur at values of the wave vector k given by $k = \pm n\pi/a$ where n is an integer (see Problem 2.6). The

Fig. 2.5(a) Relationship of E–k for electrons subjected to the potential distribution of the Kronig–Penney model and the corresponding energy band structure. The E–k relationship for free electrons is shown for comparison.

$E-k$ curve then takes the appearance shown in Fig. 2.5(a), in contrast to the smooth dashed curve which is the $E-k$ curve for a free electron ($V = 0$). The discontinuities arise from the interaction of the electrons with the periodic potential V; the corresponding energy band arrangement is shown to the right of Fig. 2.5(a).

The region in Fig. 2.5(a) for which $-\pi/a < k \leqslant \pi/a$ is called the *first Brillouin zone*. It is often convenient to redraw Fig. 2.5(a) by translating the segments of the $E-k$ curve so that they all lie within this range. This is shown in Fig. 2.5(b). (There is more to this procedure than simply pictorial convenience, as it has a theoretical justification as well, see ref. 2.1f, section 2.2.) In three dimensions the situation is obviously more complicated, but it turns out that diagrams similar to Fig. 2.5(b) can be drawn corresponding to different directions in the crystal. The first Brillouin zone in general has a complicated shape which depends upon the crystal structure being considered. However, its boundaries still lie close to π/a, although the parameter a has now to be interpreted in terms of the crystal unit cell dimensions.

2.2.1 Conductors, semiconductors and insulators

In real crystals the $E-k$ relationship is much more complicated as we can see from Fig. 2.6, which shows the relationships for silicon and gallium arsenide. It depends on the

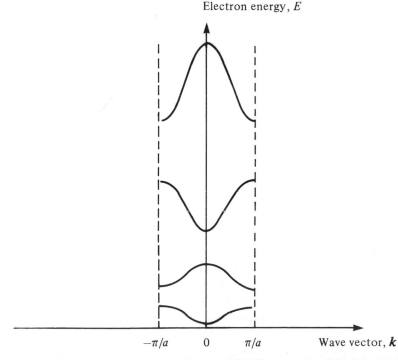

Fig. 2.5(b) Reduced zone representation of the $E-k$ relationship shown in Fig. 2.5(a). This representation is constructed by translating the segments of the $E-k$ curve so that they all lie between $k = -\pi/a$ and $k = +\pi/a$, which comprises the first Brillouin zone.

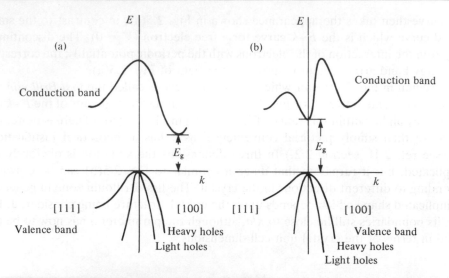

Fig. 2.6 Relationship of $E-k$ for real solids: (a) silicon (which has an indirect band-gap) and (b) gallium arsenide (which has a direct bandgap). The figure shows the conduction and valence bands and the energy gap E_g between them. Note that (i) k is specified in different crystallographic directions to the left and right and (ii) there are holes present with different effective masses (section 2.2.1 and 2.3).

orientation of the electron wave vector to the crystallographic axes, since interatomic distances and the internal potential energy distribution also depend on direction in the crystal. However, the basic effect is still energy band formation. One point that arises is that the maximum of the valence band does not always occur at the same k value as the minimum of the conduction band. We speak of a *direct* bandgap semiconductor when they do and an *indirect* bandgap semiconductor when they do not. Thus silicon has an indirect bandgap whilst gallium arsenide has a direct bandgap. For many purposes, however, it is still sufficient to regard the conduction and valence bands as being similar to those shown in Fig. 2.5(a).

The electrons occupy the allowed states in the energy bands starting from the lowest until they are all accommodated (one electron per state). Above the highest occupied energy state there may be other allowed states which are empty. In order that the electrons in a solid may respond to an applied electric field such empty levels must be readily available. This must be so, because if an electron is to be accelerated by the field it will acquire energy and move to a higher level. This it may do only if the higher levels are unoccupied and not separated by a large, forbidden energy gap.

In monovalent metallic conductors, the top occupied band is only partially filled as shown in Fig. 2.7(a) and electrons can gain energy from an external field quite easily, resulting in high conductivity. In some metals, for example those with divalent atoms, the highest two bands overlap to some extent resulting in somewhat different conducting properties (Fig. 2.7(b)).

In insulators the upper occupied band, the valence band, is completely filled with

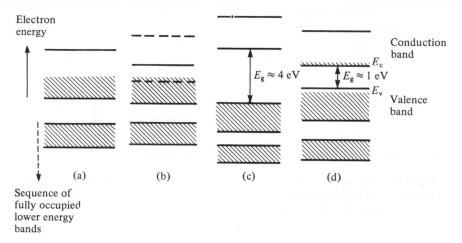

Fig. 2.7 Schematic representation of the energy bands in various materials: (a) a metal with partially filled valence band, e.g. monovalent metals; (b) a metal with two overlapping partially filled bands, e.g. divalent metals; (c) an insulator; and (d) an intrinsic semiconductor. In this, and succeeding diagrams, crosshatching is used to signify occupied electron energy levels.

electrons as shown in Fig. 2.7(c). The nearest empty states are in the conduction band, but these are separated from the valence band by a large energy gap, (large, that is, compared with the average thermal energy of the electrons, kT). At room temperature kT is about 1/40 eV, while typically $E_g \approx 4 \text{eV}$. There are, therefore, very few electrons that can respond to a field, and virtually no conduction occurs.

A similar situation arises in intrinsic semiconductors where at low temperatures the valence band is full and the conduction band is empty. In this case, however, the energy gap is sufficiently small (about 1 eV) that some electrons are excited across it at higher temperatures. Clearly, the electrons excited into the conduction band can contribute to current flow. Similarly, as there are now vacant states in the valence band (Fig. 2.7(d)), the electrons in that band can also respond to an applied field and contribute to the current flow. It is difficult to evaluate this contribution in terms of electron movement, however, as there are a very large number of electrons and a small number of unoccupied states in the band. It turns out that the contribution to the current flow from all the electrons in the nearly full valence band is the same as that which would arise from the presence of a small number of fictitious *positive* charge carriers called *holes* in an otherwise empty band. The number of holes, in fact, is simply equal to the number of empty states in the valence band. Indeed, for many purposes we may regard a hole as an unfilled state. Holes can be regarded as particles which behave in general in a similar way to electrons, apart from having a charge of opposite sign.

The concept of how holes arise in a solid is most easily illustrated by considering the energy band scheme of an intrinsic semiconductor. At any temperature above absolute zero some electrons will be excited from the valence band to the conduction band as a result of their thermal energy. When electrons make such transitions, as illustrated in

Fig. 2.8 Creation of electron–hole pairs by the thermal excitation of electrons from the valence band to the conduction band (a) and electron and hole drift in a semiconductor on application of a field $\mathscr{E}_x$ (b).

Fig. 2.8(a), empty states are left in the valence band and we say that *electron-hole* pairs have been created.

If an electric field is now applied to such a solid, electrons in the conduction band and holes in the valence band drift in opposite directions as shown in Fig. 2.8(b) and both thus contribute to the current because of their opposite charges.

2.3 Electrical conductivity

The charge carriers in a solid are in constant thermal motion. In the absence of external fields this motion is quite random resulting from the carriers being scattered by collisions with lattice atoms, impurities and crystal defects. As the motion is random there is no net displacement of charge in any direction and no net current flow.

If an electric field $\mathscr{E}_x$, is applied then the individual electrons will experience a force $-e\mathscr{E}_x$ (where we have taken e to represent the *magnitude* of the electronic charge). While this force may be insufficient to alter the random path of an individual electron appreciably, the effect when averaged over all the electrons is to generate a net motion of the electrons in the negative x direction. If the total momentum of the group of electrons in the x direction is p_x then we may write:

$$-ne\mathscr{E}_x = \frac{\mathrm{d}p_x}{\mathrm{d}t} \qquad (2.14)$$

where n is the electron density. Equation (2.14) seems to indicate a steady acceleration in the x direction. This is not the case, however, as the electrons lose momentum in collisions with defects in the crystal lattice such as lattice vibrations (or phonons) and impurities. In the steady state the acceleration due to the electric field is balanced by the rate of loss of momentum due to collisions. Consequently the group of electrons aquires a constant drift velocity v_D. Now if the electron collisions are random the probability of each electron colliding at any instant of time is constant. Thus if we have a group of n electrons at the instant of time $t = 0$, and at time t $n(t)$ of these have not undergone a collision, then we may write

$$-\frac{dn(t)}{dt} = \frac{1}{\tau}n(t) \tag{2.15}$$

where τ is a constant of proportionality.

The solution to eq. (2.15) is $n(t) = n_0 \exp(-t/\tau)$. We see that τ, which is called the *mean free time* or *relaxation time*, is the mean time between scattering events, and that the probability that an electron has a collision in the time interval dt is dt/τ. Thus the change in momentum due to collisions in time dt is $dp_x = -p_x dt/\tau$.

The rate of loss of momentum due to collisions, which may be regarded as generating a decelerating or resistive force, is then $dp_x/dt = -p_x/\tau$. Hence in the steady state, from eq. (2.14) we have

$$-\frac{p_x}{\tau} - ne\mathscr{E}_x = 0 \tag{2.16}$$

The average momentum per electron, $\bar{p}_x$, is p_x/n, which from eq. (2.16) is given by

$$\bar{p}_x = -e\tau\mathscr{E}_x \tag{2.17}$$

As expected, eq. (2.17) indicates that the electrons on average have a constant net drift velocity of

$$v_D = \frac{\bar{p}_x}{m_e^*} = \frac{-e\tau\mathscr{E}_x}{m_e^*} \tag{2.18}$$

where m_e^* is the *effective* mass of the electron (see below).

As the electron concentration is n, the current density resulting from this drift is

$$J = -nev_D \tag{2.19}$$

Hence combining eqs. (2.18) and (2.19) we have

$$J = \frac{ne^2\tau}{m_e^*}\mathscr{E}_x \tag{2.20}$$

or

$$J = \sigma\mathscr{E}_x \tag{2.21}$$

Equation (2.21) is, of course, Ohm's law and σ, the electrical conductivity, is given by

$$\sigma = \frac{ne^2\tau}{m_e^*} = ne\mu_e \tag{2.22}$$

The quantity μ_e $(= e\tau/m_e^*)$ in eq. (2.22) is called the electron *mobility*. It describes the ease with which the electrons drift in the material and it is a very important quantity in characterizing semiconductor materials.

The two basic types of scattering that influence electron and hole mobilities are known as lattice scattering and impurity scattering. In the former a carrier moving through the crystal encounters atoms which are out of their normal positions because of thermal vibrations. It is to be expected that the frequency of such events would rise as the temperature rises, leading to a decrease in the mobility with increasing temperature. On the other hand, scattering from lattice defects such as ionized impurities (including dopant atoms, see section 2.4.2), should be relatively independent of temperature. Thus we would expect lattice scattering to dominate at high temperatures and impurity scattering to dominate at low temperatures.

The mobility, as we can see from eq. (2.18), can be defined as the drift velocity per unit electric field. Equation (2.20) can be rewritten, using the mobility, as

$$J = ne\mu_e \mathscr{E}_x \tag{2.23}$$

In semiconductors we must include current flow due to both electrons and holes and the total current density is

$$J = (ne\mu_e + pe\mu_h) \mathscr{E}_x = \sigma\mathscr{E}_x \tag{2.24}$$

where μ_h is the mobility of the holes, and is given by $\mu_h = e\tau/m_h^*$, where m_h^* is the effective mass of the holes.

The concept of effective mass arises when we consider the motion of an electron through a crystal when a field is applied. The situation is not governed simply by the laws of classical physics, because the electron is influenced not only by the external field but also by an internal field, produced by the other electrons, and by the periodic potential of the atoms. Despite this we may use Newton's law to evaluate the acceleration of the electron providing we accept that the electron will exhibit an effective mass m_e^* which is different from the mass m of a free electron in vacuum. The value of the effective mass depends on the energy of the electron within an energy band and is given by (ref. 2.4)

$$m_e^* = \hbar^2 \left(\frac{\mathrm{d}^2E}{\mathrm{d}k^2}\right)^{-1} \tag{2.25}$$

For an electron at the top of an energy band, $\mathrm{d}^2E/\mathrm{d}k^2$, and hence also the effective mass, is *negative*. Such an electron will be accelerated by a field in a direction the reverse of that expected for a negatively charged electron. The existence of electrons with negative effective masses is closely linked with the concept of holes, which have positive effective masses, m_h^* (ref. 2.5).

As mentioned at the start of section 2.2.1 the $E-k$ relationships depend strongly on direction within the crystal. Hence it might appear from eq. (2.25) that we should have an anisotropic effective mass. In general, however, material parameters such as electrical conductivity that depend on effective mass are found to be isotropic. This is because there are often several minima in the $E-k$ diagram, each giving rise to an anisotropic effective mass, but each being differently orientated so that when an average is taken an isotropic result is obtained. In addition, the use of effective mass in different contexts (for example conductivity and density of states (section 2.5)) necessitates taking the average in different ways resulting in different values for the effective mass. Thus we speak of the 'conductivity effective mass' and the 'density of states effective mass'. In silicon, for example, for electrons these are $0.26m$ and $0.55m$ respectively.

To evaluate conductivities from eq. (2.22) in addition to carrier mobilities we also require values for the carrier concentrations. This is considered in more detail in section

2.5, where we shall find that carrier concentrations are strongly temperature dependent. As the mobilities, as mentioned above, are also temperature dependent, the conductivity varies with temperature in a quite complex way. This variation, however, is the basis of one method of measuring impurity excitation energies and energy gaps (ref. 2.6).

2.4 Semiconductors

2.4.1 Intrinsic semiconductors

A perfect semiconductor crystal containing no impurities or lattice defects is called an *intrinsic* semiconductor. In such a material, there are no charge carriers at absolute zero but as the temperature rises electron–hole pairs are generated as explained in section 2.2.1. As the carriers are generated in pairs the concentration n of electrons in the conduction band equals the concentration p of holes in the valence band. Thus we have

$$n = p = n_i$$

where n_i is the intrinsic carrier concentration. The value of n_i varies exponentially with temperature (section 2.5), but even at room temperature it is usually not very large. For example, in silicon $n_i \approx 1.6 \times 10^{16}$ m^{-3} at room temperature, whereas there are about 10^{29} free electrons per cubic meter in a typical metal.

Example 2.1 Electrical conductivity of metals and semiconductors

We may compare the electrical conductivity of copper and intrinsic silicon from the following data: for copper we have a density of 8.93×10^3 kg m^{-3}, an atomic mass number of 63.54, a mean free time between collisions of 2.6×10^{-14} s and we assume $m_e^* = m$; for intrinsic silicon we have $n = p = n_i = 1.6 \times 10^{16}$ m^{-3}, an electron mobility of 0.135 m^2 V^{-1} s^{-1} and a hole mobility of 0.048 m^2 V^{-1} s^{-1}.

In copper we assume that the free electron concentration is the same as the number of atoms per unit volume, which we may determine from Avagadro's number to be

$$n = \frac{6 \times 10^{26} \times 8.93 \times 10^3}{63.54} = 8.4 \times 10^{28} \text{ m}^{-3}$$

Then from eq. (2.22)

$$\sigma_{cu} = 6.4 \times 10^7 \ \Omega^{-1} \ \text{m}^{-1}$$

For intrinsic silicon, using eq. (2.24) we have

$$\sigma_{Si} = en_i(\mu_e + \mu_n) = 1.6 \times 10^{-19} \times 1.6 \times 10^{16} \times (0.135 + 0.048)$$

or $\quad \sigma_{Si} = 4.7 \times 10^{-4} \ \Omega^{-1} \ \text{m}^{-1}$

As at a given temperature there is a steady state carrier concentration, there must be a *recombination* of electron–hole pairs at the same rate as that at which the thermal generation occurs. Recombination takes place when an electron in the conduction band makes a transition into a vacant state in the valence band. The energy released in the recombination, which is about E_g, may be emitted as a photon or given up as heat to the crystal lattice in the form of quantized lattice vibrations, which are called *phonons*, depending on the nature of the recombination mechanism. When a photon is released, the process is called *radiative recombination*. The absence of photon emission indicates a *nonradiative* process, in which lattice phonons are created.

We may distinguish between two types of recombination process which we term 'band-to-band' and 'defect center' recombinations. (Some texts refer to these as 'direct' and 'indirect' transitions respectively. This terminology has not been adopted here to avoid confusion with direct and indirect bandgaps, ref. 2.7.) In the band-to-band process, which is shown in Fig. 2.9(a), an electron in the conduction band makes a transition directly to the valence band to recombine with a hole. In the defect center process, the recombination takes place via recombination centers or traps. These are energy levels E_r in the forbidden energy gap which are associated with defect states caused by the presence of impurities or lattice imperfections. Any such defect state can act as a recombination center if it is capable of trapping a carrier of one type and subsequently capturing a carrier of the opposite type, thereby enabling them to recombine.

The precise mechanism of a defect center recombination event depends on the nature and energy of the defect state. One such process is illustrated in Fig. 2.9(b). In the first step (i) an electron is trapped by the recombination center, which subsequently captures a hole (ii). When both of these events have occurred the net result is the annihilation of an electron–hole pair leaving the center ready to participate in another recombination event (iii). The energy released in the recombination is given up as heat to the lattice.

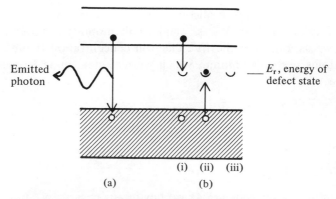

Fig. 2.9 Illustration of (a) band-to-band recombination and (b) recombination via a defect center. The first step in (b) is (i) the trapping of an electron followed by (ii) hole capture. This results in the annihilation of an electron–hole pair leaving the center ready to participate in another recombination (iii).

If the thermal generation rate is g_i and the recombination rate is r_i then, in equilibrium,

$$g_i = r_i \qquad (2.26)$$

Both rates are temperature dependent so that if the temperature is raised, g_i increases and a new value of carrier concentration n_i is established such that the higher recombination rate just balances generation.

At any temperature the probability of an electron recombining is proportional to the number of holes present; thus electrons will disappear at a rate proportional to the product of the electron and hole concentrations. Therefore we may write, in general,

$$r_i = Bnp = g_i \qquad (2.27)$$

where B is a constant of proportionality which depends on the recombination mechanism taking place (see also section 4.6). For an intrinsic material $n = p = n_i$ and $r_i = Bn_i^2$.

2.4.2 Extrinsic semiconductors

The number of charge carriers in a semiconductor can be vastly increased by introducing appropriate impurities into the crystal lattice. In this process, which is called *doping*, a crystal can be altered so that it has a predominance of either electrons or holes; that is, it can be made either *n*-type (where the *majority* carriers are negative electrons and the *minority* carriers are holes) or *p*-type (where the majority carriers are positive holes). In doped semiconductors the carrier concentrations are no longer equal and the material is said to be *extrinsic*.

In doping tetravalent elements, for example silicon, impurities from column V of the periodic table such as phosphorus and arsenic or from column III such as boron and indium are used to produce *n*-type and *p*-type semiconductors respectively. The reasons for this are as follows. When intrinsic silicon is doped with phosphorus, for example, the phosphorus atoms are found to occupy atomic sites normally occupied by silicon atoms as shown in Fig. 2.10(a). Since the silicon atoms are tetravalent only four of the five valence atoms of phosphorus are used in forming covalent bonds (ref. 2.8), leaving one electron weakly bound to its parent atom. This electron is easily freed; that is, it can easily be excited into the conduction band. Therefore in the energy band model the energy levels for the 'extra' electrons associated with these impurities lie at E_d, just beneath the conduction band as shown in Fig. 2.10(b). Such impurities are referred to as *donors*, and the energy levels at E_d as *donor levels*, since they donate electrons to the conduction band. The energy required to excite an electron from the donor levels into the conduction band is E_D, which equals $(E_c - E_d)$, where E_c is the energy of the bottom of the conduction band. If, as is frequently the case, we take the energy E_v at the top of the valence band to be zero, then $E_g = E_c$ and then E_D equals $(E_g - E_d)$. At absolute zero the donor levels are all occupied but, because E_D is so small (about 0.04 eV), even at moderately low temperatures most of the electrons are excited into the conduction band, thereby increasing the free electron concentration and the conductivity of the material (see section 2.3).

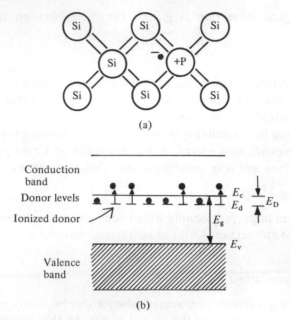

(a)

(b)

Fig. 2.10 Schematic representation of an n-type semiconductor: (a) the crystal lattice in which a silicon atom has been replaced by a phosphorus impurity atom and (b) the energy levels of the donor impurity atoms (the diagram shows that some of the impurities have 'donated' an electron to the conduction band).

We can estimate E_D as follows. If a phosphorus impurity atom loses its fifth valence electron it is left with a net positive charge of $+e$ (the impurity is said to have been ionized). It can be imagined, therefore, that this electron is bound to its parent atom in a situation which is similar to that found in the hydrogen atom, where a charge of $+e$ binds an electron to the nucleus. The ionization energy of the hydrogen atom is 13.6 eV (see Example 1.5), but in the case under discussion there are two important differences arising from the fact that the electron moves in a solid. First we must use the effective mass m_e^* rather than the free electron mass. Secondly the relative permittivity of the semiconductor must be included in the derivation of the electron energy levels. This is because the electron orbit is large enough to embrace a significant number of silicon atoms so that the electron may be considered to be moving in a dielectric medium of relative permittivity ε_r. Therefore, from eq. (1.34), the excitation energy E_D is given by

$$E_D = 13.6 \, \frac{m_e^*}{m} \left(\frac{1}{\varepsilon_r}\right)^2 \, \text{eV} \qquad (2.28)$$

Suppose, on the other hand, that silicon is doped with boron. Again it is found that the impurity atoms occupy sites normally occupied by silicon atoms as shown in Fig. 2.11(a). In this case, however, there is one electron too few to complete the covalent bonding. At temperatures above absolute zero an electron from a neighboring silicon atom can move to the impurity to complete the bonding there but, by so doing, leaves a

Example 2.2 Ionization energy of donor impurities

We may estimate the energy required to excite electrons from the donor levels to the conduction band in silicon given that $m_e^* = 0.26m$ and the relative permittivity is 11.8.

From eq. (2.28) we then have

$$E_D = 13.6 \times 0.26 \left(\frac{1}{11.8}\right)^2 = 0.025 \text{ eV}$$

(We may compare this value with the following experimental values for various donor impurities in silicon: P 0.045 eV; As 0.05 eV; and Sb 0.04 eV.)

vacant state in the valence band thereby creating an additional hole. For this reason the trivalent impurities are referred to as *acceptors* as they accept electrons excited from the valence band. It is convenient to regard this situation as a negatively ionized acceptor atom with a positive hole orbiting around it analogous to the situation described above. The energy E_A, which equals $E_a - E_v$, required to 'free' the hole from its parent impurity can be estimated as above. An average value of the effective mass of holes in

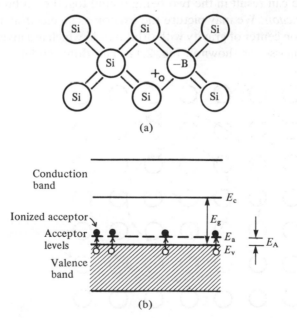

(a)

(b)

Fig. 2.11 Schematic representation of a *p*-type semiconductor: (a) the crystal lattice with trivalent impurities (e.g. boron) and (b) the energy levels of the acceptor impurity atoms (some impurities have 'accepted' electrons from the valence band).

silicon is $0.33m$ and, from eq. (2.28). $E_A = 0.032$ eV. In reality, of course, E_A is the energy required to excite an electron from the valence band to the acceptor energy levels, which lie just above the valence band as illustrated in Fig. 2.11(b).

In the compound semiconductor gallium arsenide, gallium is from group III of the periodic table, while arsenic is from group V (such semiconductors are often referred to as group III–V compounds, other members of this group are GaP, InP and AlSb). In this case n- and p-type doping can be accomplished in a number of different ways. For example, if a group II element such as zinc replaces the gallium atoms, the material becomes p-type, whereas if a group VI element such as tellurium replaces the arsenic atoms, it becomes n-type. Group IV elements such as germanium and silicon can produce either p- or n-type material depending on whether the group IV impurity replaces the arsenic or gallium ions respectively.

Similar remarks apply to the group II–VI compound semiconductors, for example CdS and ZnSe, and to the ternary and quaternary semiconductors such as GaInAs and GaInAsP.

2.4.3 Excitons

We have just seen that the introduction of suitable impurities into intrinsic semiconductor material can result in the formation of electron energy levels situated just below the bottom of the conduction band. However, electron energy levels similarly situated can also appear in *intrinsic* material. These arise because the Coulombic attraction of an electron for a hole can result in the two being bound together; a bound electron–hole pair is called an *exciton*. We may picture the exciton as an electron and a hole orbiting about their common center of gravity with orbital radii which are inversely proportional to their effective masses as shown in Fig. 2.12. The Bohr model is readily adapted to

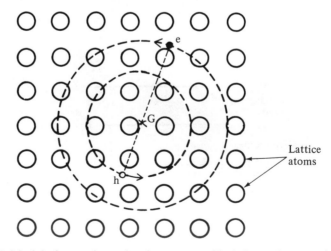

Fig. 2.12 Model of an exciton; the electron e and hole h may be regarded as being bound together and orbiting around their common center of gravity G with radii which are inversely proportional to their effective masses (we have assumed $m_h^* > m_e^*$).

this situation with the electron mass being replaced by the reduced mass m_r^* of the electron and hole; m_r^* is given by

$$\frac{1}{m_r^*} = \frac{1}{m_e^*} + \frac{1}{m_h^*}$$

The binding energy E_e, of the exciton is then given by modifying eq. (2.28) to give

$$E_e = 13.6 \frac{m_r^*}{m_e} \left(\frac{1}{\varepsilon_r}\right)^2 \text{ eV} \qquad (2.28\text{i})$$

Since m_r^*, will be the same order of magnitude as m_e^* and m_h^*, we see that exciton energy levels will be similarly placed to those of donor levels in doped semiconductors.

Excitons may move through a crystalline lattice and thus provide an important means of transferring energy from one point in the material to another. They play an important role in the luminescence of solids and will be discussed further in Chapter 4.

2.5 Carrier concentrations

In calculating semiconductor properties and in analyzing device behavior it is often necessary to know the carrier concentrations. In metals we can make a fairly good estimate of the free electron concentration by calculating the number of atoms per unit volume (from the density and atomic mass number of the metal) and multiplying by the valency. Similarly in heavily doped semiconductors we can take the majority carrier concentration to be the same as the impurity concentration. This may not be so at high temperatures when the number of electron–hole pairs generated by electron excitation across the energy gap may be greater than the number of impurities. The situation in near-intrinsic material is not so clear, however, nor is the temperature variation of carrier concentrations immediately obvious.

To calculate carrier concentrations in energy bands we need to know the following parameters:

(i) The *distribution of energy states* or levels as a function of energy within the energy band.
(ii) The *probability* of each of these states being occupied by an electron.

The first of these parameters is given by the *density of states function* $Z(E)$ which may be defined as the number of energy states per unit energy per unit volume. The form of $Z(E)$, which is shown in Fig. 2.13(a), can be derived, for example, from eq. (2.11), which gives the energy levels in a potential well. It is given by (ref. 2.9)

$$Z(E) = \frac{4\pi}{h^3} (2m_e^*)^{3/2} E^{1/2} \qquad (2.29)$$

where E is measured relative to the bottom of the band.

The second parameter depends on the fact that electrons obey the Pauli principle (page 36) and hence the probability of a particular energy level being occupied at temperature T is given by Fermi–Dirac statistics. This is in contrast to the case of atoms

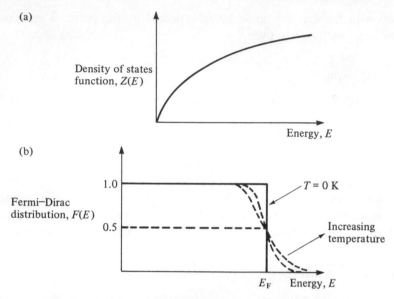

Fig. 2.13 Density of states function (a) and the Fermi–Dirac function at $T = 0\,\text{K}$ and at $T > 0\,\text{K}$ (b).

in an ideal gas where the Maxwell–Boltzmann distribution function applies (in fact, in several situations electron behavior can be approximated by Maxwell–Boltzmann statistics thereby simplifying the mathematics).

The Fermi–Dirac distribution function is found to be

$$F(E) = \frac{1}{\exp\left(\dfrac{E - E_F}{kT}\right) + 1} \tag{2.30}$$

where E_F is a characteristic energy called the *Fermi energy*.

The distribution function is shown in Fig. 2.13(b). We notice that at 0 K $F(E)$ is unity for energies less than E_F and zero for energies greater than E_F. At any temperature above absolute zero the probability of occupation of the energy level at $E = E_F$ is 0.5, as we can see from eq. (2.30). Figure 2.13(b) shows that the probability of occupation of states above E_F is finite for $T > 0$ and that there is a corresponding probability that states below E_F are empty. In fact $F(E)$ is symmetrical about E_F. This symmetry makes the Fermi level a natural reference point in calculating electron and hole concentrations. In the case of intrinsic semiconductors the 'tails' in the probability distribution extend into the conduction and valence bands respectively as shown in Fig. 2.14(b); the density of states function and carrier densities are shown in Figs. 2.14(a) and (c). As the electron and hole concentrations are equal we expect the Fermi level to lie near to the middle of the energy gap.

In n-type material there are many more electrons in the conduction band than holes in the valence band and we expect the Fermi level to lie near the donor levels. Similarly in p-type material the Fermi level lies near the acceptor levels. The Fermi level, density of states and carrier densities in n-type material are shown in Fig. 2.15. In many cases,

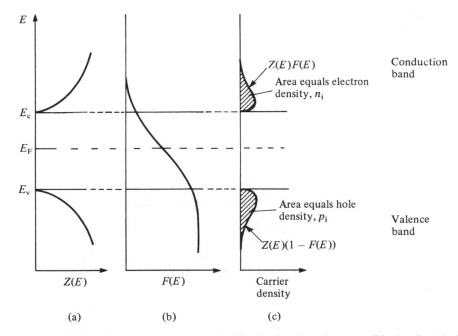

Fig. 2.14 Graphical representation of (a) the density of states, (b) the Fermi–Dirac distribution and (c) carrier densities for an intrinsic semiconductor.

for example when considering junctions between dissimilar materials, the energies of the impurity levels can be ignored if we know where the Fermi level lies.

Returning to the calculation of carrier concentrations we see that this is given by summing the product of the density of states and the occupancy probability over the energy range of interest; that is,

$$n = \int_{\substack{\text{Energy} \\ \text{band}}} F(E)\, Z(E)\, \mathrm{d}E \tag{2.31}$$

Taking firstly the case of a metal at absolute zero, where $F(E) = 1$ and the upper occupied level is E_F we have

$$n = \int_0^{E_F} \frac{4\pi}{h^3}\, (2m_e^*)^{3/2}\, E^{1/2}\, \mathrm{d}E$$

and hence

$$n = \frac{8\pi}{3h^3}\, (2m_e^*\, E_F)^{3/2} \tag{2.32}$$

In fact we often calculate n as described earlier and use eq. (2.32) to calculate E_F; typical values are 7.0 eV for copper and 11.2 eV for aluminum which are in good agreement with experimental results.

For the case of the electron concentration in the conduction band of a semiconductor we have

Fig. 2.15 Graphical representation of (a) the density of states, (b) the Fermi–Dirac distribution and (c) carrier densities for an *n*-type semi-conductor.

$$n = \frac{4\pi}{h^3}\,(2m_e^*)^{3/2} \int_{E_c}^{E_T} \frac{(E - E_c)^{1/2}}{\exp\left(\dfrac{E - E_F}{kT}\right) + 1}\, \mathrm{d}E$$

where E_T is the top of the band. Making the substitutions and approximations suggested in Problem 2.14 we have

$$\left.\begin{array}{c} n = N_c \exp\left[-\left(\dfrac{E_c - E_F}{kT}\right)\right] \\[2ex] \text{where} \\[1ex] N_c = 2\left(\dfrac{2\pi m_e^* \, kT}{h^2}\right)^{3/2} \end{array}\right\} \qquad (2.33)$$

N_c is called the *effective density of states* in the conduction band; it is constant at a constant temperature.

Example 2.3 Effective density of states in the conduction band

We may calculate the effective density of states for germanium at 300 K given that the appropriate effective mass is $0.55m$.

From eq. (2.33) we have $N_c = 1.03 \times 10^{25}$ m^{-3}.

By a similar argument, the concentration of holes in the valence band is given by

$$p = \int_{\substack{\text{Valence} \\ \text{band}}} (1 - F(E))\, Z(E)\, \mathrm{d}E$$

or

$$p = N_\mathrm{v} \exp\left(E_\mathrm{F} - E_\mathrm{v}/kT\right)$$

where

$$N_\mathrm{v} = 2\left(\frac{2\pi m_\mathrm{h}^* \, kT}{h^2}\right)^{3/2}$$

$$(2.34)$$

The carrier concentrations predicted by eqs. (2.33) and (2.34) are valid for *any* type of semiconductor provided the appropriate Fermi level is used. We may use these equations to derive an expression for E_F in intrinsic semiconductors where $n = p = n_\mathrm{i}$. It is left to the reader to show that

$$E_{\mathrm{Fi}} = \tfrac{1}{2}E_\mathrm{g} + \tfrac{3}{4}kT \ln\left(\frac{m_\mathrm{h}^*}{m_\mathrm{e}^*}\right) \qquad (2.35)$$

The second term on the right-hand side of eq. (2.35) is usually very small and, as we supposed above, $E_{\mathrm{Fi}} \simeq \tfrac{1}{2}E_\mathrm{g}$.

We also note from eqs. (2.33) and (2.34) that the product of n and p is constant for a given piece of semiconductor at constant temperature. That is,

$$np = n_\mathrm{i} p_\mathrm{i} = n_\mathrm{i}^2 = N_\mathrm{c} N_\mathrm{v} \exp\left(-E_\mathrm{g}/kT\right) \qquad (2.36)$$

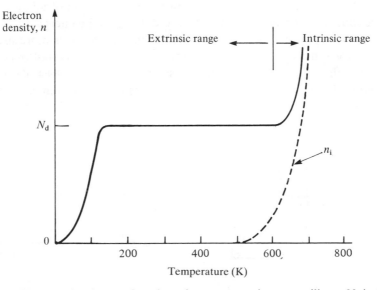

Fig. 2.16 Electron density as a function of temperature in *n*-type silicon; N_d is the density of donor impurity atoms.

Equation (2.36) is a very important relationship, for once we know n or p then the other can be determined from n_i^2; it is often referred to as the 'law of mass action' for semiconductors.

Equation (2.36) shows that the temperature variation of n_i is essentially exponential, that is $n_i \propto \exp(-E_g/2kT)$. Similarly, for an n-type semiconductor the excitation of electrons from the donor levels to the conduction band is governed by the exponential function $\exp(-E_D/kT)$. As E_D is rather small, we find that at temperatures above about 100 K nearly all the donors are ionized, while at such temperatures the number of electron–hole pairs formed by intrinsic excitation is negligible. This situation continues until the temperature rises somewhat above room temperature in the case of silicon. Thus the variation of electron concentration with temperature has the form shown in Fig. 2.16, the exact details of this variation depending on the doping level.

2.6 Work function

We saw in section 1.3 that when light of an appropriate frequency falls onto metals (and indeed onto other solids) electrons may be emitted. Similarly, when solids are heated thermionic emission of electrons may occur. The minimum energy required to enable an electron to escape from the surface of the solid, in either case, is the work function ϕ.

A simple model of this situation in the case of a metal is given in Fig. 2.17(a), which represents an electron trapped in a well of *finite* depth (that is, one from which it may escape). We see that ϕ is the energy difference between the Fermi level (which corresponds to the highest occupied energy level) and the vacuum level. We consider that when the electron reaches an energy equivalent to the vacuum level it has escaped from the metal. We might compare this situation to the ionization of an atom.

In semiconductors the work function is also defined as the energy difference between the Fermi and vacuum levels (Fig. 2.17(b)) but in this case it is more usual to use the *electron affinity* χ, defined as the energy difference between the bottom of the conduction band and the vacuum level. In some respects χ has more physical significance than ϕ as, in general, there are no electrons at the Fermi level in semiconductors.

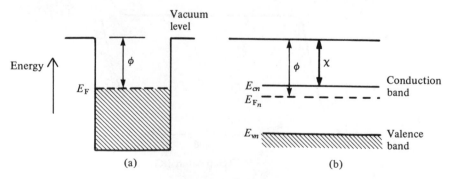

Fig. 2.17 Diagram showing the work function of (a) a metal and (b) an n-type semiconductor.

2.7 Excess carriers in semiconductors

Most semiconductor devices operate by the creation of charge carrier concentrations in excess of the thermal equilibrium values given by eqs. (2.33) and (2.34). These excess concentrations may be created by *optical excitation* or *carrier injection* via a suitable contact. In optical excitation electron–hole pairs are generated when the energy of the incident photons is sufficient to excite an electron from the valence to the conduction band, that is, $hv \geqslant E_g$. Since additional carriers are created the electrical conductivity of the material will increase. This is the basis of the photoconductive devices described in section 7.2.7.

If a carrier excess is created in this way and the exciting radiation is then suddenly cut off, the carrier concentrations will return gradually to their thermal equilibrium values due to recombination of the excess carriers. During this time, the recombination rate, being proportional to the increased carrier concentrations, will be higher than the thermal generation rate, assuming that the temperature remains constant.

If we suppose that $\Delta n(t)$ and $\Delta p(t)$ are the excess carrier concentrations at any given instant $(\Delta n(t) = \Delta p(t))$, the net rate at which the excess electron–hole pairs disappear is, from eq. (2.27),

$$- \frac{dn(t)}{dt} = - \frac{d\Delta n(t)}{dt} = B[n + \Delta n(t)][p + \Delta p(t)] - Bn_i^2 \qquad (2.37)$$

where the total concentration of electrons at time t is $n(t) = n + \Delta n(t)$, n being the equilibrium concentration (similarly for holes). As $np = n_i^2$, and $\Delta n(t) = \Delta p(t)$ we can rewrite eq. (2.37) as

$$- \frac{d\Delta n(t)}{dt} = B[(n + p)\,\Delta n(t) + (\Delta n(t))^2]$$

If the excess concentrations are not too high (that is, we have low level injection) we can ignore the term $(\Delta n(t))^2$. Furthermore, if the material is extrinsic we can usually neglect the minority carrier concentration. Thus, if the material is *p*-type we have

$$- \frac{d\Delta n(t)}{dt} = Bp\Delta n(t)$$

The solution to this equation is

$$\Delta n(t) = \Delta n(0) \exp(-Bpt) = \Delta n(0) \exp(-t/\tau_e) \qquad (2.38)$$

where $\Delta n(0)$ is the excess carrier concentration at $t = 0$ when the exciting source is switched off and the decay constant, $\tau_e = 1/Bp$, is called the *minority carrier recombination lifetime* (minority, because the calculation is made in terms of the minority carriers). Similarly, the minority carrier lifetime of holes in *n*-type material is $\tau_h = (Bn)^{-1}$. Physically τ represents the average time an excess carrier remains free before recombining. The carrier lifetime for indirect recombination is more complicated than in the above case since it is necessary to include the time required to capture each type of carrier.

2.7.1 Diffusion of carriers

Let us suppose that a concentration gradient of excess minority carriers is created in a rod of *n*-type semiconductor by injecting holes into one end of the rod via a suitable contact, as shown in Fig. 2.18. Owing to the random thermal motion of the holes at any given location it is probable that more holes will move to the right than to the left in a given time interval, and there will be a net movement or *diffusion* of holes along the rod down the concentration gradient. The net rate of flow of holes across unit area due to diffusion is found to be proportional to the gradient, that is

$$\text{hole flux} = -D_h \frac{dp(x)}{dx} \tag{2.39a}$$

Similarly for electrons we have

$$\text{electron flux} = -D_e \frac{dn(x)}{dx} \tag{2.39b}$$

D_h and D_e are the hole and electron *diffusion coefficients*. These parameters are related to the mobilities in that they are a measure of the ease of carrier motion through the crystal lattice. In fact, the Einstein relationships give the diffusion coefficients in terms of the mobilities as (ref. 2.10)

$$D_{e,h} = \mu_{e,h} \frac{kT}{e} \tag{2.40}$$

As the holes diffuse along the rod they will eventually recombine. We now show that before recombining, the holes travel a characteristic distance called the *diffusion length* L_h. Let us consider a length Δx of the rod situated a distance x from the injecting contact (as shown in Fig. 2.18); then if the cross-sectional area of the rod is A the flow of holes into the left-hand face of the element is

$$-D_h \left(\frac{d}{dx} \Delta p(x) \right) A$$

Fig. 2.18 Minority carrier injection and diffusion.

Similarly, the flow out of the right-hand face is

$$-D_h \left[\left(\frac{d}{dx} \Delta p(x) \right)_x + \frac{d}{dx} \left(\frac{d}{dx} \Delta p(x) \right)_x \Delta x \right] A$$

The difference of these two terms is the net rate at which the element gains holes, which in the steady state equals the rate of recombination of excess holes; therefore,

$$-D_h \frac{d^2(\Delta p(x))}{dx^2} \Delta x A = - \frac{\Delta p(x) \Delta x A}{\tau_h}$$

or

$$\frac{d^2 \Delta p(x)}{dx^2} - \frac{\Delta p(x)}{\tau_h D_h} = 0 \tag{2.41}$$

The solution to this equation, which is often called the steady-state diffusion equation, is

$$\Delta p(x) = B_1 \exp (x/L_h) + B_2 \exp (-x/L_h)$$

where $L_h = \sqrt{(D_h \tau_h)}$ is the hole *diffusion length*. The constants B_1 and B_2 are determined by the boundary conditions, which in the case we are describing we can express as:

$\Delta p(x) \to 0$ as $x \to \infty$, therefore B_1 must be zero; and $\Delta p(x) = \Delta p(0)$ at $x = 0$, so that $B_2 = \Delta p(0)$.

Hence we have

$$\Delta p(x) = \Delta p(0) \exp (-x/L_h) \tag{2.42}$$

Thus we see that the excess minority carrier concentration decreases exponentially with distance; we shall return to this relationship in section 2.8.2.

2.7.2 Diffusion and drift of carriers

The diffusion of charge carriers will obviously give rise to a current flow. From eqs. (2.39) we can write the electron and hole diffusion current densities as:

$$J_e(\text{diff}) = eD_e \frac{dn}{dx} \tag{2.43a}$$

$$J_h(\text{diff}) = -eD_h \frac{dp}{dx} \tag{2.43b}$$

If an electric field is present in addition to a concentration gradient, we can use eqs. (2.24) and (2.43) to write the current densities for electrons and holes as

$$J_e = e\mu_e n \mathscr{E} + eD_e \frac{dn}{dx}$$

$$J_h = e\mu_h p \mathscr{E} - eD_h \frac{dp}{dx}$$

and the *total* current density is the sum of these contributions, that is

$$J = J_e + J_h$$

2.8 Junctions

Most of the most useful electronic devices contain junctions between dissimilar materials, which may be metal–metal, metal–semiconductor or semiconductor–semiconductor combinations. We shall concentrate our attention on *p–n homojunctions*, in which a junction is formed between *p* and *n* variants of the *same* semiconductor. Such junctions, in addition to their rectifying properties, form the basis of photodiodes, light emitting diodes and photovoltaic devices, which will be discussed in later chapters. Many other devices, for example bipolar transistors and thyristors, contain two or more such junctions.

2.8.1 The p–n junction in Equilibrium

A *p–n* junction may be fabricated as a single crystal of semiconductor by a number of different techniques (ref. 2.11). Indeed the exact behavior of a junction depends to a large extent on the fabrication process used, which in turn determines the distances over which the change from *p*- to *n*-type nature occurs. For mathematical convenience we shall assume that the junction is *abrupt*, that is there is a step change in impurity type as shown in Fig. 2.19, which also shows the corresponding carrier concentrations p_p, n_p, and n_n, p_n on the *p* and *n* sides of the junction respectively. These apply only at relatively large distances from the junction; close to the junction they are modified as we shall see below.

While the assumption of a step junction may be used to establish many of the characteristics of junctions, it is not universally applicable. For example, in many cases the junction approximates to a *linearly graded* one in which there is a gradual change in doping type. One or two of the ways in which such junctions differ from abrupt ones will be pointed out in due course.

Although not the case in practice, let us assume that the junction is formed by bringing initially isolated pieces of *n*-type and *p*-type materials into intimate contact. Then, since there are many more holes in the *p*-type than in the *n*-type material, holes will diffuse from the *p* to the *n* region. The holes diffusing out of the *p*-type side leave behind 'uncovered' or ionized acceptors, thereby building up a negative space charge layer in the *p*-type side close to the junction. Similarly electrons, diffusing into the *p*-type side, leave behind a positive space charge layer of ionized donors as shown in Fig. 2.20(c). This double *space charge* layer causes an electric field to be set up across a narrow region on either side of the junction, directed from the *n*- to the *p*-type region as shown.

The direction of the junction electric field is such as to inhibit further diffusion of the majority carriers, though such diffusion is not prevented altogether. This must be so

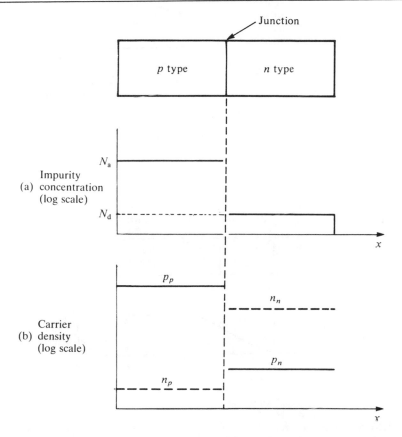

Fig. 2.19 Abrupt $p-n$ junction showing (a) the impurity concentrations and (b) the carrier densities on the two sides of the junction.

since the electric field will sweep minority carriers across the junction so that there is a drift current of electrons from the p- to the n-type side, and of holes from the n- to the p-type side, which is in the opposite direction to the diffusion current. The junction field thus builds up until these two current flows are equal, at which stage the Fermi level is constant across the junction as shown in Fig. 2.20(b) indicating that equilibrium has been reached within the crystal as a whole. Thus, as there is no *net* current flow in equilibrium,

$$J_h(\text{drift}) + J_h(\text{diff}) = 0$$

and

$$J_e(\text{drift}) + J_e(\text{diff}) = 0$$

The induced electric field establishes a *contact* or *diffusion* potential V_0 between the two regions and the energy bands of the p-type side are displaced relative to those of the n-type as shown in Fig. 2.20(b). The magnitude of the contact potential depends on the temperature and the doping levels as we shall see below. The contact potential is

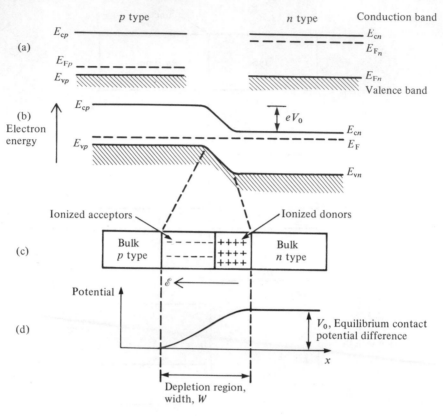

Fig. 2.20 Schematic representation of the formation of a p–n junction: (a) initially separated p-type and n-type materials, (b) the energy band distribution after the junction is formed, (c) the space charge layers of ionized impurity atoms within the depletion region W and (d) the potential distribution at the junction.

established across the space charge region, which is also referred to as the *transition* or *depletion* region, so-called because this region has been almost depleted of its majority carriers. As a consequence it is very resistive relative to the other (so-called bulk) regions of the device.

An expression relating the contact potential to the doping levels can be obtained, for example, from eq. (2.33). From this equation and adopting the notation used in Figs. 2.19 and 2.20 we can write the electron concentration in the conduction band of the p-type side as:

$$n_p = N_c \exp\left[-\left(\frac{E_{cp} - E_{Fp}}{kT}\right)\right]$$

Similarly the electron concentration in the n-type side is

$$n_n = N_c \exp\left[-\left(\frac{E_{cn} - E_{Fn}}{kT}\right)\right]$$

As we mentioned above, the Fermi level is constant everywhere in equilibrium so that $E_{Fp} = E_{Fn} = E_F$ say. Hence, eliminating N_c we have

$$E_{cp} - E_{cn} = kT \ln \left(\frac{n_n}{n_p}\right) = eV_0$$

$$\therefore V_0 = \frac{kT}{e} \ln \left(\frac{n_n}{n_p}\right) \tag{2.44}$$

At temperatures in the range $100 \text{ K} \leqslant T \leqslant 400 \text{ K}$ the majority carrier concentrations are equal to the doping levels, that is $n_n = N_d$ and $p_p = N_a$ and remembering $np = n_i^2$ we can write eq. (2.44) as

$$V_0 = \frac{kT}{e} \ln \left(\frac{N_a N_d}{n_i^2}\right) \tag{2.45}$$

Example 2.4 Equilibrium $p-n$ junction contact potential difference

The contact potential difference V_0 in a given germanium diode may be calculated from the following data: donor impurity level $N_d = 10^{22} \text{ m}^{-3}$; acceptor impurity level $N_a = 10^{24} \text{ m}^{-3}$; and intrinsic electron concentration $n_i = 2.4 \times 10^{19} \text{ m}^{-3}$.
    Then assuming room temperature ($T \approx 290 \text{ K}$) we have, from eq. (2.45), that

$$V_0 = 0.025 \ln \left[\frac{10^{22} \times 10^{24}}{(2.4 \times 10^{19})^2}\right] = 0.42 \text{ V}$$

Equation (2.44) gives us a very useful relationship between the carrier concentrations on the two sides of the junction, that is

$$n_p = n_n \exp \left(\frac{-eV_0}{kT}\right) \tag{2.46a}$$

and similarly

$$p_n = p_p \exp \left(\frac{-eV_0}{kT}\right) \tag{2.46b}$$

2.8.2 Current flow in a forward biased p–n junction

If the equilibrium situation is disturbed by connecting a voltage source externally across the junction there will be a net current flow. The junction is said to be *forward biased* if the p region is connected to the positive terminal of the voltage source as shown in Fig. 2.21(a). As we mentioned in the last section, the depletion region is very resistive in comparison to the bulk regions so that the external voltage V is dropped almost entirely across the depletion region. This has the effect of lowering the height of the

Fig. 2.21 Forward bias voltage V applied to a $p-n$ junction (a) and the resulting energy band structure (b).

potential barrier to $(V_0 - V)$ as shown in Fig. 2.21(b). Consequently, majority carriers are able to surmount the potential barrier much more easily than in the equilibrium case so that the diffusion current becomes much larger than the drift current. There is now a net current from the p to the n region in the conventional forward sense and carriers flow in from the external circuit to restore equilibrium in the bulk regions. We note that with the application of an external potential the Fermi levels are no longer aligned across the junction.

The reduction in height of the potential barrier leads to majority carriers being *injected* across the junction. On being so injected, these carriers immediately become minority carriers, and the minority carrier concentrations near to the junction rise to the new values n'_p and p'_n. This establishes excess minority carrier concentration gradients, as shown in Fig. 2.22, so that the injected carriers diffuse away from the junction. This situation is precisely the same as that described in section 2.7.1 and thus, considering the n region, the injected holes diffuse away from the junction recombining as they do so. The electrons lost in this recombination are replaced by the external voltage source so that a current flows in the external circuit. A similar argument applies

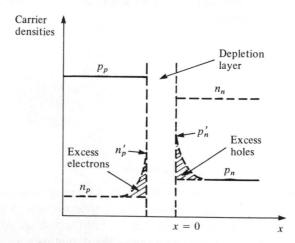

Fig. 2.22 Carrier densities in the bulk regions of a forward biased $p-n$ junction diode. Owing to carrier injection, the minority carrier densities close to the depletion layer are *greater* than the equilibrium values.

to the p region, with the roles of electrons and holes reversed. It should be noted that the majority carrier concentrations are not noticeably changed as a consequence of the injection (Fig. 2.22) unless the bias voltage is almost equal to V_0, resulting in a very large current flow.

The drift current is relatively insensitive to the height of the potential barrier since all the minority carriers generated within about a diffusion length of the edge of the depletion region may diffuse to the depletion region and be swept across it, whatever the size of the electric field there.

By the same argument that was used to give eqs. (2.46), we see that with forward bias the minority carrier concentrations in the bulk regions adjacent to the depletion layer become

$$n'_p = n_n \exp\left(\frac{-e(V_0 - V)}{kT}\right) \tag{2.47a}$$

and

$$p'_n = p_p \exp\left(\frac{-e(V_0 - V)}{kT}\right) \tag{2.47b}$$

Then taking eqs. (2.46b) and (2.47b), for example, we have

$$p'_n = p_n \exp\left(\frac{eV}{kT}\right) \tag{2.48}$$

As we noted above, the excess minority carrier concentration will decrease owing to recombination in accordance with eq. (2.42) so that we may write the excess hole concentration in the n region at a distance x from the edge of the depletion layer as

$$\Delta p(x) = \Delta p(0) \exp\left(-x/L_h\right)$$

where now

$$\Delta p(x) = p'_n(x) - p_n$$

and $\Delta p(0)$ is the value of $\Delta p(x)$ at $x = 0$, that is $(p'_n - p_n)$. Using eq. (2.48) we can therefore write

$$\Delta p(0) = p_n \left(\exp\left(eV/kT\right) - 1\right) \tag{2.49}$$

We have assumed a one-dimensional carrier flow in the x direction, which is an acceptable approximation even though in practice carrier flow occurs in three dimensions.

Now, we have argued that the electric fields in the bulk regions are very small and therefore, adjacent to the depletion layer in the n region, and in particular at $x = 0$, the total current density will be due to diffusion only. Thus the current density due to hole motion is given by eq. (2.43b). Hence differentiating eq. (2.42) and substituting into eq. (2.43b) gives

$$J_h = \frac{eD_h}{L_h} \Delta p(0) \exp\left(-\frac{x}{L_h}\right)$$

At $x = 0$, we can write this, using eq. (2.49), as

$$J_h = \frac{eD_h}{L_h} p_n \left(\exp \left(\frac{eV}{kT} \right) - 1 \right) \tag{2.50a}$$

There is a similar contribution due to electron flow, namely

$$J_e = \frac{eD_e}{L_e} n_p \left(\exp \left(\frac{eV}{kT} \right) - 1 \right) \tag{2.50b}$$

The total current density is therefore given by

$$J = J_0 \left(\exp \left(\frac{eV}{kT} \right) - 1 \right) \tag{2.51}$$

where

$$J_0 = e \left(\frac{D_h}{L_h} p_n + \frac{D_e}{L_e} n_p \right) \tag{2.51'}$$

J_0 is called the saturation current density; it represents the current flow when a few volts of reverse bias is applied to the junction (see section 2.8.3).

Example 2.5 Saturation current density

We may calculate the saturation current density in an abrupt silicon junction given the following data: $N_d = 10^{21} \text{ m}^{-3}$; $N_a = 10^{22} \text{ m}^{-3}$; $D_e = 3.4 \times 10^{-3} \text{ m}^2 \text{ s}^{-1}$; $D_h = 1.2 \times 10^{-3} \text{ m}^2 \text{ s}^{-1}$; $L_e = 7.1 \times 10^{-4} \text{ m}$; $L_h = 3.5 \times 10^{-4}$; and $n_i = 1.6 \times 10^{16} \text{ m}^{-3}$.

Assuming that all of the impurities are ionized, we have $n_n = N_d = 10^{21} \text{ m}^{-3}$ and therefore

$$p_n = \frac{2.56 \times 10^{32}}{10^{21}} = 2.56 \times 10^{11} \text{ m}^{-3}$$

Similarly $n_p = 2.56 \times 10^{10} \text{ m}^{-3}$. Therefore from eq. (2.51') $J_0 = 1.6 \times 10^{-7} \text{ A m}^{-2}$.

A typical discrete diode may have a junction area of about 10^{-6} m^2 and hence the reverse bias saturation current would be $i_0 = 1.6 \times 10^{-13} \text{ A}$.

2.8.3 Current flow in a reverse biased p–n junction

In this case the external bias is applied so that the p region is connected to the negative terminal of the voltage source as shown in Fig. 2.23(a). This has the effect of increasing the height of the potential barrier to $V_0 + V$ (Fig. 2.23(b)), thereby reducing the diffusion current to negligible proportions. The net current flow is therefore the drift current which is directed in the conventional reverse sense, that is from the n to the p

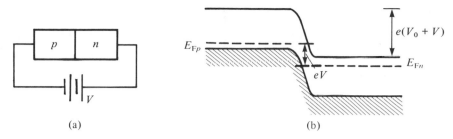

Fig. 2.23 Reverse bias voltage V applied to a $p–n$ junction (a) and the resulting energy band structure (b).

region. This results in carrier *extraction* rather than injection because the minority carriers generated near the junction diffuse to it and are swept across the depletion region. The nearer the carriers are generated to the junction the greater is the probability of this occurring so that a concentration gradient is formed towards the junction, as illustrated in Fig. 2.24. The drift of carriers across the junction is therefore 'fed' by diffusion so that we may use precisely the same arguments to derive the current–voltage relationship as we used in the previous section. Equations (2.51) also apply in this case, the only difference being, of course, that the sign of V is changed.

The current–voltage characteristic of an ideal $p–n$ junction is therefore as shown in Fig. 2.25; in forward bias the current increases exponentially with voltage, while in reverse bias the current saturates at J_0 times the junction cross-sectional area.

2.8.4 Junction geometry and depletion layer capacitance

The two space charge layers at the junction vary in width and therefore in the amount of charge they contain as the bias voltage changes, thereby giving rise to an effective

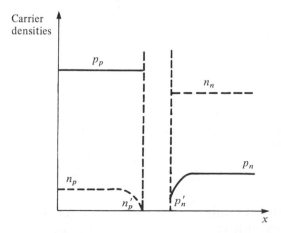

Fig. 2.24 Carrier densities in the bulk regions of a reverse biased $p–n$ junction diode. Owing to carrier extraction the minority carrier densities close to the depletion layer are *less* than the equilibrium values.

Fig. 2.25 Current–voltage characteristics of a p–n junction diode. The reverse saturation current i_0 is equal to J_0 multiplied by the junction cross-sectional area.

depletion layer (or junction) capacitance $C_j = dQ_j/dV$. This capacitance together with a charge storage capacitance (see below) is one of the factors which limits the high frequency operation of junction devices. On the other hand, the capacitance can be controlled by the bias voltage, and this effect is exploited in voltage-dependent capacitors such as the varactor diode.

The charge density within the space charge layers results simply from the charge on the ionized impurities. Thus the charge density within the p region is

$$\varrho_p = -N_a e \tag{2.52a}$$

and similarly

$$\varrho_n = N_d e \tag{2.52b}$$

Referring to Fig. 2.26, we denote the widths of the p and n space charge regions by x_p and x_n, with the origin of x at the actual junction. Using Poisson's equation we may calculate the electric field distribution within the depletion layer. For the p side, from eq. (2.52a), we have

$$\frac{d\mathscr{E}}{dx} = -\frac{eN_a}{\varepsilon_0\varepsilon_r}, \quad -x_p < x < 0 \tag{2.53}$$

where ε_r is the relative permittivity of the semiconductor. Integrating eq. (2.53) with the appropriate limits, that is,

$$\int_0^{\mathscr{E}_0} d\mathscr{E} = -\frac{eN_a}{\varepsilon_0\varepsilon_r} \int_{-x_p}^0 dx$$

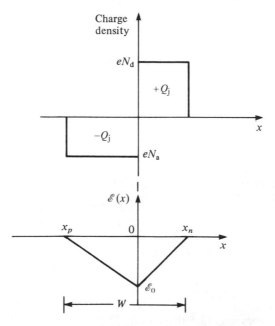

Fig. 2.26 Space charge density and variation of electric field $\mathscr{E}$ within the depletion region of a $p-n$ junction. Note that as $N_d > N_a$ in this example, then $x_n < x_p$, but the space charge on each side of the junction has the same magnitude.

gives $\mathscr{E}_0$, the maximum value of the junction electric field. We find that

$$\mathscr{E}_0 = \frac{-eN_a x_p}{\varepsilon_0 \varepsilon_r} \tag{2.54a}$$

Similarly for the n side we have

$$\mathscr{E}_0 = \frac{-eN_d x_n}{\varepsilon_0 \varepsilon_r} \tag{2.54b}$$

From eqs. (2.54a) and (2.54b) we have

$$N_a x_p = N_d x_n \tag{2.55}$$

from which we see that the depletion layer extends farthest into the least heavily doped side of the junction as we have implied in, for example, Figs. 2.20 and 2.26. This is a result of some significance in explaining the operation of many devices.

We can now relate the electric field to the junction contact potential V_0. Since in general

$$\mathscr{E}(x) = -\frac{dV(x)}{dx}$$

we may write

$$-\int_0^{V_0} dV(x) = -V_0 = \int_{-x_p}^{x_n} \mathscr{E}(x)dx$$

The right-hand integral is the area of the $\mathscr{E}(x)$ versus x triangle shown in Fig. 2.26 so that

$$-V_0 = \tfrac{1}{2} W \mathscr{E}_0 \tag{2.56}$$

where the width of the depletion layer W is

$$W = |x_p| + |x_n| \tag{2.57}$$

Substituting for $\mathscr{E}_0$ from eq. (2.54b) into eq. (2.56) gives

$$V_0 = \frac{e W N_d x_n}{2\varepsilon_0 \varepsilon_r} \tag{2.58}$$

while combining eqs. (2.55) and (2.57) gives

$$x_n = \frac{W N_a}{N_d + N_a} \tag{2.59}$$

Then substituting x_n from eq. (2.59) into eq. (2.58) yields

$$V_0 = \frac{e}{2\varepsilon_0 \varepsilon_r} \left(\frac{N_a N_d}{N_d + N_a} \right) W^2$$

If a bias voltage is applied, the above argument still holds but we must replace V_0 by $(V_0 - V)$, where V is positive for forward bias and negative for reverse bias. Therefore, finally we have

$$W = \left[\frac{2\varepsilon_0 \varepsilon_r}{e} (V_0 - V)\left(\frac{1}{N_a} + \frac{1}{N_d} \right) \right]^{\frac{1}{2}} \tag{2.60}$$

The capacitance associated with the depletion layer can be obtained as follows. The stored charge Q_j on either side of the junction is given by

$$|Q_j| = AeN_d x_n = AeN_a x_p \tag{2.61}$$

where A is the junction area. Hence using eqs. (2.59) and (2.60) we may write

$$|Q_j| = \frac{AeN_d N_a W}{N_d + N_a} = A \left[2\varepsilon_0 \varepsilon_r e (V_0 - V)\left(\frac{N_d N_a}{N_d + N_a} \right) \right]^{\frac{1}{2}} \tag{2.62}$$

If the bias voltage V changes then Q_j changes so that the junction capacitance C_j is given by differentiating eq. (2.62) with respect to V. Hence

$$C_j = \frac{dQ_j}{dV} = \frac{A}{2} \left[\frac{2e\varepsilon_0 \varepsilon_r}{(V_0 - V)} \left(\frac{N_d N_a}{N_d + N_a} \right) \right]^{\frac{1}{2}} \tag{2.63}$$

It is left as an exercise for the reader to show that $C_j = A\varepsilon_0\varepsilon_r/W$ farads, indicating that the abrupt junction behaves rather like a parallel plate capacitor of plate separation W and area A containing a medium of relative permittivity ε_r. We note, for reasonably large values of reverse bias V, that $C \propto V^{-\frac{1}{2}}$. In contrast, for the case of a graded junction it is found that $C \propto V^{-\frac{1}{3}}$.

In forward bias the junction capacitance C_j is swamped by another capacitive effect which gives rise to the charge storage or diffusion capacitance C_d. In forward bias, many carriers are injected across the junction and it is the time taken for this injected carrier density to adjust to changes in the forward bias voltage which gives rise to the charge storage capacitance. The principal mechanism whereby the injected carrier density adjusts is recombination, and consequently the expression for C_d includes the carrier lifetime τ, that is, $C_d = eI\tau/kT$ (see ref. 2.12).

Example 2.6 Junction depletion layer capacitance

We may calculate the capacitance of a silicon p^+-n junction at a reverse bias of 4 V, given that $N_d = 4 \times 10^{21}$ m^{-3}, $V_0 = 0.8$ V, $\varepsilon_r = 11.8$ and that the junction area is 4×10^{-7} m^2.

In a p^+-n junction, $N_a \gg N_d$ and hence eq. (2.63) reduces to

$$C_j = \frac{A}{2}\left[\frac{2e\varepsilon_0\varepsilon_r N_d}{V_0 - V}\right]^{\frac{1}{2}}$$

with reverse bias $V = -4$ V we find that

$$C_j = 33.4 \text{ p}F$$

2.8.5 Deviations from the simple theory

Although the theory described above and the diode characteristic shown in Fig. 2.25 is in reasonable agreement with what is observed in practice, there are several points of difference. For our purposes one of the most important deviations is that, at sufficiently large values of reverse bias, *breakdown* occurs. That is, there is a sudden and rapid increase in reverse current at a particular value of reverse bias voltage (see, for example, Fig. 2.27(b)).

Reverse breakdown occurs by two mechanisms. The first, called the *Zener effect*, is due to quantum mechanical tunneling. This takes place most readily in heavily doped junctions, which from eq. (2.60) leads to narrow depletion layers and therefore high junction fields. In effect, as we can see from Fig. 2.27(a) the energy bands on the two sides of the junction become 'crossed' so that filled states in the valence band of the p side are aligned with empty states in the conduction band of the n side. Electrons, therefore, tunnel from the p to the n side vastly increasing the reverse current.

The second mechanism, *avalanche breakdown*, occurs in lightly doped junctions with wide depletion layers. This mechanism involves impact ionization of the host atoms by energetic carriers. If carriers crossing the depletion layer acquire sufficient energy from the electric field between collisions, they may ionize lattice atoms on colliding with

p type   n type

Tunneling
electrons

(a)   (b)

Fig. 2.27 Schematic representation of electron tunneling in Zener breakdown of a heavily doped $p-n$ junction under reverse bias (a) and the resulting effect on the $i-V$ characteristic (b).

them. The electrons and holes so produced may, in turn, cause further ionizing collisions and so on to generate an avalanche of carriers. Neither breakdown mechanism is in itself destructive to the junction. If, however, the reverse current is allowed to become too large then Joule heating may cause damage to the device.

Other important deviations arise from our having ignored such factors as carrier generation and recombination within the depletion layer. Carrier recombination leads to an increase in current in the forward direction as the carriers which recombine must be replaced from the external circuit. In silicon and gallium arsenide diodes, this current mechanism may be more important than current flow due to carrier injection, especially at low currents. This and other factors lead in some cases to the $p-n$ junction having a current–voltage relationship of the form $J = J_0[\exp{(eV/\beta kT)} - 1]$, where β varies between 1 and 2 depending on the semiconductor and the temperature. Because β determines the departure from the ideal diode characteristic, it is often called the *ideality factor*.

Carrier generation in the depletion layer gives rise to a larger value of reverse bias current than the simple theory predicts. Optical generation of carriers within the depletion layer may give rise to an increase in reverse current or initiate avalanche breakdown if the reverse bias is sufficiently great. These phenomena form the bases of the photodiodes discussed in Chapter 7.

2.8.6 Other junctions

(1) Heterojunctions. In the discussion so far, the basic material has been the same on both sides of the junction. However, in some devices (for example semiconducting diode lasers), *heterojunctions* are used. In these, the material on one side of the junction differs from that on the other side. In diode lasers, some junctions are formed, for example, between GaAs and GaAlAs (see Chapter 5).

(2) Metal–semiconductor junctions. The nature of metal-to-semiconductor junctions depends primarily on the conductivity type of the semiconductor and the relative work functions of the semiconductor and the metal. In the case of an *n*-type semiconductor contacted to a metal of greater work function, simple theory predicts that the junction should be rectifying. Such rectifying junctions (called Schottky barrier diodes) find application in high frequency circuits because of their small junction capacitances. On the other hand, the prediction of simple theory that a junction between an *n*-type semiconductor and a metal of lower work function should be an ohmic contact, with a linear *i–V* characteristic in both directions, is not always borne out in practice. This is due to the presence of surface states on the semiconductor leading to stationary charges at the metal–semiconductor interface resulting in rectifying contacts (see ref. 2.13).

It is essential, of course, to be able to produce ohmic contacts to connect semiconductor devices to external circuits. A practical method for doing this is by doping the semiconductor very heavily (that is, n^+ or p^+) in the contact region. Thus if a depletion layer exists at the interface, its width will be small enough to allow carriers to tunnel through it. For example, gold containing a small percentage of antimony can be alloyed to *n*-type silicon forming an n^+ layer at the semiconductor surface and a good ohmic contact. Similarly aluminum deposited on *p*-type silicon will form a p^+ layer during a brief heat treatment, again providing a good contact. Aluminum can also be used to contact to *n*-type silicon providing an initial n^+ layer is formed at the surface.

Problems

2.1   Explain the significance of each of the four quantum numbers used to describe atomic energy states. Show how constraints on the values of these numbers leads to a system of 'shells' and 'subshells' of energy states. Draw up a table of all the possible states for $n = 4$.

2.2   Explain what is meant by the Pauli exclusion principle and show, on the basis of your answers to Problem 2.1, what is meant by the electron configurations of the elements. What are the expected electron configurations of diamond ($Z = 6$), Si ($Z = 14$), Ge ($Z = 32$), Sn ($Z = 50$) and Nd ($Z = 60$)? Why might the actual configurations be different from those that you expect?

2.3   An electron in an electron microscope is accelerated by a voltage of 25 kV; what is its de Broglie wavelength?

2.4   Show that the uncertainty principle can be expressed as $\Delta E \Delta t \geq \hbar/2\pi$, where ΔE and Δt are the uncertainties in energy and time respectively. What is the uncertainty in the velocity of an electron confined in a cube of volume of 10^{-30} m^3?

2.5   Calculate the energies of the first three levels for a cubic potential well of side length 10^{-10} m. How much energy is emitted if an electron falls from the third to the first level? If the emitted energy is in the form of a photon, what is the frequency of the associated wave?

2.6   According to the de Broglie relation we may regard a particle as having a wavelength λ. Show that when an electron is moving along a one-dimensional array of ions with separation a, then a standing wave will be set up when the electron wave vector ($2\pi/\lambda$) is equal to $n\pi/a$ where n is an integer.

From the standpoint of quantum theory, these standing waves represent electron wave functions. In fact two different wave functions can be set up corresponding to each n value since each ion may be either a node or an antinode for the standing wave. Show that these two will have different energies. (Hint: consider the coulombic energy resulting from the interaction of the electronic charge distributions with the ions). How does this energy difference manifest itself?

2.7   At 300 K the conductivity of intrinsic silicon is $5 \times 10^{-4}\ \Omega^{-1}\ m^{-1}$. If the electron and hole mobilities are 0.14 and 0.05 $m^2\ V^{-1}\ s^{-1}$ respectively what is the density of electron–hole pairs?

At 300 K the conductivity of intrinsic silicon is $5 \times 10^{-4}\ \Omega^{-1}\ m^{-1}$. If the electron and hole mobilities are 0.14 and 0.05 $m^2\ V^{-1}\ s^{-1}$ respectively what is the density of electron–hole pairs?

If the crystal is doped with $10^{22}\ m^{-3}$ phosphorus atoms, calculate the new conductivity; repeat for the case of boron doping at the same impurity level. Assume all the impurities are ionized in both cases.

2.8   Estimate the ionization energy of donors in GaAs given that $m^* = 0.07m$ and $\varepsilon_r = 10.9$.

2.9   Calculate the probabilities of finding electrons at energy levels of $E_F + 0.05$ eV and $E_F - 0.05$ eV at $T = 0$ K and 300 K.

2.10   The atomic mass number of copper is 63.54 and the density of copper is $8.9 \times 10^3\ kg\ m^{-3}$; confirm that its Fermi level is about 7.0 eV.

2.11   Calculate the intrinsic carrier concentration in GaAs at 290 K given that the electron effective mass is $0.07m$, the hole effective mass is $0.56m$ and that its energy gap is 1.43 eV.

2.12   Calculate the diffusion coefficients of electrons and holes in silicon at 290 K from the data given in Problem 2.7. If it is assumed that the electron and hole lifetimes are both 50 μs, what are their diffusion lengths?

2.13   Derive eq. (2.29) for the density of states–use eq. (2.13) to set up three-dimensional integer space in which each point (i.e. each combination of n_1, n_2, n_3) represents an energy level. Hence show, using eq. (2.11), that the number of energy states with energy less than a reference energy E_R (which equals the appropriate volume of integer space) is $4\pi/3\ (8E_R mL^2/h^2)^{3/2}$. (Remembering that only positive integers n_1, n_2, n_3 have any physical meaning and that there are two spin states per energy level, we arrive at eq. (2.29).) Show also that for two-dimensional integer space, the density of states, is independent of the energy.

2.14   Using eqs. (2.29) and (2.30), derive eq. (2.33) for the electron density in the conduction band of a semiconductor. Make, and if possible justify, the following assumptions: (1) that the energy range of the conduction band can be taken as $E_c \leqslant E \leqslant \infty$; and (2) that we can ignore the term $+1$ in the denominator of eq. (2.30) as $E - E_F \gg kT$. Furthermore, make the substitution $x = (E - E_c)/kT$ and note that

$$\int_0^\infty x^{\frac{1}{2}} \exp(-x)\ dx = \sqrt{\pi}/2$$

2.15   An electron current of 10 mA is injected into a p-type silicon rod of 1 mm^2 cross-sectional area. Assuming that the excess concentration decreases exponentially and that at 5 mm from the contact the excess concentration has fallen to 40% of its value at the contact, calculate the value at the contact. You may assume that $D_e = 3.4 \times 10^{-3}\ m^2\ s^{-1}$.

2.16   Calculate the equilibrium contact potential for a step junction in silicon if $N_a = 10^{24}\ m^{-3}$ and $N_d = 10^{24}\ m^{-3}$; take $\varepsilon_r = 12$, $n_i = 1.5 \times 10^{16}\ m^{-3}$ and $T = 290$ K. Also calculate the width of the both junction in equilibrium and with a reverse bias voltage of 50 V.

2.17   The resistivities of the p and n materials forming a p–n junction are 4.2×10^{-4} and 2.08×10^{-2} Ω m respectively. If the hole and electron mobilities are 0.15 and 0.3 m^2 V^{-1} s^{-1}, the hole and electron carrier lifetimes are 75 and 150 μs respectively and the intrinsic carrier concentration is 2.5×10^{19} m^{-3}, calculate the saturation current at 290 K given that the junction area is 10^{-6} m^2. What fraction of the current is carried by holes?

2.18   Calculate the maximum wavelengths of light which will give rise to photoeffects in intrinsic GaAs ($E_g = 1.43$ eV), CdS ($E_g = 2.4$ eV) and InSb ($E_g = 0.225$ eV).

2.19   Show that the average energy of the free electrons (which is equal to the total energy divided by the electron density) in a solid at 0 K is given by $\langle E \rangle = \frac{3}{5}E_F$. [Hint: the total energy of the free electrons can be calculated by multiplying the population of each energy state by its energy and integrating over all states.]

2.20   Show that the most probable energy of electrons in the conduction band of a semiconductor is $\frac{1}{2}kT$ above the bottom of the band.

References

2.1   (a) A. Bar-Lev, *Semiconductors and Electronic Devices*, Prentice Hall, Hemel Hempstead, 1979, section 6.1.
    (b) M. N. Rudden and J. Wilson, *Elements of Solid State Physics*, Wiley, Chichester, 1980, Chapter 1.
    (c) L. Solymar and D. Walsh, *Lectures on the Electrical Properties of Materials* (2nd edn), Oxford University Press, Oxford, 1979, Chapter 4.
    (d) B. G. Streetman, *Solid State Electronic Devices* (2nd edn), Prentice Hall, Englewood Cliffs, NJ, 1980.
    (e) C. Kittel, *Introduction to Solid State Physics* (5th edn), John Wiley, New York, 1976.
    (f) R. A. Smith, *Semiconductors* (2nd edn), Cambridge University Press, Cambridge, 1979.

2.2   R. W. Ditchburn, *Light* (2nd edn), Blackie, Glasgow, 1962, section 4.19 and Appendix IVB.

2.3   L. Solymar and D. Walsh, *op. cit.*, Chapter 7.

2.4   L. Solymar and D. Walsh, *op. cit.*, pp. 132–6.

2.5   A. Bar-Lev, *op. cit.* section 6.5.

2.6   L. Solymar and D. Walsh, *op. cit.*, pp. 172–3.

2.7   B. G. Streetman, *op. cit.*, pp. 58–61.

2.8   *Ibid*, pp. 54–5.

2.9   L. Solymar and D. Walsh, *op. cit.*, section 6.2.

2.10   A. Bar-Lev, *op. cit.*, section 4.2.

2.11   (a) M. N. Rudden and J. Wilson, *op. cit.*, Chapter 5
    (b) B. G. Streetman, *op. cit.*, pp. 126–36.

2.12   *Ibid*, pp. 185–93.

2.13   S. M. Sze, *Physics of Semiconductor Devices*, Wiley–Interscience, New York, 1969, sections 8.3 and 9.3.

Chapter 3

Modulation of light

The advent of the laser (see Chapter 5) and the increasing use of lasers in a wide variety of applications has led to a demand for devices which can modulate a beam of light. Applications of light modulators include wideband analogue optical communication systems, switching for digital information recording, information storage and processing, pulse shaping, beam deflection and scanning, and frequency stabilization and Q-switching of lasers. Some of these applications are discussed in Chapter 6. In this chapter, we have interpreted the term modulation rather broadly so that we also include sections on scanning and some aspects of laser wavelength tuning. Several of the materials, for example KDP, which are used in conventional modulators exhibit non-linear effects and consequently may be used for harmonic generation and parametric oscillation. These techniques, together with those described in section 6.5.1.6, enhance the available range of laser wavelengths.

A modulator is a device which changes the irradiance (or direction) of the light passing through it. There are several general types of modulator; namely, mechanical choppers and shutters, passive (or dye) modulators, electro-optic, magneto-optic and elasto-optic (acousto-optic) modulators. The first two types will be covered briefly in section 6.4 in which laser Q-switching is discussed. In the remaining types the refractive index and other optical characteristics of a medium are changed by the application of a force field, i.e. electrical, magnetic or mechanical (acoustical). In these cases, apart from the acousto-optical effect where the variation of refractive index creates a 'diffraction grating', the applied force field modifies the polarizing properties of the medium. This in turn may be used to modify the phase or irradiance of a beam of light propagating through the medium. Accordingly this chapter begins with a brief review of optical polarization, birefringence and optical activity in naturally occurring crystals.

3.1 Elliptical polarization

We have already described (section 1.2.1) plane polarized light in which all of the wave trains comparising a beam of light have their electric vectors lying in the same plane. In many cases of interest a beam of light may consist of two plane polarized wave trains with their planes of polarization at right angles to each other and which may also be out of phase.

Let us consider initially the special case where the amplitudes of the two wave trains arc cqual and the phase difference is $\pi/2$. In this case, if the wave trains are propagating in the z direction, we may write the component electric fields as

$$\vec{\mathscr{E}}_x = i\,\mathscr{E}_0 \cos{(kz - \omega t)}$$

and   (3.1)

$$\vec{\mathscr{E}}_y = j\,\mathscr{E}_0 \sin{(kz - \omega t)}$$

where ω is the angular frequency, k ($= 2\pi/\lambda$) is the wavenumber or propagation constant and i and j are unit vectors in the x and y directions respectively. The total electric field is the vector sum of the two components, namely

$$\vec{\mathscr{E}} = \vec{\mathscr{E}}_x + \vec{\mathscr{E}}_y$$

or

$$\vec{\mathscr{E}} = \mathscr{E}_0[i \cos{(kz - \omega t)} + j \sin{(kz - \omega t)}]$$ (3.2)

The resultant expressed by eq. (3.2) can be interpreted as a single wave in which the electric vector at a given point in space is constant in amplitude but rotates with angular frequency ω. Waves such as this are said to be *circularly polarized*. Figure 3.1 shows the electric field vector (a) at a given instant of time and (b) at a given point in space.

The signs of the terms in eq. (3.2) are such that the electric vector at a given point in space has a clockwise rotation when viewed against the direction of propagation. A wave with such an electric vector is said to be right circularly polarized.

If the sign of the second term is changed (this is equivalent to a change of π in the phase of $\mathscr{E}_y$) then the sense of rotation is counterclockwise and the wave is said to be left circularly polarized.

When the amplitudes of the electric vectors of the two waves are not the same but the

Fig. 3.1 Right circularly polarized light: (a) electric vectors at a given instant in time and (b) rotation of the vector at a given position in space. (NB: In this, and the following diagrams, for the sake of clarity only the electric vectors are shown.)

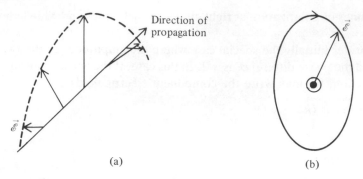

Fig. 3.2 Right elliptically polarized light; electric vectors at (a) a given instant in time and (b) a given position in space.

phase difference remains at $\pi/2$ then the resultant electric vector at any point in space rotates at frequency ω but changes in magnitude. The end of the electric vector describes an ellipse as illustrated in Fig. 3.2.

If the component waves can be represented by $\vec{\mathscr{E}}_x = \mathbf{i}\,\bar{\mathscr{E}}_0 \cos (kz - \omega t)$ and $\vec{\mathscr{E}}_y = \mathbf{j}\mathscr{E}'_0$ $\sin (kz - \omega t)$ with $\mathscr{E}_0 \neq \mathscr{E}'_0$, then the major and minor axes of the ellipse are parallel to the x and y axes. In general the electric vector amplitudes are not equal and also there is an arbitrary phase difference ϕ between the component waves. In this case the end of the resultant electric vector again describes an ellipse but with the major and minor axes inclined at an angle of $\frac{1}{2}\tan^{-1}[(\mathscr{E}_0\,\mathscr{E}'_0 \cos \phi)/(\mathscr{E}'^2_0 - \mathscr{E}^2_0)]$ to the x and y axes (see Problem 3.1). In all of these instances, the resultant wave is said to be elliptically polarized and in fact plane and circular polarization are special cases of elliptical polarization. For the purposes of this book it is not instructive to dwell on a full discussion of elliptically polarized light (ref. 3.1) but rather to describe it simply in terms of its components parallel to and perpendicular to a convenient axis or plane.

3.2 Birefringence

Ordinary glass is isotropic in its properties, but many important crystalline optical materials such as calcite ($CaCO_3$), Quartz (SiO_2) and KDP (potassium dihydrogen phosphate, KH_2PO_4) are anisotropic. This anisotropy is due to the arrangement of the atoms being different in different directions through the crystal. Thus, for example, the electric polarization **P** produced in the crystal by a given electric field $\vec{\mathscr{E}}$ is not a simple scalar multiple of the field but in fact varies in a manner that depends on the direction of the applied field in relation to the crystal lattice. One of the consequences is that the speed of propagation of a light wave in such crystals depends on the direction of propagation and the polarization of the light. In other words, the refractive index of the crystal varies with direction in the crystal. These crystals are said to be *birefringent* or *doubly refracting*; the names refer to the fact that in general there are *two* different directions of propagation that a given incident ray may take depending on the direction of polarization. The rays corresponding to these directions travel with different speeds and have mutually orthogonal planes of polarization. Consequently, when unpolarized light or light of arbitrary polarization (which, as we have seen in Chapter 1, can always be resolved into two orthogonally polarized waves) propagates through a crystal, it can be considered to consist of two independent waves which travel with different velocities.

The theory, which is not very straightforward, shows that in general crystals exhibit three different principal refractive indices and two optic axes (see Appendix 2 and refs. 3.2 for further discussion). The optic axes are *directions* in the crystal along which the velocities of the two orthogonally polarized waves are the same. In many important crystals, for example calcite, two of the principal indices are the same and there is only one optic axis. Such crystals are called *uniaxial* whereas other doubly refracting crystals, for example mica, are *biaxial*. In cubic crystals which are isotropic the principal indices are all the same.

The simplest way of demonstrating birefringence is to allow a narrow beam of unpolarized light to fall normally onto a parallel-sided calcite plate as shown in Fig. 3.3. The beam is found to divide into two parts. One, the so-called *ordinary* or O ray, passes straight through the crystal and is found to obey Snell's law; the other, the so-called *extraordinary* or E ray, diverges as it passes through the crystal and then emerges parallel to its original direction. This is found to be the case unless the direction of incidence of the original beam is parallel or perpendicular to the optic axis. The ordinary and extraordinary rays are found to have orthogonal directions of polarization.

We can explain the above and other observations on the propagation of light through anisotropic crystals quite simply using Huygens' construction. Consider a point source of light radiating uniformly inside the crystal. Then, for the case of uniaxial crystals, when a short time has elapsed there will be two wave surfaces as shown in Figs. 3.4(a) and (b).

In each case one of the wave surfaces is a sphere. It is found that the light contributing to this wave surface (represented in the diagrams by •••») is polarized with its electric vector perpendicular to the optic axis and to the *principal section*. The principal section is a plane containing the direction of propagation and the optic axis. Thus for light with this polarization the velocity of propagation of the waves is the same in all directions (that is the crystal is behaving isotropically). Such light gives rise to the ordinary rays mentioned above and the crystal has an ordinary refractive index n_O.

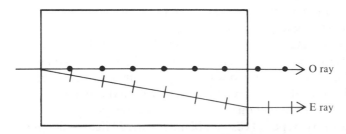

Fig. 3.3 Double refraction by a birefringent crystal.

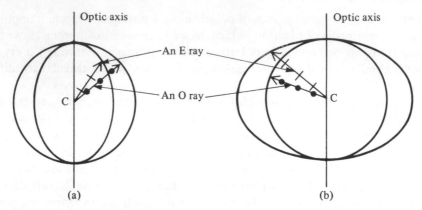

Fig. 3.4 Huygens' construction for the E- and O-wave surfaces for (a) a positive and (b) a negative uniaxial crystal.

The other wave surface is an ellipsoid of revolution which has one of its axes parallel to the optic axis. This wave surface is comprised of light waves which are polarized orthogonally to the ordinary wave and hence parallel to the principal section, and which give rise to the extraordinary rays (represented by $+\!\!+\!\!\!\rightarrow$). We see that the velocity of the extraordinary rays varies with direction. Along the optic axis, the velocity is the same as for the ordinary rays and the two wave surfaces touch. At right angles to the optic axis the extraordinary ray velocity is either a maximum, as in negative crystals (Fig. 3.4(b)), or a minimum, as in positive crystals (Fig. 3.4(a)). The refractive index n_e, of the crystal for the extraordinary rays is such that $n_o \leq n_e$ for *positive* crystals and $n_o \geq n_e$ for *negative* crystals. If we choose a coordinate system in which the z-axis is along the optic axis, the equation for this ellipsoid may be written as

$$\frac{x^2 + y^2}{v_e^2} + \frac{z^2}{v_o^2} = t^2$$

The spherical wavelets spread with velocity $v_o = c/n_o$, while the ellipsoidal wavelets spread with a velocity ranging from c/n_o to c/n_e. As the distances from the point C in Fig. 3.4 to the wave surfaces are proportional to the velocities of the E and O rays, the wave surfaces are often called *ray velocity surfaces*. They are useful for qualitative discussions but for quantitative work the *index surfaces* of the *index ellipsoid* discussed in Appendix 2 (see, for example, eq. (A2.7) and Fig. A2.1) are much more useful. The lengths of the major axes of this ellipsoid are twice the principal refractive indices mentioned on p. 79. The major axes are directed along the so-called *principal dielectric axes* (or simply the principal axes) x, y, z of the crystal. It should be noted that these directions correspond to the directions of polarization of the light waves and *not* their direction of propagation. Thus, for example, a wave propagating in the z-direction (conventionally taken to coincide with the optic axis in uniaxial crystals) can have components polarized parallel to the x- and y-directions.

Figure 3.5 shows two special cases of plane polarized light incident on the surface of a plane parallel plate cut (a) with its optic axis parallel to the surface and (b) with its optic

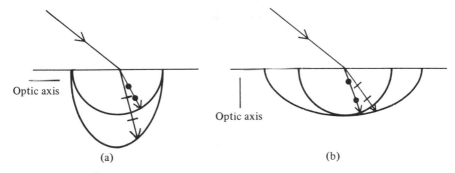

Fig. 3.5 Double refraction by a negative crystal plate in which (a) the optic axis is parallel to the crystal surface and the plane of incidence and (b) the optic axis is perpendicular to the crystal surface and parallel to the plane of incidence.

axis perpendicular to the surface. (The diagrams are for a *principal section*.) We see that for non-normal incidence there will be two diverging rays, the O ray, and the E ray. Figure 3.6 shows that for normal incidence on a crystal plate cut normal to the optic axis there will be no divergence of the E and O rays and the state of polarization of the emergent light will be the same as that of the incident light. Figure 3.7, on the other hand, shows that for a plate cut parallel to the optic axis, the O and E rays will emerge from the plate with orthogonal polarizations and different phases. As they are coherent, the perpendicular vibrations will superpose to give plane, circular or elliptically polarized light according to the relative amplitudes and the phase difference.

In general when the optic axis is at some arbitrary angle to the surface and the incident light has arbitrary polarization there will be two emergent rays, that is an O and an E ray. However, if the incident light is plane polarized either parallel to or perpendicular to the principal section, there will only be one plane polarized emergent beam.

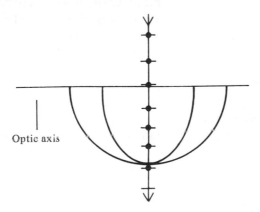

Fig. 3.6 For the case of normal incidence on a crystal plate cut with its surface normal to the optic axis, there is no double refraction; the state of polarization of the emergent light is the same as that of the incident light.

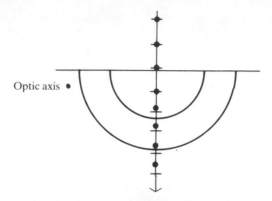

Optic axis •

Fig. 3.7 Double refraction by a negative crystal which is cut such that the optic axis is parallel to the crystal surface but normal to the plane of incidence. In this case there is no divergence of the E and O rays, but they become increasingly out of phase as they propagate through the crystal.

The two directions of vibration of the incident light which give only a single emergent beam are sometimes called the *privileged directions* or the *fast* and *slow* axes of the crystal. One direction corresponds to the vibration direction of the O ray and will be a fast axis for positive crystals, the other direction corresponds to the vibration direction of the E ray and is the slow axis. The reverse is the case for negative crystals. As we mentioned above, for light incident along the direction of the optic axis the state of polarization of the emergent beam will be the same as that of the incident beam and there are no privileged directions.

3.2.1 *Quarter-wave plate*

A crystal plate cut as in Fig. 3.7 which introduces a phase difference of $\pi/2$ between the O and E rays is called a *quarter-wave plate*. A phase difference of $\pi/2$ is equivalent to an optical path difference $|n_o d - n_e d| = \lambda/4$ where d is the plate thickness. For quartz, for example, d should be equal to 0.0164 mm for sodium light (Problem 3.2). When plane polarized light is incident on a quarter-wave plate, the emergent light is in general elliptically polarized. The axes of the ellipse are parallel to the privileged directions in the plate. If the plane of polarization of the incident beam, however, is inclined at 45° to the privileged directions then the emergent light is circularly polarized. In a similar way, one can produce half-wave and whole-wave plates. Such plates are often used in light modulation systems.

3.3 Optical activity

Certain crystals (and liquids) have the ability to rotate the plane of polarization of light passing through them; that is, they are *optically active*. Thus, for example, when a beam of plane polarized light is incident normally on a crystal plate of quartz cut perpendicular to the optic axis, it is found that the emergent beam is also plane polarized but that its electric vector vibrates in a different plane from that of the incident light. The plane of vibration may be rotated in a clockwise sense looking against the oncoming light by *right-handed* or *dextrorotatory* crystals, or in a counterclockwise sense by *left-handed* or *laevorotatory* crystals. Quartz exists in both forms. It is found that the rotation depends on the thickness of the crystal plate and the wavelength. The rotation produced by a 1 mm thick quartz plate for sodium light is 21.7° while it is 3.67° for 1 mm of sodium chloride.

Optical activity can be explained by assuming that in active crystals the velocity of propagation of circularly polarized light is different for different directions of rotation, that is the crystal has refractive indices n_r and n_l for right and left circularly polarized light. It is easy to show (see Problem 3.3) that a plane polarized wave can be resolved into two circularly polarized waves with opposite directions of rotation. If these travel through the crystal at different speeds, a phase difference will be introduced between them at different distances through the crystal. This corresponds to a rotation of the plane of the plane polarized wave which results from the recombination of the two circularly polarized waves.

3.4 Electro-optic effect

When an electric field is applied across an optical medium the distribution of electrons within it is distorted so that the polarizability and hence the refractive index of the medium changes anisotropically. The result of this electro-optic effect may be to introduce new optic axes into naturally doubly refracting crystals, for example KDP, or to make naturally isotropic crystals, for example gallium arsenide (GaAs), doubly refracting.

The change in refractive index as a function of the applied field can be obtained from an equation of the form (see ref. 3.3 and Appendix 2)

$$\Delta\left(\frac{1}{n^2}\right) = r\mathscr{E} + P\mathscr{E}^2 \tag{3.3}$$

where r is the *linear* electro-optic coefficient and P is the *quadratic* electro-optic coefficient. In solids, the linear variation in the refractive index associated with $r\mathscr{E}$ is known as the Pockels effect while the variation arising from the quadratic term is called the Kerr effect (not to be confused with the magneto-optic effect also named after Kerr).

In the case of the Pockels effect, the precise effects of the applied electric field depends on the crystal structure and symmetry of the material under consideration. With KDP, for example, if the electric field is applied along the z-direction then the x and y principal axes are rotated through 45° into new principal axes x' and y' and the refractive indices in these new directions become (see Appendix 2)

$$n_{x'} = n_o + \frac{n_o^3}{2} r_{63} \mathscr{E}_z$$

and

$$n_{y'} = n_o - \frac{n_o^3}{2} r_{63} \mathscr{E}_z \qquad (3.4)$$

in agreement with eq. (3.3) assuming that the Kerr constant P is very small, i.e.

$$\Delta \left(\frac{1}{n^2} \right) = - \frac{2\Delta n}{n^3} = r_{63} \mathscr{E}_z \qquad (3.5)$$

where r_{63} is the appropriate electro-optic coefficient for KDP. In general, the change in refractive index is of this form and we shall therefore drop the subscripts from r.

Example 3.1 Change in refractive index due to the Pockels effect

We may calculate the change in refractive index for a 10 mm wide crystal of deuterium substituted KDP (that is KD*P) for an applied voltage of 4000 V.

From eq. (3.5) and using the data in Table 3.1 (page 88) we have

$$n - n_o = \pm \tfrac{1}{2} \times 26.4 \times 10^{-12} \times 1.51^3 \times \frac{4000}{10^{-2}}$$

or

$$|n - n_o| = 1.8 \times 10^{-5}$$

Let us consider a beam of plane polarized light propagating in the z-direction through a crystal such as KDP (or GaAs in which the field may induce principal axes) with its plane of polarization at 45° to the induced axes x' and y' as shown in Fig. 3.8. If the incident beam is given by $\mathscr{E} = \mathscr{E}_0 \cos(\omega t - kz)$, then the components along the x' and y' directions will be

$$\mathscr{E}_{x'} = \frac{\mathscr{E}_0}{\sqrt{2}} \cos(\omega t - kz)$$

and

$$\mathscr{E}_{y'} = \frac{\mathscr{E}_0}{\sqrt{2}} \cos(\omega t - kz) \qquad (3.6)$$

In view of the fact that these components experience the refractive indices given by eq. (3.4), they will become increasingly out of phase as they propagate through the crystal. Thus if the crystal is of thickness L the phase changes of the two components will be

Fig. 3.8 Beam of plane polarized light incident on an electro-optic crystal plate subject to a voltage V will be resolved into components along x' and y', the induced principal directions. The induced birefringence occurs in a plane normal to the applied electric field.

$$\phi_{x'} = \frac{2\pi}{\lambda}\, n_{x'}\, L$$

and

$$\phi_{y'} = \frac{2\pi}{\lambda}\, n_{y'}\, L$$

Using eqs. (3.4) which relate the refractive index to the electric field, we see that

$$\phi_{x'} = \frac{2\pi}{\lambda}\, Ln_{\mathrm{o}}(1 + \tfrac{1}{2}r\, n_{\mathrm{o}}^2 \mathscr{E}_z)$$

or

$$\phi_{x'} = \phi_0 + \Delta\phi \tag{3.7a}$$

and

$$\phi_{y'} = \frac{2\pi}{\lambda}\, Ln_{\mathrm{o}}(1 - \tfrac{1}{2}r\, n_{\mathrm{o}}^2 \mathscr{E}_z)$$

or

$$\phi_{y'} = \phi_0 - \Delta\phi \tag{3.7b}$$

where

$$\Delta\phi = \frac{\pi}{\lambda}\, Lrn_{\mathrm{o}}^3 \mathscr{E}_z = \frac{\pi}{\lambda}\, rn_{\mathrm{o}}^3 V \tag{3.8}$$

In eq. (3.8) we have taken $\mathscr{E}_z$ to equal V/L, where V is the applied voltage.

The net phase shift, or total *retardation*, between the two waves resulting from the application of the voltage V is seen to be

$$\Phi = \phi_{x'} - \phi_{y'} = 2\Delta\phi = \frac{2\pi}{\lambda} r n_o^3 V \tag{3.9}$$

and the emergent light will in general be elliptically polarized.

From eqs. (3.6) and (3.7), the components of the wave emerging from the electro-optic crystal can now be written as (omitting common phase factors)

$$\mathscr{E}_{x'} = \frac{\mathscr{E}_0}{\sqrt{2}} \cos{(\omega t + \Delta\phi)} \tag{3.10a}$$

and

$$\mathscr{E}_{y'} = \frac{\mathscr{E}_0}{\sqrt{2}} \cos{(\omega t - \Delta\phi)} \tag{3.10b}$$

The phase shift $\Delta\phi$ for each component depends directly on the applied voltage V (eq. (3.8)) so that we can vary the phase shift by varying the voltage applied to a given crystal. Suppose that we now insert a plane polarizing element orientated at right angles to the polarizing element producing the original plane polarized beam after the electro-optic crystal, as shown in Fig. 3.9. Then, as we can see from Fig. 3.9, the transmitted electric field components will be $-\mathscr{E}_{x'}/\sqrt{2}$ and $\mathscr{E}_{y'}/\sqrt{2}$, that is, using eqs. (3.10) we can write the transmitted electric field as:

$$\mathscr{E} = \frac{\mathscr{E}_0}{2} [\cos{(\omega t + \Delta\phi)} - \cos{(\omega t - \Delta\phi)}]$$

or

$$\mathscr{E} = -\mathscr{E}_0 \sin{\Delta\phi} \sin{\omega t}$$

Thus the irradiance of the transmitted beam, which is given by averaging $\mathscr{E}^2$ over a complete period $T = 2\pi/\omega$, can be written as

$$I = \frac{\omega}{2\pi} \int_0^{2\pi/\omega} \mathscr{E}^2 dt$$

or

$$I = I_0 \sin^2{\Delta\phi} = I_0 \sin^2{(\Phi/2)} \tag{3.11}$$

where I_0 is the irradiance of the light incident on the electro-optic crystal. As the phase retardation in the Pockels effect is proportional to the voltage, we can see from eqs. (3.11) and (3.9) that the transmittance as a function of applied voltage is given by

$$\frac{I}{I_0} = \sin^2{\left(\frac{\pi}{\lambda} r n_o^3 V\right)}$$

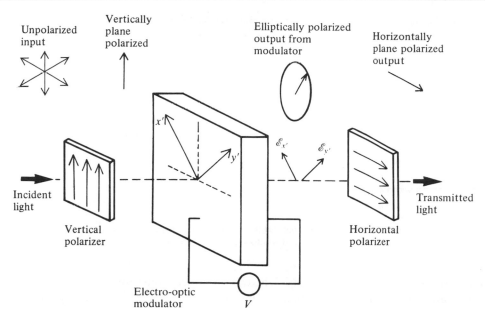

Fig. 3.9 Arrangement of the components of a Pockels electro-optic modulator in which an electro-optic crystal is placed between crossed polarizers. The state of polarization at various positions within the device is also indicated; the components transmitted by the horizontal polarizer are $-\mathscr{E}_{x'}/\sqrt{2}$ and $\mathscr{E}_{y'}/\sqrt{2}$.

which we may write as

$$\frac{I}{I_0} = \sin^2 \left(\frac{\pi}{2} \frac{V}{V_\pi} \right) \tag{3.12}$$

where $V_\pi [= \lambda/(2rn_0^3)]$ is the voltage required for maximum transmission, i.e. $I = I_0$. V_π is often called the *half-wave voltage* since it causes the two waves polarized parallel to the principal axes to acquire a relative spatial displacement of $\lambda/2$, which is equivalent to a phase difference of π. Thus a beam of plane polarized light incident on the modulator would have its plane of polarization rotated by 90° when a voltage V_π is applied to the modulator (see Example 3.2). The value of V_π depends on the electro-optic material and the wavelength (see Problem 3.4 and Table 3.1).

Thus we see that the transmittance of the system shown in Fig. 3.9 can be altered by the application of a voltage along the direction of propagation as illustrated in Fig. 3.10. Such systems are referred to as Pockels electro-optic modulators.

It is obvious that the modulation is not linear; indeed from eq. (3.12) for small voltages V, the transmitted irradiance is proportional to V^2. The effectiveness and ease of operation of a Pockels modulator can be enhanced by including a quarter-wave plate in the beam between the initial polarizer and the modulator as shown in Fig. 3.11(a). This introduces a phase difference of $\pi/2$ between the two polarized components before they enter the voltage-sensitive modulator. A bias is therefore introduced into the transmission curve so that the transmission is varied about the point Q, as illustrated in

Table 3.1 Characteristics of some electro-optic materials used in Pockels cells

Material	Linear electro–optic coefficient, r (pm V^{-1})	$n_o^\dagger$	$n_e^\dagger$	Relative permittivity,[‡] ε_r
KH_2PO_4 (KDP)	10.6	1.51	1.47	42
KD_2PO_4 (KD*P)	26.4	1.51	1.47	50
AH_2PO_4 (ADP)	8.5	1.52	1.48	12
Cadmium telluride (CdTe)	6.8	2.6		7.3
Lithium tantalate ($LiTaO_3$)	30.3	2.175	2.180	43
Lithium niobate ($LiNbO_3$)	30.8	2.29	2.20	18
Gallium arsenide (GaAs)	1.6	3.6		11.5
Zinc sulfide (ZnS)	2.1	2.32		16

[†] Values near a wavelength of 550 nm.
[‡] Low frequency values.
Note: Several of the materials listed have more than one linear electro-optic coefficient; we have quoted the one which is relevant for use in Pockels cell modulators. (See ref. 3.3b, p. 112, for further details).

Fig. 3.11(b), rather than about zero. The change in transmission in the vicinity of Q is more nearly linear with voltage than at the origin.

Example 3.2 Half-wave voltage

We may calculate the half-wave voltage for KDP, for example, at a wavelength of 1.06 μm, using Table 3.1. We have

$$V_\pi = \frac{\lambda}{2rn_o^3} = \frac{1.06 \times 10^{-6}}{2 \times 10.6 \times 10^{-12} \times (1.51)^3}$$

$$V_\pi = 14.5 \text{ kV}$$

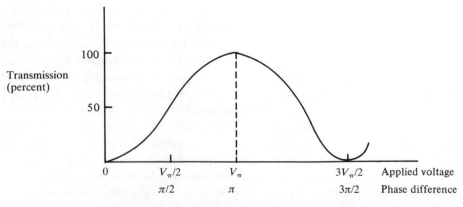

Fig. 3.10 Transmission curve for the system shown in Fig. 3.9 as a function of the applied voltage.

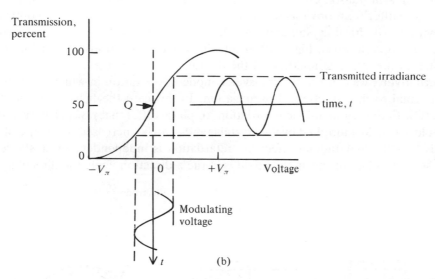

Fig. 3.11 Arrangement of the components of a Pockels electro-optic cell biased with a quarter-wave plate (a) and the resulting transmission as a function of applied voltage (b). The bias results in a 50% irradiance transmission; in the vicinity of this point the variation of transmission with applied voltage is almost linear.

With a quarter-wave plate bias, we see that the phase difference between the components is

$$\Phi = \frac{\pi}{2} + 2\Delta\phi = \frac{\pi}{2} + \pi\frac{V}{V_\pi}$$

and therefore from eq. (3.12)

$$\frac{I}{I_0} = \sin^2\left(\frac{\pi}{4} + \frac{\pi}{2}\frac{V}{V_\pi}\right) = \tfrac{1}{2}\left(1 + \sin\frac{\pi V}{V_\pi}\right)$$

For small V (up to about 5% of V_π) the change in irradiance is therefore nearly linear with V. If a small sinusoidally varying voltage of amplitude m and frequency f is applied to the modulator, then the irradiance of the transmitted beam will also vary at frequency f, as illustrated in Fig. 3.11(b). That is, we may write

$$\frac{I}{I_0} = 0.5 + \frac{\pi m}{2}\sin 2\pi ft \tag{3.13}$$

where $m\sin 2\pi ft = V/V_\pi$ should be very much less than unity otherwise the irradiance variation will be distorted and contain an appreciable amount of higher order odd harmonics.

The modulator described above is called a longitudinal effect device as the electric field is applied in the direction of propagation of the beam. This can be done by using electrodes with small apertures in them on either side of the electro-optic crystal or by evaporating semitransparent conducting films onto the crystal surfaces. Both of these techniques suffer from obvious disadvantages. To avoid these, an electro-optic modulator with a cylindrical crystal and ring electrode geometry has been developed. This device, which is shown in Fig. 3.12, results in very uniform transmission (or polarization) across the effective aperture of the device.

Alternatively, we can use the transverse mode of operation in which the field is applied normal to the direction of propagation. In this case the field electrodes do not interfere with the beam and the retardation (or phase difference), which is proportional to the electric field multiplied by the crystal length, can be increased by the use of longer crystals. (In the longitudinal effect the retardation is independent of crystal length.) Suppose that as before the applied field is in the z-direction while the direction of pro-

Fig. 3.12 Cylindrical, ring-electrode longitudinal electro-optic cell. Typical cell dimensions are length 25 mm, radius 6 mm and electrode width 8 mm.

pagation is along the y' induced principal axis, as shown in Fig. 3.13. Then, if the incident light is polarized in the $x'-z$ plane at 45° to the x' principal axis, the retardation, using eqs. (A2.15) and (A2.16) of Appendix 2, is

$$\Delta\phi = \phi_{x'} - \phi_z = \frac{2\pi}{\lambda} L(n_{x'} - n_o)$$

$$= \frac{2\pi}{\lambda} L\left((n_o - n_e) + \frac{n_o^3}{2}r\mathscr{E}_z\right)$$

$$= \frac{2\pi L}{\lambda} (n_o - n_e) + \frac{\pi}{\lambda} rn_o^3 \frac{VL}{D} \tag{3.14}$$

where L is the length of the crystal, D is the crystal dimension in the direction of the applied voltage V and n_o, n_e are the refractive indices for light polarized parallel to the principal directions. The voltage-independent term will bias the irradiance transmission curve. The half-wave voltage may be reduced by having a long, thin cell. Therefore the frequency response of transverse cells is better than in longitudinal cells as it is easier to change small voltages. However, transverse modulators suffer from having very small apertures.

In many practical situations the modulation signal is at very high frequencies and may occupy a large bandwidth so that the wide frequency spectrum available with laser sources may be fully utilized. The capacitance of the modulator and finite transit time of the light through it give rise to limitations in the bandwidth and maximum modulation frequency. Let C be the capacitance due to the electro-optic crystal and its electrodes and R_s be the internal resistance of the modulating source. Then if R_s is greater than $(2\pi f_0 C)^{-1}$, where f_0 is the average modulation frequency, most of the modulation potential drop will be across R_s and therefore will be wasted as it will not contribute to the electro-optic retardation. This problem can be overcome by connecting the crystal in a resonant circuit as shown in Fig. 3.14. The value of the inductance $\mathscr{L}$ is such that $4\pi^2 f_0^2 = 1/\mathscr{L}C$ so that at resonance $(f = f_0)$ the impedance of the circuit is simply $R_{\mathscr{L}}$, which is chosen to be greater than R_s and hence most of the modulation voltage

Fig. 3.13 Transverse electro-optic modulator. The electric field is applied normal to the direction of propagation.

Fig. 3.14 Electro-optic crystal represented by a parallel plate capacitor C connected in a resonant circuit.

appears across the crystal. The resonant circuit has a finite bandwidth, that is its impedance is high only over the frequency range $\Delta f = (2\pi R_\mathscr{L} C)^{-1}$ (centered on f_0). Therefore the maximum modulation bandwidth must be less than Δf for the modulated signal to be a faithful representation of the applied modulating voltage.

In practice the bandwidth Δf is governed by the specific application, though band-widths in the region of $10^8 - 10^9$ Hz are readily obtained. In addition, if a peak phase difference or retardation is required we can evaluate the power we need to apply to the crystal. The peak retardation $\Phi_\mathrm{m} = (2\pi/\lambda) r n_\mathrm{o}^3 V_\mathrm{m}$ (eq. (3.9)) corresponds to a peak modulating voltage $V_\mathrm{m} = (\mathscr{E}_z)_\mathrm{m} L$. The power $P = V_\mathrm{m}^2/2R_\mathscr{L}$ needed to obtain the peak retardation is therefore related to the modulation bandwidth by

$$P = \frac{\Phi_\mathrm{m}^2 \lambda^2 C 2\pi \Delta f}{2(4\pi^2 r^2 n_\mathrm{o}^6)}$$

or

$$P = \frac{\Phi_\mathrm{m}^2 \lambda^2 A \varepsilon_\mathrm{r} \varepsilon_0 \Delta f}{4\pi r^2 n_\mathrm{o}^6 L} \tag{3.15}$$

We have taken the capacitance of the crystal at the modulation frequency f_0, to be $C = A \varepsilon_\mathrm{r} \varepsilon_0 / L$, where A is the cross-sectional area of the crystal normal to the direction of propagation, in which the crystal length is L.

Example 3.3 Power requirement for modulation using a Pockels cell

We may estimate the power required to give a phase retardation of $\pi/30$ at a frequency bandwidth of 10^9 Hz using a KD*P Pockels cell with a circular aperture of 25 mm diameter and 30 mm length at a wavelength of 633 nm.

From eq. (3.15) and using Table 3.1 we find that

$$P = \frac{\pi^2 \times (633 \times 10^{-9})^2 \times \pi \times (12.5 \times 10^{-3})^2 \times 50 \times 8.85 \times 10^{-12} \times 10^9}{30^2 \times 4 \times \pi \times (26.4 \times 10^{-12})^2 \times (1.51)^6 \times 30 \times 10^{-3}}$$

$$= 306.4 \text{ W}$$

The maximum modulation frequency f_m should be such that the electric field applied to the crystal does not change substantially in a time equal to the transit time t_t of the light through the crystal; that is,

$$t_t = \frac{Ln}{c} \ll \frac{1}{f_m}$$

Typically $L = 10$ mm and hence f_m must be substantially less than 2×10^{10} Hz in KDP where the refractive index $n \simeq 1.5$.

To overcome this restriction the modulating signal can be applied transversely in the form of a wave traveling along the electrodes with a velocity equal to the phase velocity of the optical signal propagating through the modulating crystal. The optical wave then experiences a constant refractive index as it passes through the modulator and much higher modulation frequencies are possible. Although in principle it is quite easy to arrange synchronization of the electrical and optical waves, in practice it is more difficult because of the limitations of the optical materials available. Ideally we need a material for which $n = \sqrt{\varepsilon_r}$, where ε_r is the relative permittivity (or dielectric constant) of the medium. In most materials this is not the case ($n < \sqrt{\varepsilon_r}$) and the desired synchronization must be obtained by reducing ε_r by including air gaps in the electrical waveguide cross section. Alternatively, a traveling wave modulator may be realized by, in effect, slowing down the optical wave. This may be achieved, for example, by letting it propagate through the modulator along the zigzag path shown in Fig. 3.15.

3.4.1 Materials

Any transparent crystal lacking a center of symmetry exhibits a first order electro-optic effect. To be useful, however, such crystals must have a sizable electro-optic coefficient r and be available in reasonably sized, good quality crystals at modest cost. Some of the properties of technologically useful materials are listed in Table 3.1 (see page 88).

Fig. 3.15 Traveling wave modulator. The light beam follows the zigzag path shown so that in effect it propagates along the modulator at the same speed as the modulating signal and thereby experiences a constant modulating field.

Table 3.2 Typical values of the Kerr constant K for $\lambda = 589.3$ nm at about 20°C

Material	$K (\times 10^{-14}\,\mathrm{mV^{-2}})$
Water	5.2
Nitrobenzene	244
Nitrotoluene	137
Glasses – various	0.03–0.17

Potassium dihydrogen phosphate and ammonium dihydrogen phosphate crystals (KDP and ADP) are available in large sizes at relatively low cost but are soluble in water and fragile. They also have rather large half-wave voltages V_π; however, if deuterium is substituted for hydrogen (i.e. KD*P) the electro-optic properties are greatly enhanced. Other materials such as lithium tantalate and lithium niobate (LiTaO$_3$ and LiNbO$_3$) have much smaller half-wave voltages but reasonably sized crystals are rather expensive. Cadmium telluride and gallium arsenide are used in the infrared in the spectral ranges 1–28 µm and 1–14 µm respectively.

3.5 Kerr modulators

Many isotropic media, both solids and liquids, when placed in an electric field behave as uniaxial crystals with the optic axis parallel to the electric field. In this opto-electric effect, which was discovered in glass by J. Kerr (1875), the change in refractive index is proportional to the square of the applied field as we indicated in section 3.4. The difference in refractive indices for light polarized parallel to and perpendicular to the induced optic axis is given by:

$$\Delta n = n_\mathrm{p} - n_\mathrm{s} = K\lambda\mathscr{E}^2 \tag{3.16}$$

where K is the Kerr constant (see Table 3.2) and λ is the vacuum wavelength.

The electric field induces an electric moment in nonpolar molecules and changes the moment of polar molecules. There is then a reorientation of the molecules by the field which causes the medium as a whole to become anisotropic. This explains the delay that occurs between the application of the field and the appearance of the maximum effect. The delay can be several seconds, but for nonpolar liquids the delay is very small and probably less than 10^{-11} s. A Kerr cell filled with one of these liquids and placed between crossed polaroids can be used as an optical switch or modulator instead of a Pockels cell. Modulation at frequencies up to 10^{10} Hz has been obtained.

Although liquid Kerr cells containing nitrobenzene have been used extensively for many years, for example in the accurate measurement of the velocity of light, they suffer from the disadvantage of requiring a large power to operate them. A more promising approach is to use mixed ferroelectric crystals operating at a temperature near to the Curie points where a greatly enhanced opto-electric effect is observed. (Ferroelectric materials exhibit a spontaneous electric polarization, similar to the spontaneous magnetization of ferromagnetic materials, below a certain temperature. This is the

Curie point, above which the crystal structure changes and the ferroelectricity disappears.) Potassium tantalate niobate (KTN), which is used in Kerr-effect devices, is a mixture of two crystals with high and low Curie points giving a Curie point for the mixture near to room temperature. The crystal has to be 'poled' by the application of a large bias voltage. This has the effect of causing the ferroelectric domains with electric polarization in the direction of the applied field to grow at the expense of the other domains until the whole crystal is polarized in one direction. This then reduces the a.c. voltage required for 100% modulation to about 50 V peak and the half-wave voltage is very much less than for other materials, being about 250 V in KTN and barium titanate ($BaTiO_3$). Nevertheless, most practical electro-optic modulators make use of the Pockels effect.

3.5.1 Optical frequency Kerr effect

In the optical frequency Kerr effect, as the name implies, the refractive index change is brought about by an applied optical frequency field. This effect offers the interesting possibility of one beam of light being used to switch another if the refractive index change can be used to eliminate the second beam. Such optical beam switching can be achieved using a Fabry–Perot interferometer with an electro-optic material as the spacer between the reflecting plates. Intense maxima in the interference pattern formed by light passing through the interferometer occur when $p\pi = (2\pi/\lambda)nd \cos \theta$, where p is the order of interference, d is the plate separation and θ the angle between an internally reflected ray and the surface normal. A small change in the refractive index n of the spacer material induced by an optical beam would detune the interferometer and switch off the transmitted beam.

3.6 Scanning and switching

We saw in section 3.4 that the application of a voltage V_π to a Pockels cell will in effect rotate the plane of polarization of the transmitted optical beam through 90°. Thus if a block of birefringent material is placed after the cell the beam can be switched from one position to another as shown in Fig. 3.16. An array of m such combinations in sequence

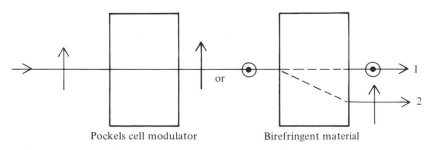

Pockels cell modulator   Birefringent material

Fig. 3.16 Beam switching using a Pockels cell modulator. As the applied voltage is changed from zero to V_π the beam is switched from position 1 to position 2.

can obviously be used to address 2^m different locations. Such a system may be used, for example, in bit-oriented optical memories (ref. 6.4d, Chapter 21).

Alternatively the arrangement shown in Fig. 3.17(a) can be used for beam switching. Here we have two similar prisms of KDP, for example, but with opposite orientations. Thus, if an electric field is applied in the z direction and the optical beam travels in the direction of one of the induced principal axes with its polarization parallel to the other principal axis, then the beam will 'see' different refractive indices in the two prisms. The difference in refractive indices will be $n_o^3 r \mathscr{E}_z$, that is, from eqs. (3.4), we see that a ray entirely in the upper prism travels in a medium of refractive index

$$n_2 = n_o - \frac{n_o^3}{2} r \mathscr{E}_z$$

while a ray entirely in the lower prism travels in a medium where the effect of the applied field is reversed so that

$$n_1 = n_o + \frac{n_o^3}{2} r \mathscr{E}_z$$

Fig. 3.17 Schematic diagram of a beam deflector: (a) the double-prism KDP beam deflector and (b) the principle of beam deflection in a medium where the refractive index varies linearly in a direction normal to the direction of propagation.

We can see that this arrangement leads to a deflection of the beam by considering Fig. 3.17(b). Here we show a crystal whose refractive index and hence optical path length varies with the transverse distance x across the crystal. If we assume that the variation of n with x is uniform then ray 1 'sees' a refractive index n while ray 2 'sees' an index $n + \Delta n$. The rays 1 and 2 will traverse the crystal in times t_1 and t_2 where

$$t_1 = \frac{Ln}{c} \text{ and } t_2 = \frac{L(n + \Delta n)}{c}$$

The difference in transit times results in ray 2 lagging behind ray 1 by a distance $\Delta y = L \Delta n / n$; this is equivalent to a deflection of the wavefront by an angle θ' measured inside the crystal just before the beam emerges. We can see from Fig. 3.17(b) that $\theta' = \Delta y / W$, where W is the width of the crystal.

Using Snell's law, the angle of beam deflection θ measured outside the crystal is given by (assuming θ is small)

$$\theta = n\theta' = n \frac{\Delta y}{W}$$

or

$$\theta = \frac{L \Delta n}{W}$$

Thus, using the arrangement shown in Fig. 3.17(a), the deflection θ is given by

$$\theta = (L/W)n_o^3 r \mathscr{E}_z \tag{3.17}$$

Light beams can also be deflected by means of diffraction gratings which are electro-optically induced in a crystal by evaporating a periodic metallic grating electrode onto the crystal. A voltage applied to this electrode induces a periodic variation in the refractive index thereby creating an efficient phase diffraction grating. This technique is especially useful in the integrated optical devices discussed in Chapter 9. Beam deflectors and scanners are used in laser displays, printers and scribers, for optical data storage systems and in optical character recognition.

3.7 Magneto-optic devices

The presence of magnetic fields may also affect the optical properties of some substances thereby giving rise to a number of useful devices. In general, however, as electric fields are easier to generate than magnetic fields, electro-optic devices are usually preferred to magneto-optic devices.

3.7.1 Faraday effect

This is the simplest magneto-optic effect and the only one of real interest for optical modulators; it concerns the change in refractive index of a material subjected to a steady

Table 3.3 Typical values of the Verdet constant V for $\lambda = 589.3\,\text{nm}$

Material	$V\,(\text{rd}\,\text{m}^{-1}\,\text{T}^{-1})$
Quartz (SiO$_2$)	4.0
Zinc sulfide (ZnS)	82
Crown glass	6.4
Flint glass	23
Sodium chloride (NaCl)	9.6

magnetic field. Faraday found (1845) that when a beam of plane polarized light passes through a substance subjected to a magnetic field, its plane of polarization is observed to rotate by an amount proportional to the magnetic field component parallel to the direction of propagation. This is very similar to optical activity which, as we saw in section 3.3, results from certain materials having different refractive indices n_r and n_l for right and left circularly polarized light. There is one important difference in the two effects. In the Faraday effect the sense of rotation of the plane of polarization is independent of the direction of propagation. This is in contrast to optical activity where the sense of rotation is related to the direction of propagation. Thus in the case under discussion the rotation can be doubled by reflecting the light back through the Faraday effect device.

The rotation of the plane of polarization is given by

$$\theta = VBL \tag{3.18}$$

where V is the Verdet constant (see Table 3.3 for some representative values), B is the magnetic flux parallel to the direction of propagation and L is the path length in the material. The Faraday effect is small and wavelength dependent; the rotation for dense flint glass is $\theta \simeq 1.6^0\,\text{mm}^{-1}\,\text{T}^{-1}$ at $\lambda = 589.3\,\text{nm}$.

We can also express θ in terms of the refractive indices n_r and n_l, i.e.

$$\theta = \frac{2\pi}{\lambda}\,(n_r - n_l)L$$

A Faraday rotator used in conjunction with a pair of polarizers acts as an optical isolator which allows a light beam to travel through it in one direction but not in the opposite one. It may therefore be used in laser amplifying chains to eliminate reflected, backward traveling waves, which are potentially damaging. The construction of a typical isolator is shown in Fig. 3.18.

Light passing from left to right is polarized in the vertical plane by polarizer P_1. The Faraday rotator is adjusted to produce a rotation of $45°$ in the clockwise sense. The second polarizer P_2 is set at $45°$ to P_1 so that it will transmit light emerging from the rotator. However, a beam entering from the right will be plane polarized at $45°$ to the vertical by P_2 and then have its plane rotated by $45°$ in the clockwise sense by the rotator. It will therefore be incident on P_1 with its plane of polarization at right angles to the plane of transmission and be eliminated. The device thus isolates the components on its left from light incident from the right.

Fig. 3.18 Optical isolator based on the Faraday effect. The reflected ray is shown displaced for clarity.

One potential application of magneto-optics currently receiving attention is large capacity computer memories. Such memories must be capable of storing very large amounts of information in a relatively small area and permit very rapid readout and, preferably, random access. The usual magnetic memories have a number of limitations of size and reading speed. Optical techniques can overcome both these constraints (refs. 3.4).

The magneto-optic memories developed so far are read via the Faraday effect or the magnetic Kerr effect, which relates to the rotation of a beam of plane polarized light reflected from the surface of a material subjected to a magnetic field. In either case a magnetized ferro- or ferrimagnetic material rotates the plane of polarization of laser light incident on it.

Writing may be achieved by heating the memory elements on the storage medium to a temperature above the Curie point using a laser beam. The element is then allowed to cool down in the presence of an external magnetic field thereby acquiring a magnetization in a given direction. Magnetizations of the elements in one direction may represent 'ones', in the opposite direction 'zeros'. To read the information the irradiance of the laser beam is reduced and then directed to the memory elements. The direction of the change in the polarization of the laser beam on passing through or being reflected from the memory elements depends on the directions of magnetization; therefore we can decide if a given element is storing a 'one' or 'zero'.

Prototype systems, incorporating, for example, a 50 mW He–Ne laser with a Pockels modulator and manganese bismuth (MnBi) thin film storage elements, have enabled information to be stored, read and erased at rates in excess of 1 Mb s[-1].

3.8 Acousto-optic effect

The acousto-optic effect is the change in the refractive index of a medium caused by the mechanical strains accompanying the passage of an acoustic (strain) wave through the medium. The strain and hence the refractive index varies periodically with a wavelength Λ equal to that of the acoustic wave. The refractive index changes are caused by the photoelastic effect which occurs in all materials on the application of a mechanical stress. It can be shown that the change in refractive index is proportional to the square root of the total acoustic power (ref. 3.5).

In general, the relationships between changes of refractive index and mechanical strain, and between the strain and stress are rather complicated (ref. 3.6). However, for simplicity we can consider the case of a monochromatic light wave, wavelength λ, incident upon a medium in which an acoustic wave has produced sinusoidal variations of wavelength Λ in the refractive index. The situation is shown in Fig. 3.19, where the solid horizontal lines represent acoustic wave peaks (pressure maxima) and the dashed horizontal lines represent acoustic wave troughs (pressure minima). As the light enters the medium, the portions of the wavefront near to a pressure peak will encounter a higher refractive index and therefore advance with a lower velocity than those portions of the wavefront which encounter pressure minima. The wavefront in the medium therefore soon acquires the wavy appearance shown by the dashed curve in Fig. 3.19. The acoustic wave velocity is very much less than the light wave velocity, so we may ignore it and consider the variation in refractice index to be stationary in the medium.

As elements of the light wave propagate in a direction normal to the local wavefront, almost all the wave elements will suffer a change in direction leading to a re-distribution of the light flux, which tends to concentrate near regions of compression. In effect, the

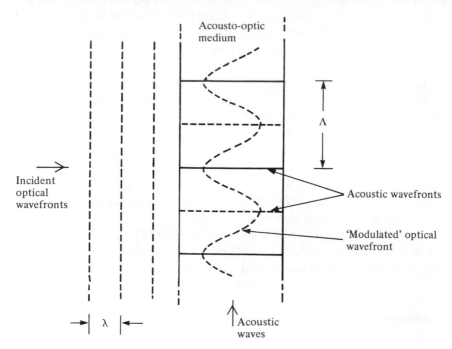

Fig. 3.19 Schematic illustration of acousto-optic modulation. The acoustic waves change the refractive index of the medium in a periodic way so that the plane optical wavefronts take on the 'wavy' appearance shown (very much exaggerated) as they propagate through the medium.

acoustic wave sets up a diffraction grating within the medium so that optical energy is diffracted out of the incident beam into the various orders. There are two main cases of interest, namely (a) the *Raman–Nath regime* and (b) the *Bragg regime*.

In the Raman–Nath regime, the acoustic diffraction grating is so 'thin' that the diffracted light suffers no further redistribution before leaving the modulator. The light is diffracted as from a simple plane grating such that

$$m\lambda_0 = \Lambda \sin \theta_m \tag{3.19}$$

where $m = 0, \pm1, \pm2, \ldots$, is the order and θ_m is the corresponding angle of diffraction, as illustrated in Fig. 3.20.

The irradiance I of the light in these orders depends on the 'ruling depth' of the acoustic grating, which is related to the amplitude of the acoustic grating. This, in turn, is related to the amplitude of the acoustic modulating wave (that is, the stress produced). The fraction of light removed from the zero order beam is $\eta = (I_0 - I)/I_0$, where I_0 is the transmitted irradiance in the absence of the acoustic wave. Thus amplitude variations of the acoustic wave are transformed into irradiance variations of the optical beam.

The physical basis of the Bragg regime is that light diffracted from the incident beam is extensively re-diffracted before leaving the acoustic field. Under these conditions, the acoustic field acts very much like a 'thick' diffraction grating, that is, a grating made up of planes rather than lines. The situation is then very similar to that of Bragg diffraction (or 'reflection') of X rays from planes of atoms in a crystal. Consider a plane wavefront incident on the grating planes at an angle of incidence θ_i as shown in Fig. 3.21(a); significant amounts of light will emerge only in those directions in which constructive

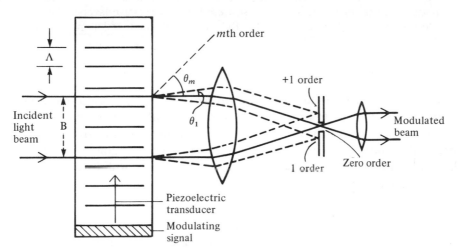

Fig. 3.20 Geometry for Raman–Nath (or transmission-type) acousto-optic diffraction grating modulation. The amount of light diffracted into the orders $m \geqslant 1$ from the incident beam, and hence the modulation of the transmitted beam, depends on the amplitude of the modulating signal.

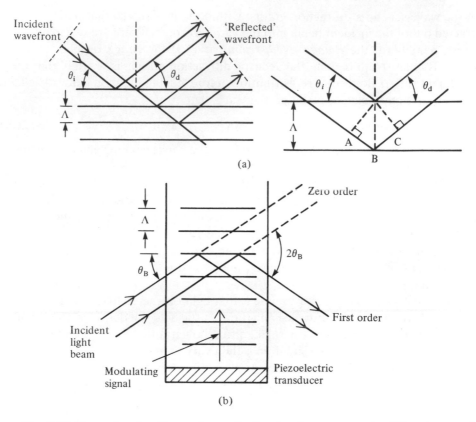

Fig. 3.21 Geometry for Bragg (or reflection-type) acousto-optic diffraction grating modulation: (a) incident rays being scattered from successive layers – for constructive interference the path difference AB + BC must equal an integral number of wavelengths $m\lambda$ and (b) the amount of light 'reflected' into the first order depends on the amplitude of the modulating signal.

interference occurs. The conditions to be satisfied are: (a) light scattered from a given grating plane must arrive in phase at the new wavefront and (b) light scattered from successive grating planes must also arrive in phase at the new wavefront, implying that the path difference must be an integral number of wavelengths. The first of these conditions is satisfied when $\theta_d = \theta_i$, where θ_d is the angle of diffraction. The second condition requires that

$$\sin \theta_i + \sin \theta_d = \frac{m\lambda}{\Lambda}$$

with $m = 0, 1, 2 \ldots$. The two conditions are simultaneously fulfilled when

$$\sin \theta_i = \sin \theta_d = \frac{m\lambda}{2\Lambda} \tag{3.20}$$

The diffraction is similar to that obtained with a plane grating, but only for special angles of incidence; the angle of incidence must equal the angle of diffraction.

Although in the simplified theory given above strong scattering can take place when m is equal to any positive integer, a more rigorous treatment, taking into account the fact that scattering is not from discrete planes but from a continuous medium, shows that scattering only takes place when $m = 1$. This is shown in Fig. 3.21(b), the equation for the so-called Bragg angle θ_B then becomes $\sin \theta_B = \lambda/2\Lambda$. The modulation depth $(I_0 - I)/I_0$ (or diffraction efficiency, η) in this case can theoretically equal 100% in contrast to about 34% for the Raman–Nath case. At the Bragg angle it is given by $\eta = \sin^2 \phi/2$, where $\phi = (2\pi/\lambda)(\Delta n\, L/\cos \theta_B)$, Δn being the amplitude of the refractive index fluctuation and L the length of the modulator (ref. 3.7).

The acoustic waves, which create the diffraction grating, are of course moving through the medium and, as a consequence, the diffracted wave behaves as if it had been reflected from a 'mirror' moving with the same velocity as the grating, and therefore appears to originate from a source moving at *twice* the mirror (or grating) velocity. Thus the frequency of the reflected beam is changed by the Doppler effect (see Section 5.7) and is given by

$$\nu' = \nu_0 \left(1 \pm 2\upsilon_a\,(c/n)\right)$$

where $\pm\upsilon_a$ is the component of the velocity of the acoustic wave along (or away from) the original beam direction and n is the medium refractive index. The frequency shift is thus:

$$\Delta\nu = \nu' - \nu_0 = \frac{\pm 2\nu_0\upsilon_a n}{c}$$

If the light is incident at an angle $\theta_i = \theta_d$ to the acoustic wave as shown in Fig. 3.21a, then $\upsilon_x = \upsilon_a \sin \theta_d$ where υ_a is the acoustic wave velocity. The frequency shift is then:

$$\Delta\nu = \frac{\pm 2\nu_0\upsilon_a \sin \theta_d\, n}{c} \tag{3.21}$$

Combining eqs. (3.21) and (3.20) and taking $m = 1$ gives a frequency shift of $\pm\upsilon_a/\Lambda$ or $\pm f_0$, where f_0 is the acoustic wave frequency. This change in frequency can be used as the basis of a frequency modulator.

The minimum time required to move from a condition where the acoustic wave interacts with the light beam and 'turns off' the undiffracted light to a condition where there is no diffraction is the transit time of the acoustic wave across the optical beam. This is simply, from Fig. 3.20, $t_{min} = B/\upsilon_a$, where B is the optical beam width. Hence the bandwidth of the modulator is limited to about υ_a/B. Commercial modulators have bandwidths of up to 50 MHz. This limitation is partly due to the frequency dependence of the acoustic losses of available acousto-optic materials. At the present time only $LiNbO_3$ and $PbMoO_4$ appear to have sufficiently low loss to have a reasonable prospect of being operated at appreciably higher frequencies.

Acousto-optic modulators can in general be used for similar applications to electro-optic modulators, though they are not so fast. On the other hand, because the electro-optic effect usually requires voltages in the kilovolt range, the drive circuitry for modulators based on this effect is much more expensive than for acousto-optic modulators, which operate with a few volts.

Example 3.4 Acousto–optic modulator

Given the following data for a PbMO$_4$ acousto-optic modulator, we may calculate the Bragg angle; the maximum change in refractive index of the material and the maximum width of the optical beam of wavelength 633 nm that may be modulated with a bandwidth of 5 MHz.

The modulator length is 50 mm, diffraction efficiency 70%, while the acoustic wavelength is 4.3×10^{-5} m and the acoustic velocity is 3500 m s^{-1}.

The angle of diffraction (from eq. 3.20) is

$$\theta_B = \sin^{-1}\left(\frac{633 \times 10^{-9}}{2 \times 4.3 \times 10^{-5}}\right) = 7.4 \text{ m rad (or } 0.42°)$$

The value of ϕ is given by

$$\phi = 2 \sin^{-1} \sqrt{\eta} = 2 \sin^{-1} \sqrt{0.7}$$

$$\phi = 113.6°$$

$$\therefore \Delta n = \frac{\phi\lambda \cos \theta_B}{2\pi L} = 1.27 \times 10^{-5}$$

The bandwidth is v_a/B and hence the maximum optical beamwidth B is

$$\frac{3500}{5 \times 10^6} = 0.7 \text{ mm}$$

3.9 Nonlinear optics

Practical applications of nonlinear optical effects have arisen as a direct consequence of the invention of the laser. The very high power densities made available by lasers have enabled several phenomena, which were previously regarded as theoretical curiosities, to be observed and exploited.

The explanation of nonlinear effects lies in the way in which a beam of light propagates through a solid. The nuclei and associated electrons of the atoms in the solid form electric dipoles. The electromagnetic radiation interacts with these dipoles causing them to oscillate which, by the classical laws of electromagnetism, results in the dipoles themselves acting as sources of electromagnetic radiation. If the amplitude of vibration is small, the dipoles emit radiation of the same frequency as the incident radiation. As the irradiance of the radiation increases, however, the relationship between irradiance and amplitude of vibration becomes nonlinear resulting in the generation of harmonics of the frequency of the radiation emitted by the oscillating dipoles. Thus frequency doubling or second harmonic generation and indeed higher order frequency effects occur as the incident irradiance is increased. The electric polarization (or dipole moment per unit volume) P can be expressed as a power series expansion in the applied electric field $\mathscr{E}$ by

$$P = \varepsilon_0 \left(\chi \mathscr{E} + \chi_2 \mathscr{E}^2 + \chi_3 \mathscr{E}^3 + \ldots \right) \tag{3.22}$$

where χ is the linear susceptibility and χ_x, χ_3, ... are the nonlinear optical coefficients. The resulting (nonlinear) relationship between P and $\mathscr{E}$ is shown in Fig. 3.22.

If the applied field is of the form $\mathscr{E} = \mathscr{E}_0 \sin \omega t$ such as produced by an electromagnetic wave, then substitution into eq. (3.22) gives:

$$P = \varepsilon_0 \left(\chi \mathscr{E}_0 \sin \omega t + \chi_2 \mathscr{E}_0^2 \sin^2 \omega t + \chi_3 \mathscr{E}_0^3 \sin^3 \omega t + \ldots \right)$$

$$= \varepsilon_0 \left(\chi \mathscr{E}_0 \sin \omega t + \tfrac{1}{2} \chi_2 \mathscr{E}_0^2 \left(1 - \cos 2\omega t \right) + \ldots \right) \tag{3.23}$$

Equation (3.23) contains a term in 2ω which corresponds to an electromagnetic wave having twice the frequency of the incident wave. The magnitude of the term in 2ω, however, does not approach that of the first term $\varepsilon_0 \chi \mathscr{E}_0$ until the electric field is about 10^6 V m^{-1} (which is not entirely negligible in comparison to the internal fields of crystals, i.e. $\mathscr{E}_{int} \approx 10^{11}$ V m^{-1}). A field of 10^6 V m^{-1} corresponds, at optical wavelengths, to a power density of about 10^9 W m^{-2}, while the electric fields and power density of sunlight are of the order of 100 V m^{-1} and 20 W m^{-2} respectively, so it is not too surprising that the observation of nonlinear effects had to await the advent of the laser. (Large nonlinear effects have been observed in some semiconductors with power densities of only about 5×10^4 W m^{-2} due to free carrier effects.) Harmonic generation is only observed in those solids that do not possess a center of symmetry. In symmetric materials, an applied electric field produces polarizations of the same magnitude but of opposite sign according to whether the electric field is positive or negative and there is no net polarization. Consequently the coefficients of even powers of $\mathscr{E}$ in eq. (3.22) are zero. In anisotropic media such as quartz, ADP and KDP, however, harmonics are generated as indicated in Fig. 3.23(a) where we see that the symmetrical optical field produces an asymmetrical polarization. A Fourier analysis of the polarization (Fig. 3.23(b)) shows that it consists of components having frequencies ω and 2ω as well as a d.c. component.

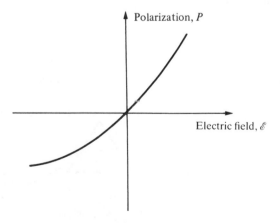

Fig. 3.22 Curve showing electric polarization versus electric field for a nonlinear material (that is, a material lacking a center of symmetry), assuming χ_2 is negative.

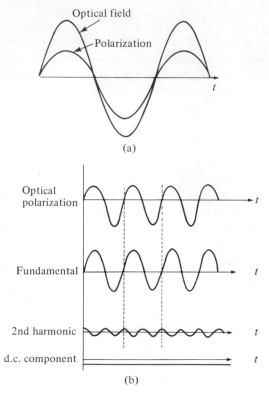

Fig. 3.23 Applied sinusoidal optical (i.e. electrical) field and the resulting polarization for a nonlinear material (a) and Fourier analysis of the asymmetrical polarization wave (b) into (i) a fundamental wave oscillating at the same angular frequency (ω) as the wave inducing it, (ii) a second harmonic of twice that frequency (2ω) and (iii) an average (d.c.) negative component.

Second harmonic generation was first observed in 1961 by Franken and his co-workers (ref. 3.8), who focused the 694.3 nm output from a ruby laser onto a quartz crystal as shown in Fig. 3.24 and obtained a very low intensity output at a wavelength of 347.15 nm. In these experiments the conversion efficiency from the lower to the higher frequency was typically $10^{-6}\%$ to $10^{-4}\%$. The reason for this is that wavelength dispersion

Fig. 3.24 Simplified diagram showing the arrangement for optical frequency doubling. The KDP crystal is mounted in the correct orientation for index matching.

within the crystal causes the frequency doubled light to travel at a different speed from that of the fundamental. As the latter is generating the former throughout its passage through the crystal, the two waves periodically get out of phase and destructive interference occurs. The irradiance of the frequency doubled light thus undergoes fluctuations through the crystal with a periodicity of l_c, which is called the *coherence length* and is typically only a few microns.

We can derive an expression for l_c as follows. Let us consider a plane wave propagating through an anisotropic crystal; the fundamental wave has a space–time variation of the form $\exp[i(k_1 z - \omega t)]$, whereas that of the second harmonic is $\exp[i(k_2 z - 2\omega t)]$. The amplitude of the second harmonic as it emerges from the crystal can be found by summing the contributions for the conversion which occurs in each element dz within the crystal, that is

$$\mathcal{E}(2\omega, L) \propto \int_0^L \mathcal{E}^2(\omega, z)\, dz \tag{3.24}$$

where L is the thickness of the crystal. If we let the time taken for the optical disturbance of frequency 2ω to travel from each point z to L be τ, then we can write eq. (3.24) as

$$\mathcal{E}(2\omega, L) \propto \int_0^L \exp 2i[k_1 z - \omega(t - \tau)]\, dz$$

where

$$\tau = \frac{L - z}{c_{2\omega}} = \frac{(L - z)k_2}{2\omega}.$$

$c_{2\omega}$ being the speed of the second harmonic and k_2 being the corresponding wave vector. Substituting for τ we have:

$$\mathcal{E}(2\omega, L) \propto \int_0^L \exp 2i\left[\left(k_1 - \frac{k_2}{2}\right)z + \frac{k_2 L}{2} - \omega t\right] dz$$

Integrating and squaring we find that the irradiance of the second harmonic is given by

$$|\mathcal{E}(2\omega, L)|^2 \propto \left[\frac{\sin\left(k_1 - \frac{k_2}{2}\right)L}{\left(k_1 - \frac{k_2}{2}\right)}\right]^2 \tag{3.25}$$

Equation (3.25) indicates that the irradiance of the second harmonic reaches a maximum after the waves have propagated a distance $L = l_c = \pi/(2k_1 - k_2)$ into the crystal. Thereafter, the energy in the second harmonic is returned to the fundamental wave and after two, or indeed any even number of coherence lengths, the irradiance of the second harmonic falls to zero.

This difficulty can be overcome by a technique known as *index* or *phase* matching. The commonest method uses the birefringent properties of the nonlinear medium,

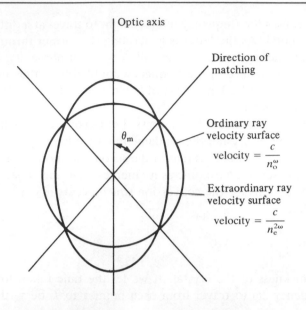

Fig. 3.25 Index matching in a negative uniaxial crystal. The condition $n_e^{2\omega} = n_o^{\omega}$ is satisfied for propagation at an angle θ_m to the optic axis ($\theta_m \simeq 50°$ in KDP). The eccentricities and velocity differences are greatly exaggerated for clarity – similarly the E ray velocity surface at frequency ω and the O ray velocity surface at frequency 2ω have been omitted.

which of course must be anisotropic if harmonic generation is to occur at all. As we have illustrated in Fig. 3.25, it is possible to choose a direction through the crystal such that the velocity of the fundamental (corresponding to the O ray of frequency ω and refractive index n_o^{ω}) is the same as that of the second harmonic (corresponding to the E ray of frequency 2ω and refractive index $n_e^{2\omega}$); that is, we choose a direction such that $n_o^{\omega} = n_e^{2\omega}$. With this technique, known as index matching, the coherence length increases from a few micrometers to a few centimeters. The conversion efficiency can be increased by orders of magnitude as the fundamental and harmonic waves remain in phase and there is a continuous build up in the irradiance of the second harmonic. The conversion may be about 20% for a single pass through a KDP crystal a few centimeters long. However, care must be taken in a number of respects of maximize the efficiency. For example, as the refractive indices are temperature dependent it may be necessary to control the crystal temperature. Furthermore, as only one direction of propagation is perfectly index matched, the laser divergence must be minimized and similarly lasers with broad linewidths may have lower conversion efficiencies.

Second harmonic generation enables us to extend the range of laser wavelengths into the blue and ultraviolet parts of the spectrum, which are not rich in naturally occurring laser lines. One very important laser application which may benefit from second harmonic generation is laser induced nuclear fusion, which appears to be more efficient at higher optical frequencies (see section 6.8).

Nonlinear processes can be described in terms of a photon model, according to which we can view second harmonic generation as the annihilation of two photons of angular frequency ω and the simultaneous creation of one photon of frequency 2ω. That is, conservation of energy requires that

$$\hbar\omega + \hbar\omega = \hbar(2\omega) \tag{3.26}$$

while conservation of momentum for the photons similarly requires that

$$\hbar k^{\omega} + \hbar k^{\omega} = \hbar k^{2\omega}$$

or

$$2k^{\omega} = k^{2\omega} \tag{3.27}$$

It is left to the reader to show that this equation is equivalent to the above refractive index matching criterion. The photon model forms a useful basis for the discussion of a related nonlinear phenomenon, that of parametric amplification and oscillation.

3.9.1 Parametric oscillation

Second harmonic generation can be regarded as a special case of sum frequency conversion, whereby power from a 'pump' wave at angular frequency ω_3 is transferred to waves at frequencies ω_1 and ω_2; that is,

$$\omega_3 \rightleftharpoons \omega_1 + \omega_2 \tag{3.28}$$

The 'reaction' represented by eq. (3.28) goes to the left in odd-numbered coherence lengths and to the right in even-numbered ones. If the frequency ω_3 alone is applied to a suitable nonlinear material such as lithium niobate, then the two smaller frequencies can build up from noise; ω_1 and ω_2 are known as the 'signal' and idler' frequencies respectively. The particular subdivision into ω_1 and ω_2 is determined by the index-matching criterion, which conservation of momentum gives as

$$\hbar k_1 + \hbar k_2 = \hbar k_3 \tag{3.29}$$

Example 3.5 Coherence length in second harmonic generation

Given that n^{ω} at 0.8 µm is 1.5019 and $n^{2\omega}$ at 0.4 µm is 1.4802 in KDP, we can calculate the coherence length as follows: the coherence length is given by $l_c = \pi/(2k_1 - k_2)$, which we can show can be written as

$$I_c = \frac{\lambda_0}{4(n^{\omega} - n^{2\omega})}$$

where λ_0 is the vacuum wavelength at the fundamental frequency. Therefore, using the data provided

$$I_c = \frac{0.8 \times 10^{-6}}{4 \times 0.0217} \simeq 10^{-5} \text{ m}$$

Fig. 3.26 Arrangement of the components in Giordmaine and Miller's observation of parametric oscillation in lithium niobate.

It can be seen from eq. (3.28) that if ω_3 and ω_1 are fixed, then ω_2 is also fixed ($\omega_2 = |\omega_3 - \omega_1|$). However, if only frequency ω_3 is fixed, then the other two are free to range over many values; this effect is known as *parametric amplification*. The index-matching criterion (eq. (3.29)) is very severe and the process is usually carried out in an optical cavity (see Chapter 5) with mirrors that are highly reflecting at ω_1 or ω_2, but not at ω_3, whence we have a parametric oscillator. Tuning can be achieved simply by varying the index-matching conditions through, for example, mechanical or temperature control of the length of the cavity. A schematic diagram of the system used by Giordmaine and Miller (ref. 3.9), who first achieved parametric oscillation in 1965, is shown in Fig. 3.26. In this case, the output was tuned by changing the temperature of the lithium niobate crystal; a typical tuning curve is shown in Fig. 3.27. A temperature change of about 11°C produced output frequencies in the range 3.1×10^{14} to 2.6×10^{14} Hz (which corresponds to the wavelength range 968–1154 nm). While the conversion efficiency in

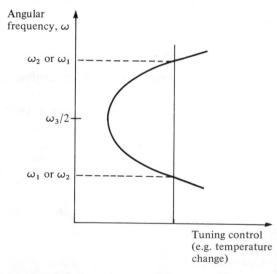

Fig. 3.27 Schematic tuning curve for a parametric oscillator.

these experiments was only 1%, efficiencies of about 50% have now been achieved, with coherent wavelengths ranging from the infrared to the ultraviolet being produced.

As we mentioned above, the production of coherent radiation at frequencies in the ultraviolet may be an essential requirement for laser induced nuclear fusion. The additional laser wavelengths made available by harmonic generation, parametric oscillation and other tuning techniques (section 6.5.1.6) lead to applications in photochemistry, high resolution spectroscopy and the remote detection and identification of atmospheric pollutants by optical radar or lidar (see also section 6.6.3). One example of photochemistry which may become important is isotope separation (ref. 6.8); others include the selective excitation of reactions and techniques for the study of the kinetics of ultrafast chemical reactions (see the references on pages 511–3 of ref. 6.4d).

Problems

3.1   Prove that the angle of inclination of the axes of the ellipse (which represents elliptically polarized light at a fixed position in space) to the horizontal and vertical directions is given by $1/2 \tan^{-1} [(\mathscr{E}_0 \mathscr{E}_0' \cos \phi)/(\mathscr{E}_0'^2 - \mathscr{E}_0^2)]$ (the symbols are as defined in section 3.1).

3.2   The principal refractive indices for quartz, n_e and n_o, are 1.55336 and 1.54425 respectively; calculate the thickness of a quarter-wave plate for sodium D light $\lambda = 589.3$ nm. How may such a plate be made more robust? Calculate the thickness of a calcite $\lambda/2$ plate for which $n_o = 1.658$ and $n_e = 1.486$ at the same wavelength.

3.3   Show how a beam of plane polarized light may be considered as consisting of two oppositely directed circularly polarized beams of light. Hence show that the rotation of the plane of polarization in an optically active medium of thickness d is $(\pi d/\lambda) (n_r - n_l)$. The specific rotation of quartz is 29.73° mm^{-1} at $\lambda = 508.6$ nm; calculate the difference in refractive indices $(n_r - n_l)$.

3.4   Using the data given in Table 3.1 calculate some typical half-wave voltages. Why would you expect the half-wave voltage to vary with wavelength?

3.5   Suggest the design of a double-prism KD*P beam deflector to give a beam deflection of 2° (the voltage applied to the crystal should not exceed 20 kV).

3.6   Using the data provided in Table 3.1 calculate the modulation power required for a longitudinal GaAs modulator to give a retardation of π. Take the modulation bandwidth to be 0.1 MHz, the length of the crystal to be 30 mm and the cross-sectional area to be 100 mm^2. Verify that the same power would be required if the crystal were used in the transverse mode.

3.7   Design an optical isolator using zinc sulfide (see Table 3.3), take the permeability of zinc sulfide to be unity and assume that the magnetic field is produced by a solenoid wound directly onto the zinc sulfide at the rate of 5 turns per mm.

References

3.1   (a) G. R. Fowles, *Introduction to Modern Optics* (2nd edn) Holt, Rinehart & Winston, New York, 1975, Chapter 6.

    (b) E. Hecht, *Optics* (2nd edn) Addison-Wesley, Mass., 1987, Chapter 8.

3.2   (a) See ref. 3.1a, Chapter 9.

    (b) A. Yariv, *Optical Electronics* (3rd edn) Holt-Saunders, Japan, 1985.

3.3   (a) J. F. Nye, *Physical Properties of Crystals*, Oxford University Press, Oxford, 1957, Chapter 13.

    (b) I. P. Kaminow, *An Introduction to Electro-optic Devices*, Academic Press, New York, 1974.

3.4   (a) D. Chen, 'Magnetic materials for optical recording', *Appl. Opt.*, **13**, 1974, 767.

    (b) D. Chen and J. D. Zook, 'An overview of optical data storage technology', *Proc. IEEE*, **63**, 1975, 1207.

3.5   D. A. Pinnow, *IEE J. Quant. Electronics*, **QE. 6**, 1970, 223.

3.6   (a) A. Yariv, *Quantum Electronics*, Wiley, New York, 1975, sections 14.8–14.11.

    (b) See ref. 3.2b, Chapter 12.

3.7   L. Levi, *Applied Optics*, Vol. II, Wiley, New York, 1980, Chapter 14.

3.8   P. A. Franken, A. E. Hill, C. W. Peters and G. Weinreich, 'Generation of optical harmonics', *Phys. Rev. Letters*, **7**, 1961, 118.

3.9   J. A, Giordmaine and R. C. Miller, 'Tunable optical parametric oscillation in $LiNbO_3$ at optical frequencies', *Phys. Rev. Letters*, **14**, 1965, 973.

Chapter 4

Display devices

Introduction

We may divide display devices into two broad categories: (a) those that emit their own radiation (active devices) and (b) those that in some way modulate incident radiation to provide the display information (passive devices). However, before discussing the devices themselves, we consider the circumstances under which matter can be induced to emit radiation.

4.1 Luminescence

Luminescence is the general term used to describe the emission of radiation from a solid when it is supplied with some form of energy. We may distinguish between the various types of luminescence by the method of excitation. For example:

Photoluminescence   excitation arises from the absorption of photons
Cathodoluminescence excitation is by bombardment with a beam of electrons
Electroluminescence   excitation results from the application of an electric field (which may be either a.c. or d.c.)

Whatever the form of energy input to the luminescing material, the final stage in the process is an electronic transition between two energy levels, E_1 and E_2 $(E_2 > E_1)$, with the emission of radiation of wavelength λ where (see section 1.3)

$$\frac{hc}{\lambda} = E_2 - E_1 \qquad (4.1)$$

Invariably E_1 and E_2 are part of two *groups* of energy levels, so that, instead of a single emission wavelength, a *band* of wavelengths is usually observed.

When the excitation mechanism is switched off, we would expect the luminescence to persist for a time equal to the lifetime of the transition between the two energy levels E_1 and E_2. When this is so, we speak of *fluorescence*. Often, however, the luminescence persists for much longer than expected, a phenomenon called *phosphorescence*.

Phosphorescence is often attributable to the presence of metastable (or very long life-time) states with energies less than E_2. Electrons can fall into these states and remain trapped there until thermal excitation releases them some time later. Materials exhibit-ing phosphorescence are known as *phosphors*. Generally speaking, phosphor materials depend for their action on the presence within the material of impurity ions called *activators*. These replace certain of the host ions in the crystal lattice. Unless the charge on the activator ion is identical with that of the host ion it replaces, the charge balance will be upset and few will be able to enter the lattice. Improved solubility of the activator in these circumstances may result from the introduction of further impurity atoms with different ionic charge. These are known as *co-activators*.

We may distinguish between two main types of energy level system. In the first the energy levels are those of the activator ion itself, whilst in the second they are those of the host lattice modified by the presence of the activator ions. We refer to these two types as 'characteristic' and 'non-characteristic' respectively.

In characteristic luminescence, the excitation energy is usually transferred rapidly (i.e. in a time very much less than 10^{-8} s) to the activator ion. The persistence of the luminescence is then entirely due to the lifetime of the excited state level of the activa-tor. It should be noted that, whilst for atomic electric dipole transitions this is of the order of 10^{-8} s, it can be much longer if such transitions are forbidden (see ref. 4.1). Hence fluorescence cannot be unambiguously associated with characteristic luminescence.

In noncharacteristic luminescent materials, both activators and co-activators are usually present. These create acceptor and donor energy levels in the material (see section 2.4.2), although in phosphors these levels are usually referred to as hole and electron *traps* respectively. Energy absorption within the solid creates excess electron–hole pairs and, as the hole trapping probability is usually much greater than the electron trapping probability, most of the excess holes quickly become trapped. Any electron that then finds itself in the vicinity of a trapped hole can recombine with it and generate luminescence. As the electrons migrate through the crystal, however, they themselves

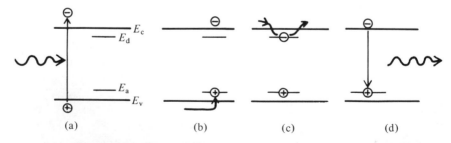

Fig. 4.1 Electron–hole generation and recombination processes in non-characteristic luminescent materials. Electron–hole pairs are generated by photon absorption (a), and the holes are quickly trapped at acceptor sites (b). Electrons may then recombine with these trapped holes, thereby giving rise to luminescent emission (d). However, before such a recombination can take place, the electron itself may spend some time trapped at a donor site (c).

are subject to trapping. The electron traps could, of course, act as recombination centers were there appreciable numbers of free holes present, but the difference in trapping probability prevents this. Instead an electron may remain in its trap for some time before subsequently being released by thermal excitation. It may then go on to be retrapped or to recombine with a trapped hole. This process is illustrated in Fig. 4.1.

The time that an electron spends in a trap depends on the depth of the trap below the conduction band ($E_c - E_d$) and also the temperature T. It is generally found that the probability of escape per unit time can be written in the form Q exp $- [(E_c - E_d)/kT]$ where Q is a constant approximately equal to $10^8 s^{-1}$ (ref. 4.2). Thus with fairly 'deep' traps (i.e. large $E_c - E_d$) and at low temperatures the time spent in a trap may be comparatively large, which will cause a long persistence of the luminescence after the cessation of excitation. In some cases this may amount to hours or even days.

Example 4.1 Luminescence lifetime due to traps

If we take a value of 0.4 eV for $E_c - E_d$, the probability of escape per second of a trapped electron at room temperature (where $kT = 0.025$ eV) is $\approx 10^8$ exp $(-0.4/0.025) \approx 10 \ s^{-1}$.

Hence we would expect a luminescence lifetime of ≈ 0.1 s.

We turn now to a more detailed discussion of photoluminescence, cathodoluminescence and electroluminescence.

4.2 Photoluminescence

As we have seen, in photoluminescence energy is transferred to the crystal by the absorption of a photon. In characteristic luminescent materials the activator ion itself absorbs the photon directly. It might be expected, therefore, that since the same energy levels are involved in absorption as in emission then the wavelengths for absorption and emission would be identical. In fact it is found that the peak emission wavelength is invariably shifted towards the red end of the spectrum compared to the peak of the absorption spectrum. This phenomenon is known as the *Stokes shift*, and it may be understood by taking account of the effect of the vibrations of the surrounding crystal lattice on the energy levels of the activator ions. The latter are often positively charged, typical examples being Cr^{3+} and Mn^{2+}, although the exact charge state depends on the host. For the sake of argument, we assume here that each activator ion is positively charged and is surrounded by six equidistant negatively charged ions at a distance R from the activator as shown in Fig. 4.2. We further assume that in the most important vibrational mode of the group, the activator ion remains at rest whilst the six surrounding negative ions all vibrate radially and in phase. Because of electrostatic interactions,

Fig. 4.2 Assumed impurity site structure in a characteristic luminescent material. The impurity ion (I) is surrounded by six host ions (H) each at a distance R from it.

the positions of the energy levels of the activator ion will depend on R. A schematic diagram illustrating this variation for two energy levels is shown in Fig. 4.3. We take these to represent the energy levels between which luminescent transitions take place. The most important feature of this diagram is that the minima of the two curves do not occur at the same values of R. This is perhaps not too suprising, since the quilibrium distribution of charge round the activator ion will be different when it is in the two states (corresponding as they do to different charge distributions round the ion).

Consider now the absorption of a photon when the activator is in its ground state, the most probable value for R is R_0 (the position of minimum energy). Photon absorption is a very rapid process and takes place virtually instantaneously as far as the vibration of the surrounding ions is concerned. On the energy level diagram, therefore, the process may be represented by a vertical transition (i.e. with R remaining constant). Immediately after the transition to the excited state the surrounding ions will not be at their equilibrium positions for this state and subsequently they will relax to their new equilibrium positions (at $R = R_1$ in Fig. 4.3). When a downward transition takes place, R again remains constant and an inspection of Fig. 4.3 shows that the emitted photon will then have less energy than that of the absorbed photon.

We have of course been talking about the *most probable* transition. The ions surrounding the activator will always be in a state of oscillation, and hence at the instant of absorption the value of R may well differ from R_0. The same will be true for the excited state. Thus instead of a single absorption (or emission) wavelength a *band* of absorption (or emission) wavelengths is seen. Further, since the amplitude of the oscillation will increase with increasing temperature, we would expect the width of the absorption and emission bands also to increase with increasing temperature. Figure 4.4 shows a curve of absorption and emission in KCl:Tl and illustrates the Stokes shift seen in this material at room temperature.

The Stokes shift finds commercial application in fluorescent lamps, in which an elec-

Fig. 4.3 Schematic diagram of the variation in the energy of two electron energy levels of an impurity ion in a characteristic luminescent material as a function of the nearest neighbor ion separation R. If the ion is initially at the point A on the diagram (where $R = R_0$), then photon absorption can take place, and the energy of the ion will change to that of the point B. The surrounding ions then relax to a new equilibrium position (R_1), and the impurity ion moves from B to C, losing energy by phonon emission. The impurity ion may then make a transition to the point D and emit a photon. Once at D the surrounding ions relax back to R_0, and the impurity ion returns to the point A, again losing energy by phonon emission.

trical discharge is passed through a mixture of argon and mercury vapor. The emitted radiation has a bluish color and an appreciable amount of the radiant energy is in the ultraviolet. If the walls of the discharge tube are coated with a suitable luminescing material, the ultraviolet radiation may be converted to useful visible radiation, thereby increasing the luminous efficiency of the lamp.

4.3   Cathodoluminescence

In cathodoluminescence, although the emission processes are the same as those outlined above for photoluminescence, the excitation mechanisms are somewhat more involved. When a beam of energetic electrons (say with energy greater than 1 keV) hits a solid, a fraction (about 10%) is backscattered. The remainder penetrate into the solid

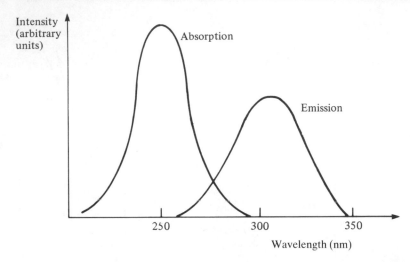

Fig. 4.4 Absorption and emission spectra for thallium activated potassium chloride (KCl:Tl) at room temperature. The emission peak occurs at a higher wavelength than that of the absorption curve. This is an example of the Stokes shift.

where they rapidly lose energy, mainly by causing bound electrons to be ejected from their parent ions. These secondary electrons in turn may generate further secondary electrons, provided they have sufficient energy. The final stages in the secondary generation process consist of the excitation of electrons from states at the top of the valence band (with energy E_v) to those at the bottom of the conduction band (with energy E_c). Energy conservation considerations alone would dictate that the exciting electrons must have an energy of at least $E_c - E_v = E_g$ above E_c if they are to create electron–hole pairs and still remain in the conduction band. In fact the additional constraint of momentum conservation requires a somewhat higher minimum energy of approximately $E_c + 3E_g/2$ (ref. 4.3). When electrons have energies between this value and E_c, they can only lose energy by exciting lattice vibrations (phonons). In addition, of course, even when electrons have higher energies than $E_c + 3E_g/2$, energy may still be wasted in phonon generation. It has been found empirically for a range of semiconductor materials that the total number of electron–hole pairs generated may be written as $E_B/\beta E_g$, where E_B is the total electron beam energy and $\beta \approx 3$. This inefficiency in the generation of electron–hole pairs is a major factor in causing cathodoluminescence to be a considerably less efficient process than photoluminescence.

In noncharacteristic materials, electron–hole recombination and luminescent emission then takes place as for photoluminescence. In characteristic materials, on the other hand, it is thought that the next step is the formation of excitons (bound electron–hole pairs, see section 2.4.3). These migrate through the lattice and may subsequently transfer their recombination energy to the activator ions.

As the primary electrons rapidly lose energy they penetrate only a little way into the solid they are exciting. It has been found experimentally that the penetration depth, or range, R_e of an electron beam of energy E_B is given by (ref. 4.4):

$$R_c = KE_B^b \tag{4.2}$$

where the parameters K and b depend on the material. For ZnS, for example, the range is in microns when $K = 1.2 \times 10^{-4}$ and $b = 1.75$. Thus a 10 keV electron beam has a range in ZnS of $1.2 \times 10^{-4} \times 10^{1.75}$ or 0.7 μm.

It is often found that cathodoluminescent efficiency increases with increasing beam voltage. This may be attributed mainly to the fact that at low beam voltages most of the electron–hole pairs are generated close to the surface of the luminescent material, where there is often a relatively high concentration of nonluminescent recombination centers. (Another instance of a similar deleterious surface effect is the reduction in efficiency of photoconductive detectors with decreasing wavelength, as discussed in section 7.2.7.)

4.4 Cathode ray tube

The relative ease with which a beam of electrons can be directed and focused led to the early development and retention of the cathode ray tube (CRT) as an important tool for the analysis of rapidly varying electrical signals as well as providing a versatile optical display device. Only a very brief discussion of the CRT can be attempted here and the reader is referred to ref. 4.5 for more details. Figure 4.5 shows the basic construction. Electrons are generated by thermionic emission (see section 2.6) by heating a specially impregnated cathode surface (usually based on oxides of barium and strontium) and then focused onto the viewing screen by a series of metal electrodes held at various potentials. A grid for the control of the electron flow is usually also included. The whole

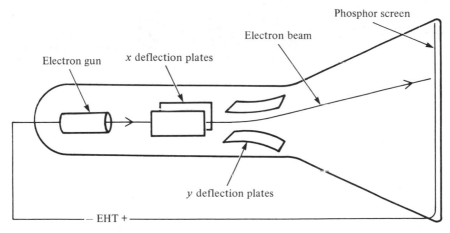

Fig. 4.5 Schematic diagram of a CRT. An electron beam originating at an electron gun passes through an electrostatic deflection system (which uses two sets of plates at right angles, one for the 'x' deflection and the other for the 'y' deflection) and then falls onto a phosphor screen. Details of the electron gun and its associated focusing system are not shown.

assembly is known as an 'electron gun'. The electron beam is scanned across the viewing screen in a series of lines; when one line scan is completed the beam is rapidly switched to the start of the line below. Beam deflection is controlled by electrostatic or electromagnetic fields acting at right angles to the beam direction. Electrostatic deflection enables the highest beam deflection rates to be achieved, whilst electromagnetic deflection enables higher beam accelerating potentials to be employed, which results in a smaller spot size and higher screen brightness. When the beam strikes the viewing screen, radiation is generated by cathodoluminescence. The screen consists of a thin layer of small (dimensions ≈ 5 µm) phosphor granules, with a layer of aluminum (≈ 0.1 µm thick) evaporated onto the gun side (Fig. 4.6). This layer serves two purposes; firstly, it prevents charge build-up on the phosphor granules (which generally have low conductivities) and secondly it helps to reflect light emitted in a direction away from the observer back towards him.

The thicknesses of both the aluminum and phosphor layers are fairly critical. If the aluminum is too thick, an appreciable fraction of the electron beam energy will be absorbed within it, whilst if it is too thin its reflectivity will be poor. If the phosphor layer is too thick, on the other hand, then scattering and absorption reduces the light output, whilst too thin a layer can result in incomplete coverage of the screen area.

For normal display operations (e.g. television) the beam is scanned line by line over the viewing area. In video applications the display consists of some 625 lines in Europe and 525 in America. To avoid an image that 'flickers' the picture must be renewed at a rate greater than about 45 Hz. However, it is possible to avoid having to renew the

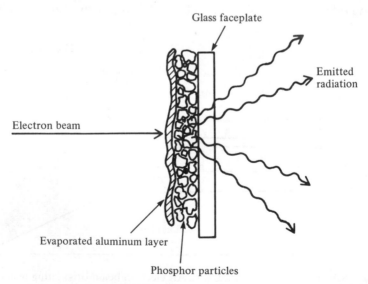

Fig. 4.6 Cross section of a CRT screen. A layer of phosphor is sandwiched between a glass faceplate and an evaporated aluminum layer. High energy electrons penetrate the aluminum and exite cathodoluminescence in the phosphor particles. The aluminum layer reduces charge build-up and helps to reflect light back out through the faceplate.

entire picture at this rate by using a raster scan that splits the picture up into two inter-laced halves. Thus if a complete picture scan takes t_s seconds then we may arrange that during the first $t_s/2$ seconds lines 1, 3, 5, 7 etc. are scanned, whilst during the second $t_s/2$ seconds lines 2, 4, 6, 8 etc. are scanned. Because the two images are effectively super-imposed, the eye treats the picture repetition rate as if it were $2/t_s$ Hz rather than $1/t_s$ Hz. This reduction in the rate at which picture information is required before flicker be-comes troublesome is very useful because it halves the transmission frequency band-width that would otherwise be required. In Europe the entire picture is scanned in 1/25 s, whilst in America this time is 1/30s. Varying light irradiances are obtained by varying the beam current. Ideally, the phosphor used should have a luminescent decay time shorter than the picture cycle time, otherwise streaking effects due to image per-sistence are obtained. CRT displays can be made sufficiently bright for them to be visible under nearly all ambient lighting conditions. The brightness limits are usually reached when the phosphor screen rapidly deteriorates under high beam currents.

Color displays for home video viewing are obtained using the 'shadowmask' prin-ciple, in which three electron guns are used that are slightly inclined to each other so that their beams coincide at the plane of the shadowmask. The latter is a metal screen, with holes in it, placed just in front of the phosphor screen. Having passed through one of the holes in the shadowmask the three beams diverge, and on striking the phosphor screen are again physically distinct. The phosphor screen consists of groups of three phosphor dots, placed so that when the three beams pass through a hole in the shadow-mask they each hit a different dot. Figure 4.7 illustrates the basic geometry. Each of the three phosphor dots emits one of the primary colors (i.e. blue, green and red), so that any desired color can be generated by varying the relative excitation intensities. Some

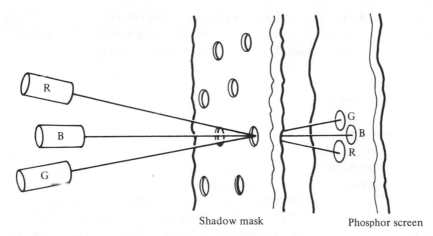

Shadow mask

Phosphor screen

Fig. 4.7 Use of the shadowmask for obtaining color displays. Three separate guns are used; these are inclined slightly to each other so that their beams will all pass through a single hole in the shadowmask. After passing through the hole, the beams diverge so that each falls on one of the three circular areas composed of phosphors each of which emits one of the three primary colors (blue, green and red).

of the more commonly used phosphors are zinc sulfide doped with silver: $ZnS:Ag$ (blue); zinc cadmium sulfide doped with copper: $Zn_xCd_{1-x}S:Cu$ (green); and yttrium oxysulfide doped with europium and terbium: $Y_2O_2S:Eu,Tb$ (red). The first two are noncharacteristic phosphor materials whilst the latter is a characteristic material. There is obviously some loss in resolution capability over a monochrome display since the colored dots are physically displaced on the screen. Furthermore, the alignment of the shadowmask with the guns and the phosphor screen is critical, and can be spoilt by fairly harsh environmental conditions such as stray magnetic fields. Several modifications have been made to the basic shadowmask principle to try and rectify some of these disadvantages; the interested reader is referred to ref. 4.5 for further information.

A somewhat different method of CRT color display has been developed within the last few years using so-called 'penetration phosphors'. These utilize the variation in electron penetration depth that can be obtained by varying the beam voltage (see eq. (4.2)). For example, a two-color display may be obtained by using a mixture of two different kinds of phosphor particle (which we take to be red and green emitting). One, the red emitting, is an ordinary phosphor granule whilst the other, the green emitting, has a non-luminescent coating. At low beam voltages the display will only show red, since the electrons will have insufficient energy to penetrate the non-luminescent coating of green phosphor. At higher voltages, electron penetration into the green phosphor will give rise to both red and green emission. If the red phosphor is less efficient than the green, then at high voltages green will predominate. As can be imagined, however, there are considerable difficulties in obtaining any desired intermediate color at any desired level. Further, the necessary changes in beam potential cannot be achieved rapidly enough for video applications. For a relatively static display that requires only a few fixed colors, however, this technique offers good resolution capability and relative freedom from magnetic interference.

In conclusion, the CRT provides a very versatile display which can readily cope with complex displays. Its main disadvantages are its comparatively low screen area to volume ratio and the necessity for a high voltage power supply.

4.5 Electroluminescence

We are concerned here with what might be termed 'classical electroluminescence' as opposed to 'injection electroluminescence', which uses fabricated $p-n$ junctions and which will be dealt with in section 4.6. Four main types of device may be distinguished, depending on the type of drive (a.c. or d.c.) and the character of the active layer (powder or thin film). The first electroluminescent device to be extensively studied was the a.c. powder device, proposed in 1936. In this a phosphor powder (usually $ZnS:Cu$) is suspended in a transparent insulating binding medium of high dielectric constant and is sandwiched between two electrodes (one of which is transparent) as shown in Fig. 4.8(a). Usually there is no complete conducting path between the electrodes, so that d.c. excitation is not possible. When an alternating voltage, $V_0 \cos(2\pi f t)$, is applied across

(a)

Glass cover

Transparent electrode
(usually SnO$_2$)

Phosphor particles
in nonconducting
binding medium

Metal electrode

(b)

Anode
(transparent)

Copper depleted
surface

ZnS : Cu, Mn

Cu$_x$S coating

Cathode

Fig. 4.8 Construction of an a.c. electroluminescent device (a). Phosphor particles are suspended within a transparent insulating medium and sandwiched between two electrodes, one of which is transparent. When an alternating voltage is applied to the electrodes the phosphor particles emit light. (b) The construction of a d.c. electroluminescent device. The phosphor particles have a coating of Cu$_x$S. This coating is removed from the anode side of the particles in contact with the anode by the application of an initial high current pulse. Under normal conditions, light is emitted only from the Cu$_x$S depleted particles.

the cell, however, light is emitted in the form of short bursts which last about 10^{-3} s and occur once every half cycle. It is found that the integrated light output power P can be written in the form (ref. 4.6):

$$P = P_0(f) \exp\left[-\left(\frac{V_1}{V_0}\right)^{\frac{1}{2}}\right] \tag{4.3}$$

where V_1 is a constant and $P_0(f)$ is a function of frequency. The strongest emission from within the phosphor grain is found to take place from the side temporarily facing the cathode. Several possible emission mechanisms have been proposed (see ref. 4.6 for further details); it is generally agreed, however, that there will be a high electric field within the phosphor particle. It is then possible that this field is sufficiently strong to enable electrons from occupied acceptor levels to 'tunnel' to states of the same energy

Fig. 4.9 Possible mechanism for electroluminescence emission involving quantum mechanical tunneling. In (a) an electron in an acceptor state 'tunnels' through the forbidden gap region into states of the same energy. It is only able to do this if there is a considerable electric field present, thus causing the energy bands to be tilted. An electron in the conduction band may now fall into the vacated level resulting in radiative emission (b).

in the conduction band, as illustrated in Fig. 4.9(a). (Quantum mechanical tunneling was briefly mentioned in section 2.8.5.) Other electrons in the conduction band are then able to fall into these vacated levels and emit radiation (Fig. 4.9(b)).

Another possibility is that an electron moving in the electric field may acquire sufficient energy to enable it to excite an electron from the valence band to the conduction

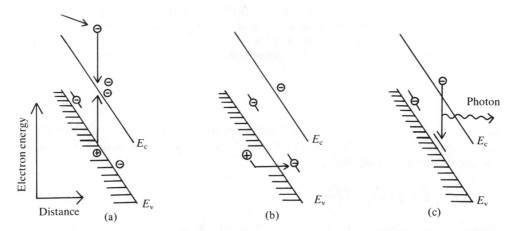

Fig. 4.10 Possible mechanism for electroluminescence emission involving an avalanche process. In (a) an electron moving in the high electric fields present may acquire sufficient energy to excite an electron from the valence band into the conduction band. The hole left behind then moves up into an acceptor state effectively emptying it of an electron (b). Finally, an electron in the conduction band may then make a radiative transfer into the empty acceptor level (c).

band. The resulting hole quickly becomes trapped at an impurity acceptor site, thereby effectively emptying it of an electron. An electron in the conduction band can then make a radiative transition by falling into the empty acceptor level. The sequence of events is illustrated in Fig. 4.10. In phosphors containing manganese, there is evidence that the Mn^{2+} ions themselves may be directly excited by the high energy electrons, radiation being emitted when the ion subsequently undergoes de-excitation. A.c. powder devices usually require several hundred volts (r.m.s.) to drive them. They exhibit luminances of about 40 nits (for a definition of the nit, see section 4.8), have power efficiencies (i.e. the ratio of optical power out to electrical power in) of about 1% and have lifetimes of about 1000 h. By using different phosphor powders, red, green, yellow and blue displays are possible.

A more recent development is the d.c. powder display. These have a structure basically similar to that of the a.c. device. However, the phosphor particles ($ZnS : Cu,Mn$) are coated with a conducting layer of Cu_xS. Provided the phosphor particles are not too widely dispersed within the binder there will be a conducting path from the anode to the cathode. Before normal operation, the cell must be 'formed' by applying a high voltage across it for a short time. This causes copper ions to migrate away from the phosphor surfaces next to the anode. A thin high resistance layer of ZnS is then created next to the anode, across which most of the applied voltage appears, and from which light

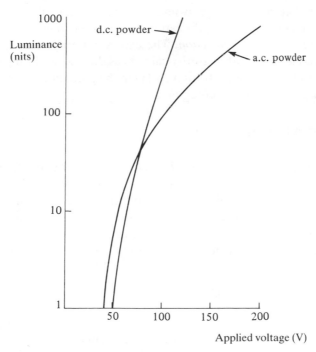

Fig. 4.11 Typical luminances obtained for a.c. and d.c. electroluminescent powder devices as a function of the applied voltage (r.m.s. volts in the case of the a.c. device).

emission takes place during subsequent operation at lower voltages (Fig. 4.8(b)). Luminances of about 300 nits are possible at voltages of around 100 V d.c., although the power conversion efficiencies are low at approximately 0.1%. The luminance versus drive voltage characteristics for both a.c. and d.c. powder electroluminescent devices are shown in Fig. 4.11.

In addition, a.c. and d.c. devices have been made where the active layer is a vacuum deposited thin film of phosphor material (usually based on ZnS), but both types tend to have rather poor lifetimes. Although a considerable amount of research effort has been put into the development of electroluminescent displays, they have yet to make any significant commercial impact. They do offer the possibility, however, of large area displays with a high surface area to volume ratio, and their output-voltage characteristics, particularly of the d.c. powder devices, are suitable for matrix addressing, as discussed in section 4.10.

4.6 Injection luminescence and the light emitting diode

The basic structure giving rise to injection luminescence is that of a $p-n$ junction diode operated under forward bias which was discussed in Chapter 2. Under forward bias, majority carriers from both sides of the junction cross the internal potential barrier and enter the material at the other side, where they are then the minority type of carrier and cause the local minority carrier population to be larger than normal. This situation is described as *minority carrier injection*. The excess minority carriers diffuse away from the junction recombining with majority carriers as they do so. Using eq. (2.42), we may write the excess electron concentration $\Delta n(x)$ in the p material as a function of distance x from the edge of the depletion region as:

$$\Delta n(x) = \Delta n(0) \exp\left(-\frac{x}{L_e}\right)$$

The process is illustrated in Fig. 4.12. Ideally, in a light emitting diode (LED) every injected electron takes part in a radiative recombination and hence gives rise to an emitted photon. In practice this is not so, and the efficiency of the device may be described in terms of the *quantum efficiency*, which is defined as the rate of emission of photons divided by the rate of supply of electrons. In reverse bias, no carrier injection takes place and consequently no light is emitted. The current–voltage (i, V) relationship for a diode can usually be written (see eq. (2.51) and section 2.8.5) as

$$i = i_0\left[\exp\left(\frac{eV}{\beta kT}\right) - 1\right] \tag{4.4}$$

where i_0 is a constant (the reverse saturation current).

The number of radiative recombinations that take place is usually proportional to the carrier injection rate and hence to the total current flowing. If the transitions take place

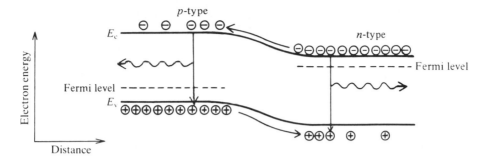

Fig. 4.12 Injection of minority carriers and subsequent radiative recombination with the majority carriers in a forward biased *p–n* junction.

directly between states at the bottom of the conduction band and at the top of the valence band, then the emission wavelength λ_g is given by (see eq. (4.1)):

$$\frac{hc}{\lambda_g} = E_c - E_v = E_g$$

$$\therefore \lambda_g = \frac{hc}{E_g} \tag{4.5}$$

For example, GaAs has an energy band gap of 1.43 eV, which corresponds to a value for λ_g of 0.86 μm. In fact, because of thermal excitation, electrons in the conduction band have a most probable energy which is $kT/2$ above the bottom of the conduction band (see Problem 2.20). Band-to-band transitions therefore result in a slightly shorter emission wavelength than given by eq. (4.5), and self-absorption can further distort the situation. However, as we shall see later, most transitions involve energy levels within the energy gap, and for these eq. (4.5) represents the shorter wavelength limit. We now consider the transmission process in more detail.

4.6.1 *Radiative recombination processes*

Radiative recombination in semiconductors occurs predominantly via three different processes, namely (1) interband transitions, (2) recombination via impurity centers and (3) exciton recombination.

4.6.1.1 INTERBAND TRANSITIONS

This recombination process is illustrated schematically on an energy level diagram for both direct and indirect bandgap materials in Figs. 4.13(a) and (b) respectively. (*E–k* diagrams for silicon and gallium arsenide are shown in Fig. 2.6.) It is important to realize that the transition must conserve the total wave vector of the system. The photon wave vector is given by $2\pi/\lambda$; whilst the electron wave vectors involved range

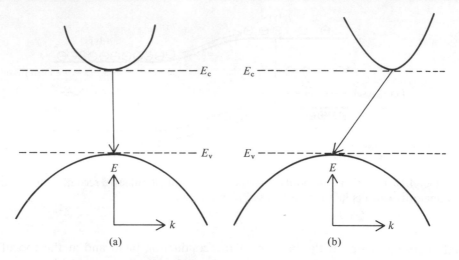

Fig. 4.13 Interband transitions for (a) a direct band gap semiconductor and (b) for an indirect band gap semiconductor. In the former cases, there is no change in the electron k value, whilst in the latter case there is.

between approximately $-\pi/a$ and $+\pi/a$, where a is the crystal lattice spacing (i.e. the k values at the first Brillouin zone boundaries; see Fig. 2.5(b)). For visible radiation, $\lambda \approx 0.5 \times 10^{-6}$ m whereas crystal lattice spacings are approximately 10^{-10} m; hence $2\pi/\lambda \ll \pi/a$. The photon wave vector is thus much smaller than the possible electron wave vectors. Consequently if the only particles involved are an electron and a photon, then the electron must make a transition between states having virtually the same wave vector. On an $E-k$ diagram, therefore, only *vertical* transitions are allowed. It is possible to have nonvertical transitions (as illustrated in Fig. 4.13(b)) but to conserve the wave vector, a phonon must either be created or destroyed at the same time. The equation for the wavelength of the emitted photon is then given by:

$$\frac{hc}{\lambda} = E_g \pm E_p \tag{4.6}$$

where E_p is the phonon energy. The $+$ and $-$ signs correspond to phonon *annihilation* and *creation* respectively. Phonon energies are of the order of 0.01 eV, and hence the photon wavelength in fact differs little from λ_g. However, because now *three* particles are involved instead of just two for the indirect transition process, the transition is much less probable. We know from the discussion in section 2.4 that the interband recombination rate r may be written:

$$r = Bnp \tag{4.7}$$

where B is a constant. Calculated values for B in various semiconductor materials are shown in Table 4.1 (see ref. 4.7), from which we see that the values of B for indirect bandgap materials are some 10^6 times smaller than for direct bandgap materials. We conclude that band-to-band radiative transitions in indirect bandgap semiconductors

Table 4.1

Group(s)	Element/ compound	Direct/ indirect	E_g (eV)	Readily doped n or p type	$B(\mathrm{m^3 s^{-1}})$	λ_g (nm)
IV	C	i	5.47			227
	Si	i	1.12	yes	$1.79\ 10^{-21}$	1106
	Ge	i	0.67	yes	$5.25\ 10^{-20}$	1880
IV–VI	SiC (hex. α)	i	3.00	yes		413
III–V	AlP	i	2.45			506
	AlN	i	5.90	no		210
	AlSb	i	1.50			826
	AlAs	i	2.16			574
	GaN	d	3.40	no		365
	GaP	i	2.26	yes	$5.37\ 10^{-20}$	549
	GaAs	d	1.43	yes	$7.21\ 10^{-16}$	861
	InN	d	2.40			516
	InP	d	1.35	yes	$1.26\ 10^{-15}$	918
	InAs	d	0.35		$8.50\ 10^{-17}$	3540
	InSb	d	0.18		$4.58\ 10^{-17}$	6870
II–VI	ZnO	d	3.20	no		387
	ZnS(α)	d	3.80	no		326
	ZnS(β)	d	3.60	no		344
	ZnSe	d	2.28	no		480
	ZnTe	d	2.58	no		544
	CdS	d	2.53	no		490
	CdSe	d	1.74	no		712
	CdTe	d	1.50	yes		826

are relatively rare, and unless some other radiative transition mechanisms are possible these materials will not be suitable for LEDs.

One disadvantage of radiation derived from direct band gap recombination is that the probability of the emitted radiation being re-absorbed in band-to-band transitions can be high when the radiation has to traverse an appreciable thickness of the semiconductor material.

4.6.1.2 IMPURITY CENTER RECOMBINATION

Three types of recombination involving impurity energy levels are shown in Fig. 4.14. Thus we may have (a) conduction band–acceptor level transitions, (b) donor level–valence band transitions and (c) if in addition a pair of donor and acceptor states are close together, then donor–acceptor transitions are possible. When the electron is in either type of impurity state, then it will be fairly strongly localized. This spatial localization implies that the electron can have a range of momentum values, since uncertainties in both position (Δx) and momentum (Δp) are related via the Heisenberg uncertainty relation (eq. 2.3)

$$\Delta x\ \Delta p \geqslant \frac{\hbar}{2}$$

Because the electron wave vector k and the momentum p are related by $k = p/\hbar$ we may write:

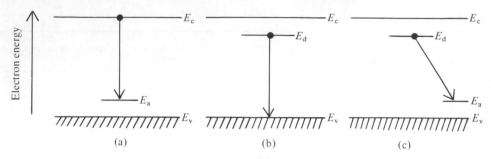

Fig. 4.14 Three types of recombination involving impurity energy levels are shown. In (a) an electron moves from the conduction band into an empty acceptor level. In (b) an electron in a donor level recombines with a hole in the valence band. In (c) an electron in a donor level falls into an empty acceptor level. This latter process requires that the donor and acceptor levels are physically close together.

$$\Delta k \geqslant \frac{1}{2\Delta x} \tag{4.8}$$

We express the electron localisation by putting $\Delta x = Na$, where a is the lattice spacing and N a number which is not expected to be much larger than unity: whence $\Delta k \geqslant 1/(2\,Na)$. In our discussion at the start of section 4.6.1.1, we noted that electrons in the conduction and valence bands have a range of k values extending approximately from $-\pi/a$ to $+\pi/a$. Hence we see that there may be sufficient spread in the k values of the impurity states to allow a significant number of transitions between them and the band extrema without calling on the assistance of phonons. Transitions via impurity states therefore provide a possible mechanism whereby indirect bandgap semiconductors can increase their radiative efficiency.

Typical values for $E_c - E_d$ and $E_a - E_v$ are of the order of 0.02 eV, so that the emission wavelength will be slightly longer than that given by eq. (4.5). However, thermal excitation within the bands themselves will tend to decrease the magnitude of this effect (see the discussion following eq. (4.5)). Since the radiation has a lower photon energy than that required to excite an electron across the bandgap, it is not subject to re-absorption to the same extent as that derived from band-to-band recombinations.

4.6.1.3 EXCITON RECOMBINATION

Exciton states exist within the energy gaps of even pure semiconductor materials (they must not be confused with impurity donor or acceptor states). We may visualize such states as being akin to Bohr-like states in which an electron and hole circle round their common center of gravity at relatively large distances (Fig. 2.12). The electron and hole are relatively weakly bound and the exciton states are situated just below the bottom of the conduction band. In Chapter 2 we showed (eq. (2.28i)) that we may write the exciton binding energy E_e as:

$$E_e = 13.6 \frac{m_r^*}{m_e}\left(\frac{1}{\varepsilon_r}\right)^2 \text{eV} \tag{4.9}$$

where m_r^* is the reduced mass. For example, in GaAs we have $\varepsilon_r = 11.5$, $m_e^* = 0.068\,m$ and $m_h^* = 0.47\,m$, whence $m_r^* = 0.06\,m$ and $E_e = 5.9$ meV. Observed experimental values are in reasonable agreement with this simple model calculation; in GaAs the exciton binding energy is found to be 4.8 meV. The exciton is capable of movement through the lattice, although, since exciton energies can be affected by the presence of impurities, in some circumstances an exciton may remain 'bound' in the vicinity of the impurity. If the impurities are neutral donors or acceptors then the exciton binding energy is usually about one tenth of that of the centers to which they are bound. (The binding energy may be much larger than this at isoelectronic traps, which are discussed later.) Bound exciton states may be sufficiently well localized so that electron–hole recombination can take place in indirect bandgap semiconductors via these states without the need for phonon intervention to conserve the wave vector.

4.6.2 LED materials

We may summarize the main requirements for a suitable LED material as follows: first, it must have an energy gap of appropriate width; secondly, both p and n types must exist, preferably with low resistivities; and finally, efficient radiative pathways must be present. Equation (4.5) indicates that, in order to obtain visible radiation, energy gaps greater than or equal to about 2 eV are required. Unfortunately, materials with such large gaps tend to have high resistivities even when doped. Furthermore, in most cases the wider the energy gap, the greater are the difficulties met in material preparation. This is often because the materials have high melting temperatures and low structural stability.

Table 4.1 lists a number of relevant properties of several elemental and binary semiconductors. There are no suitable elemental semiconductors as they all have indirect energy gaps. Diamond, the only element with a large enough energy gap to enable visible emission to take place, is too intractable and expensive a material. Both silicon and germanium can be readily fabricated into diodes (hence their outstanding success in purely electronic components), but they have extremely low radiative efficiencies, and in any case would emit in the infrared spectral region.

Silicon carbide is a promising material, but its high melting point and the consequent growth difficulties prevented its early use in LEDs. It can be doped as both n and p type, and commercially available blue emitting diodes of SiC have become available recently. Doping with B, Al, Sc and Be gives rise to yellow, blue, green and red emission respectively. Unfortunately, SiC is an indirect bandgap material and radiative transitions via donors and acceptors are rather inefficient leading to a low overall brightness.

GaN is another candidate for a blue LED material, although it has not proved possible to make it p type. The devices made so far have been based on an $n^+ - i$ structure, where the electrons are injected from the n^+ region into the i region. Here they can combine radiatively with holes via zinc acceptors. Voltages of between 5 and 10 V are required and maximum allowed currents are only of the order of 10 mA, brightness is again relatively low. At present, the prospects for blue LEDs with high brightnesses are rather poor.

The most important of the III–V compounds are undoubtedly GaAs, GaP and the ternary 'alloy' $GaAs_{1-x}P_x$ formed from them. These will be discussed more fully in the next section.

None of the II–VI compounds have so far proved very practicable, and only in CdTe has it been possible to make p–n junctions. In principle, heterojunction could be made between ZnTe (which can be made p type) and one of the other II–VI compounds, but the high internal resistances of these devices result in efficiencies too low, as yet, for commercial applications.

4.6.3 Commercial LED materials

Gallium arsenide (GaAs). This is a direct bandgap semiconductor with $E_g = 1.443$ eV ($\lambda_g = 860$ nm); p–n junctions are readily formed and have high luminescent efficiencies. One of the easier methods of making such junctions is to diffuse zinc into pulled crystals of n–GaAs. The radiation emitted arises from band-to-band transitions, and is consequently subject to heavy re-absorption. Because of this, the peak emission wavelength is shifted to about 870 nm.

More efficient diodes are made by doping with silicon. In solution-grown GaAs, silicon can act either as a donor (replacing Ga), or as an acceptor (replacing As), depending on the growth conditions. Low temperature growth ($\lesssim 820°C$) favours acceptors whilst high temperature growth favours donors. In addition to these, a complex acceptor center may be formed with $E_a - E_v \simeq 0.1$ eV (the exact energy depends on the growth conditions). In silicon-doped diodes, the main luminescent transition is between the conduction band and this complex acceptor level. The peak wavelength of the transition can lie between 910 and 1020 nm. This radiation has insufficient energy to cause band-to-band transitions, and hence re-absorption is low. In consequence, such diodes are highly efficient, and external quantum efficiencies of up to 10% can be obtained. Typical emission spectra for both Zn-diffused and Si-doped diodes are shown in Fig. 4.15. Si doped diodes have now become the 'standard' near-IR emitting devices, although recently more efficient diodes based on GaAlAs have become available (see later).

Gallium phosphide (GaP). This is an indirect bandgap semiconductor with $E_g = 2.26$ eV ($\lambda_g = 549$ nm), and hence band-to-band transitions are not normally observed. Group V elements, such as N and Bi, are commonly used as dopants to assist radiative recombination. Since these have the same valency as the atoms which they replace (P), they do not form normal donor or acceptor states. The sites do act as recombination centers, however, and are termed *isoelectronic traps*. They are believed to act as follows. First, a free carrier is trapped at the center. Trapping is found to take place only when the impurity atom differs considerably from the atom it replaces both in its size and its tendency to acquire negative charge (electro-negativity). Once a carrier is trapped, the resulting Coulomb potential attracts a carrier of the opposite charge and a bound exciton is formed. Finally the exciton recombines and emits radiation.

The nitrogen isoelectronic trap has an 8 meV binding energy for electrons and the

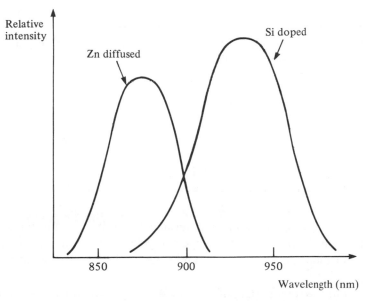

Fig. 4.15 Spectral output of an Si-doped and a Zn-diffused GaAs LED.

subsequent exciton is only fairly weakly bound; consequently, the resulting emission has a wavelength fairly close to λ_g. (Typical peak emission is at about 565 nm, i.e. green radiation.) External quantum efficiencies are fairly low ($\approx 0.1\%$), mainly because the exciton is weakly bound and readily dissociates at room temperature into a free electron and hole.

At high nitrogen concentrations ($> 10^{25}$ m^{-3}) the room temperature emission shifts to 590 nm (i.e. yellow). This is possibly due to the formation of energy levels within the bandgap arising from nearest neighbor N–N molecular complexes.

Red emission is also possible from GaP by using (Zn,O) double doping. Here, the Zn dopant atoms replace the Ga, whilst O replaces the P. When the two different dopant atoms lie on nearest neighbor sites, it is believed that again an isoelectronic trap is formed. Here, however, the binding energies are much larger (≈ 0.3 eV) than for the N trap, and consequently exciton recombination occurs at higher wavelengths (≈ 690 nm, i.e. red).

Such LEDs have quite high quantum efficiencies; however, they are not widely used since the emission wavelength is in a region where the human eye is not very sensitive, and consequently they do not appear very bright. In addition their efficiency drops with increasing drive current.

Gallium arsenide phosphide (GaAs$_{1-x}$P$_x$). Another useful LED material is the ternary alloy GaAs$_{1-x}$P$_x$. This material is interesting in that it changes from being a direct bandgap material (when $x < 0.45$) to being an indirect bandgap material (when $x > 0.45$). At the changeover point, the bandgap is approximately equal to 2.1 eV. Using diodes of composition GaAs$_{0.6}$P$_{0.4}$, red emission is obtained, which arises from direct band-to-

Table 4.2

Material	Dopant	Peak emission (typical values) (nm)	Color	External quantum efficiencies (commercial diodes) (%)
GaAs	Si	$910 \rightarrow 1020$	Infrared	10
GaP	N	570	Green	0.1
GaP	N, N	590	Yellow	0.1
GaP	Zn, O	700	Red	4
$GaAs_{0.6}P_{0.4}$		650	Red	0.2
$GaAs_{0.35}P_{0.65}$	N	632	Orange	0.2
$GaAs_{0.15}P_{0.85}$	N	589	Yellow	0.05
$Ga_{0.6}Al_{0.4}As$	Zn	650	Red	3
$Ga_xAl_{1-x}As$ $(1 < x < 0.7)$	Si	$870 \rightarrow 890$	Infrared	15

band transitions The indirect bandgap material can also be made to emit visible radiation by doping with N; for example, orange and yellow emission may be obtained from diodes of composition $GaAs_{0.35}P_{0.65}$: N and $GaAs_{0.15}P_{0.85}$: N respectively.

Gallium aluminum arsenide ($Ga_xAl_{1-x}As$). A fairly recent newcomer to the LED scene is $Ga_xAl_{1-x}As$, which is used in highly efficient red and near-infrared LEDs. In the former, a heterojunction is formed between n-type $Ga_{0.3}Al_{0.7}As$ and p-type $Ga_{0.6}Al_{0.4}As$. The n layer is the surface layer and, in operation, electrons are injected into the p material with efficiencies that approach 100%. Radiative recombination (via acceptor levels) gives rise to radiation with a wavelength of 650 nm. This radiation passes through the surface n-type layer with little attenuation, because its bandgap is larger than that of the p-layer. The high efficiencies of such 'super-bright' red LEDs may well cause them to replace the 'standard' types based on GaAsP. As far as near-infrared diodes based on GaAlAs are concerned, the basic structure is again a $p-n$ junction with the n layer uppermost. As with GaAs, doping with Si is used to form both donor and acceptor states. The bandgap is varied through the device by varying the Al concentration from 30% at the surface to 0% some way beyond the junction into the p material. Peak emission occurs between 870 and 890 nm. Again, the efficiencies of these LEDs are better than those made from the 'standard' material, GaAs. However, the GaAs diodes do have the advantage that they can withstand higher peak currents.

Table 4.2 summarizes the characteristics of the most commonly used LED materials.

4.6.4 LED construction

A typical LED construction is shown in Fig. 4.16. It is obviously advantageous if most of the radiative recombinations take place from the side of the junction nearest the surface, since then the chances of re-absorption are lessened. We may ensure this by arranging that most of the current flowing across the diode is carried by those carriers that are injected into the surface layer. We assume that, as in Fig. 4.16, the surface layer is p type. The fraction of the total diode current that is carried by electrons being injected into the p side of the junction (η_e) is then given by:

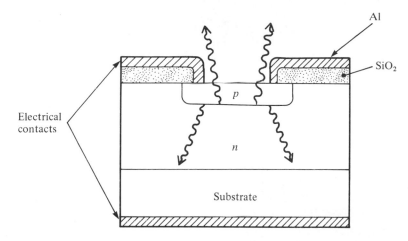

Fig. 4.16 Construction of a typical LED. A shallow $p-n$ junction is formed and electrical contacts made to both regions whilst leaving as much of the upper surface of the p material uncovered so that the flow of radiation from the device is impeded as little as possible.

$$\eta_e = \frac{\dfrac{D_e n_p}{L_e}}{\dfrac{D_e n_p}{L_e} + \dfrac{D_h p_n}{L_h}}$$

or $\quad \eta_e = \left(1 + \dfrac{D_h L_e p_n}{D_e L_h n_p}\right)^{-1}$ $\hspace{3cm}$ (4.10)

This equation is readily derived from the results of section 2.8.2. By using the Einstein relation (eq. (2.40)), i.e. $D_{e,h} = (kT/e)\mu_{e,h}$, and the relation $n_p p_p = p_n n_n = n_i^2$ (eq. (2.36)), eq. (4.10) becomes

$$\eta_e = \left(1 + \frac{\mu_h p_p L_e}{\mu_e n_n L_h}\right)^{-1} \hspace{3cm} (4.11)$$

In III–V compounds $\mu_e \gg \mu_h$, and so, assuming that $L_e \simeq L_h$, we see that there is a natural tendency for η_e to be close to unity. We may reinforce this tendency by arranging that $n_n \gg p_p$ (i.e. by making an n^+-p diode).

Although the internal quantum efficiencies of some LED materials can approach 100%, the external efficiencies are much lower. The main reason is that most of the emitted radiation strikes the material interface at greater than the critical angle and so remains trapped. Unfortunately, the high refractive indices of the III–V materials discussed here give rise to small critical angles. Consider, for example, radiation from a point source within a medium of refractive index n_1 impinging on a plane interface with another medium of refractive index n_2, where $n_2 < n_1$, as shown in Fig. 4.17.

Light from a point source impinging on an interface between two media of refractive indices n_1 and n_2 ($n_1 < n_2$). Only those rays (e.g. (1)) that have an angle to the normal

Fig. 4.17 Phenomenon of total internal reflection. When a beam is incident at an angle θ_1 onto an interface between two media of refractive indices n_1 and n_2 ($n_2 < n_1$) then the refracted beam (beam 1) makes an angle θ_2 with the normal to the surface where $n_1 \sin \theta_1 = n_2 \sin \theta_2$ (Snell's law). At the critical angle ($\theta_1 = \theta_c$), $\theta_2 = 90°$ (beam 2) and the refracted beam emerges along the interface. At angles of incidence greater than θ_c (beam 3), the beam is totally reflected back into the first medium.

less than the critical angle (θ_c) enter the second medium. Those with angles greater than θ_c (e.g. (3)) are reflected back into the first medium.

From eq. (1.14), we have that the critical angle θ_c is given by

$$\theta_c = \sin^{-1}\left(\frac{n_2}{n_1}\right) \tag{4.12}$$

Light originating at recombination centers near the p–n junction will be radiated isotropically, whereas only that within a cone of semi-angle θ_c will escape. In Problem 4.1, it is shown that the fraction F of the total generated radiation that is actually transmitted into the second medium is

$$F \simeq \frac{1}{4}\left(\frac{n_2}{n_1}\right)^2 \left[1 - \left(\frac{n_1 - n_2}{n_1 + n_2}\right)^2\right] \tag{4.13}$$

There are two obvious ways to increase F; the first is to ensure that most rays strike the surface at less than the critical angle. This may be achieved by shaping the semiconductor/air interface into a hemisphere, as shown in Fig. 4.18(a). However, although this technique is used occasionally in high power diodes, it is too difficult and expensive for most situations. The second, and much commoner, technique is to encapsulate the junction in a transparent medium of high refractive index. This is usually a plastic

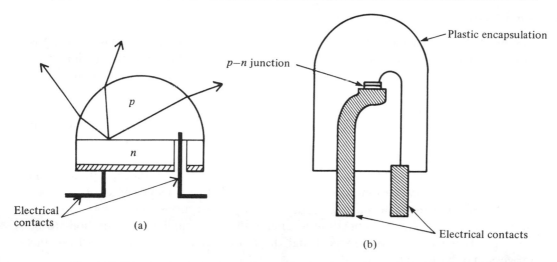

Fig. 4.18 Two methods used to reduce reflection losses in LEDs. In (a) the p material is made into a hemispherical dome. More radiation then strikes the semiconductor/air interface at less than the critical angle than would otherwise be the case. In (b) the $p-n$ junction is surrounded by plastic encapsulation. Losses at the plane semiconductor/plastic interface are then less than for a corresponding semiconductor/air interface.

material with a refractive index of about 1.5. Using eq. (4.13) with $n_1 = 3.6$ and $n_2 = 1.5$ we obtain $F = 0.036$, giving a nearly threefold increase in light output over the simple semiconductor/air interface (Example 4.2). Of course there will be some losses at the plastic/air interface, but these are easily minimized by molding the plastic into an approximately hemispherical shape (see Fig. 4.18(b)).

The diodes themselves may be fabricated using either vapor or liquid phase expitaxy, for more details the reader is referred to ref. 4.8.

Example 4.2 Transmission efficiency of a plane GaAs LED surface

If we take a GaAs/air interface where $n_1 = 3.6$ and $n_2 = 1$, then the fractional transmission for isotropic radiation originating inside the GaAs is given by eq. (4.13) as

$$F = \frac{1}{4}\left(\frac{1}{3.6}\right)^2 \left[1 - \left(\frac{2.6}{4.6}\right)^2\right]$$
$$\simeq 0.013$$

The basic reason for this low value for a GaAs/air interface may be better appreciated if we calculate the critical angle. From eq. (4.12) we have $\theta_c = \sin^{-1}(1/3.6) = 16°$.

4.6.5 *Response times of LEDs*

For most display purposes the fast response time obtainable from LEDs ($<$ 1 µs) is not essential. However, because of their importance in optical communications (see Chapter 9), it is convenient to pursue the matter here. There are two main factors which limit the speed with which an LED can respond to changes in the drive current. One of these is due to the effects of junction capacitance C_j which arises from the variation in charge stored in the depletion region when the external voltage is varied. It was shown in section 2.8.4 that C_j varies as the reciprocal of the square root of the applied voltage (assuming an 'abrupt' junction).

The other limitation is due to what is sometimes termed 'diffusion capacitance', and results from the storage of mobile carriers within a diffusion length or so of the junction. When the external voltage is reduced, charge must diffuse away from the junction and disappear by recombination to enable the new equilibrium conditions to be attained. It is shown in Appendix 3 that the frequency response resulting from this process may be written:

$$R(f) = \frac{R(0)}{(1 + 4\pi^2 f^2 \tau_e^2)^{\frac{1}{2}}} \tag{4.14}$$

Here $R(f)$ is the response at frequency f and τ_e is the minority carrier lifetime provided we have conditions of low level injection. For high level injection the concept of a constant lifetime no longer applies (see the arguments leading to eq. (2.38)), and we must assume some average value of τ_e. However, in practice eq. (4.14) is found to be a good representation of the frequency response of most LEDs. It is evident that for a good frequency response we require τ_e to be as small as possible. From eq. (2.38) we have for *p*-type material that:

$$\tau_e = (Bp)^{-1} \tag{4.15}$$

where *p* is the majority carrier population (here assumed to be holes) and *B* is the constant tabulated in Table 4.1. We see that τ_e may be reduced by using highly doped material. Unfortunately if compounds such as GaAs are doped at near the solubility limit for acceptor impurities, then nonradiative centers are formed. Germanium is a fairly commonly used acceptor impurity in GaAs and above a concentration of some 10^{24} atoms m^{-3} the external quantum efficiency starts to decline. At this concentration, the electron lifetime is given by eq. (4.15):

$$\tau_e = [7 \times 10^{-16} \times 10^{24}]^{-1} = 1.4 \times 10^{-9}\,\text{s}$$

An alternative approach to attaining short response times is to use lightly doped material with a narrow active region and to operate the diode under conditions of heavy forward injection. The injected electron and hole densities are then much greater than they are in equilibrium. If Δn and Δp are the injected carrier concentrations of electrons and holes, then eq. (4.15) can be written:

$$\tau_e = (B\Delta p)^{-1} \tag{4.16}$$

If the injection current density is J and the active region width is t, then in equilibrium the number of recombinations per second per unit volume must be J/te. Since an excess population density Δp gives rise to $\Delta p/\tau_e$ recombinations per second per unit volume, we must have

$$\Delta p = \frac{J\tau_e}{et} \tag{4.17}$$

Eliminating Δp from eqs. (4.16) and (4.17), we obtain

$$\tau_e = \left(\frac{et}{JB}\right)^{\frac{1}{2}} \tag{4.18}$$

We see that in this case τ_e may be reduced by reducing t and increasing J. However, since the lifetime is now current dependent, this approach can lead to signal distortion.

4.6.6 LED drive circuitry

As we have seen, the electrical characteristics of LEDs are essentially those of ordinary rectifying diodes. Typical operating currents are between 20 mA and 100 mA, whilst the forward voltages vary from 1.2 V for GaAs to 2 V for GaP. (The operating voltage is approximately equal to the built-in diode potential, which in turn is slightly less than the energy gap expressed in eV.) Simple drive circuits for d.c. and a.c. voltage operation are shown in Figs. 4.19(a) and (b) respectively. The current through the diode is limited by a series resistance R_s whose value may be calculated from

$$R_s = \frac{V_b - V_d}{i_d} \tag{4.19}$$

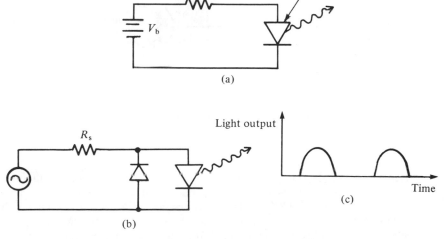

Fig. 4.19 Simple LED drive circuits for (a) d.c. operation and (b) a.c. operation. In both cases a series resistor R_s limits the maximum current flow. In the a.c. circuit a diode is placed with reversed polarity across the LED to prevent damage from excessive reverse bias voltages. (c) shows the light output obtained with the circuit of (b).

where V_b is the power source voltage, V_d the diode operating voltage and i_d the desired diode current. In the a.c. circuit, a rectifying diode is placed across the LED to protect it against reverse bias breakdown.

These two circuits provide for continuous 'on' operation. If it is desired to switch the diode on or off, or to modulate the output, then the circuits shown in Figs. 4.20(a) and (b) respectively may be used. In Fig. 4.20(a) the transistor is used as a simple switch.

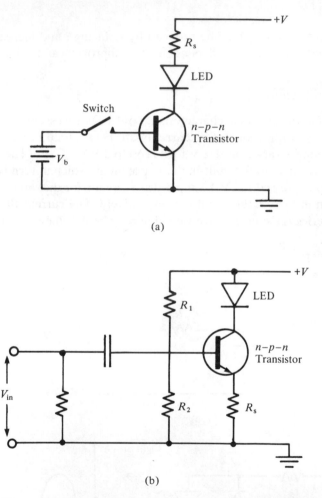

(a)

(b)

Fig. 4.20 LED modulation circuits. (a) provides for a simple on/off modulation via a switch. The voltage V_b is sufficient to switch the transistor on. There is then a low impedance path between the collector and emitter terminals, and the current flowing through the LED is determined by the voltage V and the series resistance R_s. In (b) the diode output may be modulated by voltage V_{in}. The resistors R_1 and R_2 bias the transistor so that the average current through the transistor, and hence through the LED, is about half the maximum value. Both the transistor and the LED are then biased well into their linear regions.

With no voltage applied to the base, the transistor has a very high impedance between the collector and emitter and hence no current flows through the LED. If a large enough base voltage is then applied so that the emitter base junction becomes heavily forward biased, the transistor has a relatively low impedance between emitter and collector and a substantial current can flow, resulting in the LED being turned on. In Fig. 4.20(b) the transistor is biased so that the quiescent diode current is about half its peak value and both the transistor and the LED are biased well into their linear regions. Changes in the current flowing through the LED are then directly proportional to changes in the input voltage.

4.7 Plasma displays

Plasma displays rely on the glow produced when an electrical current is passed through a gas (usually neon). Free electrons and ionized gas atoms are present during the discharge. Under the influence of the external field the electrons acquire a high kinetic energy and when they collide with the gas atoms (or ions) they transfer this energy to the atoms, thereby exciting them into energy levels above the ground state. The atom may then lose energy radiatively and return to the ground state.

Either a.c. or d.c. excitation may be used, although the former is the most common. Figure 4.21 shows the basic construction of an a.c. plasma display element in which the electrodes are external to the gas cavity. (When d.c. excitation is used, the electrodes must be inside the gas cavity.) Typical cavity widths and gas pressures are 100 μm and 400 torr respectively. The discharge is initiated by applying a firing voltage V_f of some 150 V; however, once the discharge has started, it may be sustained with a reduced voltage V_s of about 90 V. A suitable voltage waveform for this situation is shown in Fig. 4.22.

If the initial voltage pulse is relatively wide then an appreciable amount of charge accumulates on the interior of the dielectric layer. This charge build-up causes an internal electric field to be set up which, being in opposition to the external field, may

Fig. 4.21 Construction of an a.c. plasma element. The gas cavity is some 10^{-4} m in width with transparent electrodes on the outside of the containing dielectric layers. The main constituent of the gas is usually neon at a pressure of some 400 Torr.

Fig. 4.22 Voltage waveform used for driving an a.c. plasma display. After an initial firing pulse V_f the voltage may be reduced to V_s for the remaining time the display is required to be on.

be strong enough to extinguish the discharge if the pulse width is too long. Such accumulated charge takes a long time to dissipate and this effect can be used to enable the device to have a memory. Thus, if at some time subsequent to the first ('writing') pulse an external voltage pulse of opposite polarity is applied, the field from the stored charge will add to the external field and the discharge will again commence, but now at a lower voltage than V_f.

4.8 Display brightness

The basic photometric units of brightness were introduced in section 1.4, and we briefly review the topic again here. The sensitivity of the human eye to different wavelengths was shown in Fig. 1.17, which enables us to convert objective radiometric units into subjective photometric units. Thus, if the radiant power in the wavelength interval λ to $\lambda + d\lambda$ (nm) is $W(\lambda)d\lambda$ (watts), then the total luminous flux Φ (lumens) is given by:

$$\Phi = 680 \int_{380}^{780} W(\lambda)V_\lambda d\lambda \tag{4.20}$$

where V_λ is the relative response of the eye as shown in Fig. 1.17. This function has its maximum at a wavelength of 555 nm, where the eye's absolute sensitivity is 680 lm W^{-1}.

A calculation of Φ enables the conversion efficiency (in lm W^{-1}) of electrical power into effective visible radiation to be determined. However, the apparent brightness of a display element depends on the emission area and the angle from which it is observed. If the element is sufficiently small for it to appear as a point source (which is the case if it subtends an angle at the eye of some 2 minutes of arc or less), then a meaningful measure of brightness is the flux per unit solid angle in the direction of the observer. This is called the *luminous intensity* of a source and the units are *lumens per steradian* or *candela*. For sources with an apparently finite area, the brightness is more properly referred to as the *luminance* and the appropriate units are *lumens per steradian per unit projected area of emitting surface*, i.e. *candelas* m^{-2} or *nits*.

The variation in luminance with viewing angle is something that must be determined experimentally for each type of display. It is often assumed that emitters either show isotropic emission or Lambertian emission. If the luminance of a surface viewed at an angle θ to the normal is $B(\theta)$, then for isotropic emission $B(\theta) = \text{constant} = B(0)$. For Lambertian emission, on the other hand, $B(\theta) = B(0) \cos(\theta)$. It is readily shown (see Problem 4.5) that for an isotropic surface of area A the total flux Φ_i is given by

$$\Phi_i = 2\pi A B(0) \tag{4.21a}$$

whilst for a corresponding Lambertian surface

$$\Phi_L = \pi A B(0) \tag{4.21b}$$

Quite a number of displays are Lambertian, but there are some, for example GaP red LEDs, that approximate more to isotropic emitters.

Example 4.3 Brightness of an LED

We take an LED with a chip diameter of 0.2 mm which is viewed from a distance of 1 m. It emits at a wavelength of 550 nm and has an external quantum efficiency of 0.1%. We further assume that the emission is isotropic and that the diode is operated at 2 V and 50 mA. We must first decide whether the diode acts as a point or as an extended source. At the eye, the emitting area subtends an angle θ where:

$$\tan \frac{\theta}{2} = \frac{2 \times 10^{-4}}{2 \times 1}$$

whence $\theta < 1$ minute of arc, and the LED acts as a point source.

The total radiant power output W is given by:

$$W = hc/\lambda \times \text{quantum efficiency} \times \text{photon emission rate}$$

$$= \frac{hc}{550 \times 10^{-9}} \times (0.001) \times \frac{(50 \times 10^{-3})}{e}$$

$$\therefore W = 1.13 \times 10^{-4} \text{ watts}$$

Using Fig. 1.17, we take an average luminosity at 550 nm of 600 lm W^{-1}; hence the luminous flux from the source is $1.13 \times 10^{-4} \times 600$, that is 6.8×10^{-2} lm. Isotropic emission means the flux is uniformly distributed over the solid angle 2π, and so the luminous intensity at normal incidence is $6.8 \times 10^{-2}/2\pi = 1.1 \times 10^{-2}$ candela.

Another brightness unit sometimes encountered is the foot-Lambert (ft-L). This is obtained by dividing the total luminous flux (in lumens) by the area of the device in square feet. This definition assumes a Lambertian emission pattern. To convert nits or candelas m^{-2} into foot-Lamberts (for a Lambertian surface) we multiply by the factor 0.292 (this factor may be derived from eq. (4.21b).

The brightness level required of a display will depend on the ambient lighting conditions. In a dimly lit room, a brightness of about 10 nits may suffice, whilst in a well-lit environment about 1000 nits may be needed, although the difference between the display on and off brightnesses also plays an important role. A more thorough discussion is given in ref. 4.9.

4.9 Liquid crystal displays

We turn now to the most important of the 'passive' types of display (and indeed the only one of this type we shall mention), namely liquid crystal displays (LCDs). These have come into prominence in the last few years mainly as display elements for digital watches and pocket calculators. Here one of the prime requisites is for low power consumption, particularly so for the digital watch because of the necessarily low capacity of the power source. LCDs consume the least power of all common display devices because no light generation is required. This, as we have seen in the LED, for example, is a very inefficient process; at most the efficiency is only some 10%. There are two basic types of LCD available. These are (a) reflective, which requires front illumination, and (b) transmissive, which requires rear illumination. Most reflective types utilize ambient light for illumination with provision for secondary illumination via a small incandescent lamp or LED if ambient levels become too low. At the heart of all LCD devices is a cell formed between two glass plates each with a conductive coating. The cell has a thickness of about 10 μm (sometimes less) and is filled with a liquid crystal material.

The liquid crystal state is a phase of matter which is exhibited by a large number of organic materials over a restricted temperature range. At the lower end of the temperature range, the material becomes a crystalline solid, whilst at the upper end it changes into a clear liquid. Within this range it has a milky yellowish appearance and combines some of the optical properties of solids with the fluidity of liquids. A major characteristic of all liquid crystal compounds is the rod-like shape of their molecules. When they are in the liquid crystal phase, these molecules can take up certain orientations relative both to each other and to the liquid crystal surface. It is usual to describe this orientation in terms of a *director*, i.e. a unit vector pointing along the time-averaged preferred orientation of the molecules in any small volume.

There are three basic types of ordering in liquid crystals, which are termed *nematic, cholesteric* and *smectic*. Only the first two of these are of importance in display devices at present and are illustrated in Fig. 4.23. In nematic ordering, the molecules (or, rather, the directors) are aligned parallel to each other, but apart from remaining parallel the molecules are free to move relative to each other so that the phase has liquid properties. A nematic liquid crystal molecule usually consists of two benzene rings linked with

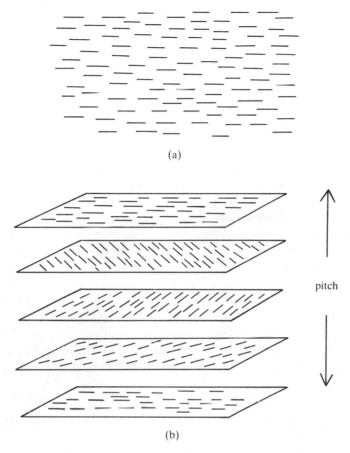

(a)

pitch

(b)

Fig. 4.23 (a) Nematic ordering and (b) cholesteric ordering. In (a) the directors all line up parallel to each other. In (b) a large number of planes of nematic order are formed where the directors rotate as we move along a direction perpendicular to the planes.

a central group. A typical example is 4-methoxybenzylidene-4-butylanaline (MBBA), which has the chemical formula CH_3—O—⟨⟩—CH=N—⟨⟩—C_6H_9. MBBA shows liquid crystal behavior over the temperature range 20°C to 47°C. For more general details concerning liquid crystal materials the reader is referred to ref. 4.10.

In the cholesteric phase (Fig. 4.23(b)), we may regard the material as being made up from a large number of planes each having a nematic like structure, but with each plane showing a progressive change in the director direction from the one below. The director directions thus display a helical twist through the material. The distance between planes having the same director direction is called the *pitch*, *p*. Cholesteric liquid crystals exhibit some interesting color effects. If, for example, light of wavelength λ is incident normally on the director planes, then strong Bragg reflection will occur when $p = n\lambda$ (*n* an integer), but not otherwise. Thus if white light is shone onto a cholesteric liquid crystal, it can appear strongly colored. Furthermore, the pitch is usually temperature dependent, so that the colour of the reflected light will also be temperature dependent. Obviously this can form the basis of a thermometer. Most liquid displays, however, are of the so-called twisted nematic type, and we shall spend the rest of this section discussing them.

When a nematic liquid crystal material comes into contact with a solid surface, the directors often become aligned either perpendicular to the surface (*homeotropic* ordering) or parallel to the surface (*homogeneous* ordering). These two forms can be produced by suitable treatment of the surface. In the case of homogeneous ordering, this can often be achieved by rubbing the surface once or twice along a particular direction with a soft fabric (e.g. cotton) before it comes into contact with the liquid crystal material. The liquid crystal molecules then take up an orientation parallel to the direction of rubbing.

One of the most important electrical characteristics of liquid crystal materials is that they show different dielectric constants $\varepsilon_{\parallel}$ and $\varepsilon_{\perp}$, depending on whether the external field is parallel to, or perpendicular to, the molecular axis. If $\varepsilon_{\parallel} > \varepsilon_{\perp}$ we refer to it as a *positive* material. The application of an external electric field to a positive material will tend to make the molecules lie along the electric field, since this will tend to minimize their energy (see Problem 4.7). We see that there is thus a possibility of changing the homogeneous type of ordering into a homeotropic type by the application of a field which is perpendicular to the surface (assuming a positive material). This transition is found to take place above a critical field ($\mathscr{E}_c$) and is illustrated in Fig. 4.24.

The most common liquid crystal display uses a 'twisted nematic' cell. In this, the op-

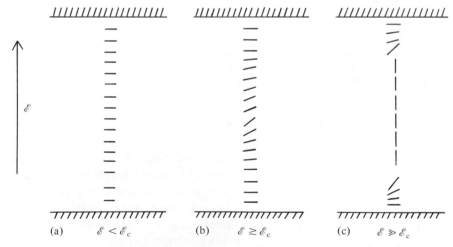

Fig. 4.24 Behavior of molecules in an initially homogeneously ordered liquid crystal material as an increasing electric field of magnitude $\mathscr{E}$ is applied in a direction perpendicular to the liquid crystal/solid interface. If $\mathscr{E}$ is less than a critical value ($\mathscr{E}_c$) the ordering is not affected (a). If $\mathscr{E} \gtrsim \mathscr{E}_c$ the molecules furthest away from the interface begin to align along the field direction (b). If $\mathscr{E} \gg \mathscr{E}_c$ then most of the molecules are aligned along the field direction (c).

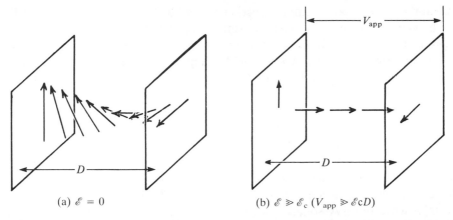

(a) $\mathscr{E} = 0$ (b) $\mathscr{E} \gg \mathscr{E}_c$ ($V_{app} \gg \mathscr{E}cD$)

Fig. 4.25 Behavior of the molecules in a liquid crystal cell of thickness D with (a) no applied voltage ($\mathscr{E} = 0$) and (b) with a voltage applied such that $\mathscr{E} \gg \mathscr{E}_c$. In the former, the molecules undergo a 90° rotation across the cell, in the latter they are mostly ordered with their axes parallel to the applied field.

posite walls of the cell are treated to produce a homogeneous arrangement in which the molecular alignment directions at the walls are at right angles to each other. Thus the molecules undergo a 90° rotation across the cell as shown in Fig. 4.25(a). When a beam of polarized light is incident on the cell the strong optical anisotropy of the liquid causes the polarization to undergo a 90° rotation. With a strong enough electric field across the cell, however, (i.e. $\mathscr{E} \gg \mathscr{E}_c$) the molecular alignments will become as shown in Fig. 4.25(b) and in this state the molecular alignments will have no effect on an incident polarized light beam.

In operation, the cell is sandwiched between two pieces of polaroid whose polarizing directions correspond to the molecular ordering direction of the particular cell surfaces they are next to. In the reflective mode a reflector is placed behind the back sheet of polaroid. Figure 4.26 shows the arrangement and traces the behavior of a polarized beam as it traverses the system. With no applied voltage, the incident light is first

Fig. 4.26 Illustration of the action of an LCD display device.

polarized, then has its polarization direction rotated by 90° as it traverses the cell, then passes through the second polarizer and is then reflected back along its path where the same process is repeated. With no field applied, therefore, the device reflects incident radiation and appears bright. When a field is applied the direction of polarization of light traversing the cell is not rotated and hence cannot pass through the second polarizer. Little light will be reflected from the device and it will appear dark.

The amount of light reflected from an LCD as a function of applied voltage is shown schematically in Fig. 4.27. The reflectance, initially constant, falls rapidly beyond a critical voltage (where $V_c = \mathscr{E}_c D$; D = cell thickness) and again becomes constant beyond a voltage V_{sat}. A typical value for V_{sat} is 3 V. D.c. operation tends to shorten the operating lifetime of the device due to electromechanical reactions taking place, and hence a.c. waveforms are invariably used. The cell responds to the r.m.s. value of the voltage waveform. A square waveform is often used which has a frequency of between 25 Hz and 1 kHz.

Transmission LCD displays do not have the reflector, and must be provided with rear illumination, but otherwise they operate in a very similar fashion to the reflective displays.

The use of polarizers in the twisted nematic cell substantially reduces the maximum amount of light that can be reflected from it (see Problem 4.8). In addition, the angle of viewing is found to be restricted to about ±45°. A greater image contrast over a wider range of viewing angles is claimed for the recently developed 'supertwist' liquid crystal displays. These are still basically twisted nematic displays, the main difference being that the directors undergo a 270° rather than a 90° rotation across the cell. This is achieved by a combination of a special treatment of the surfaces as well as the addition of some cholesteric material to the nematic material.

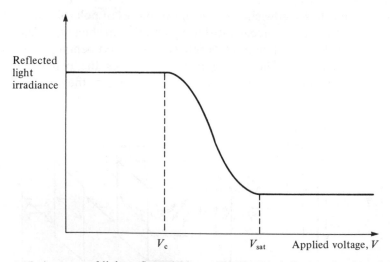

Fig. 4.27 Amount of light reflected from a liquid crystal display as a function of applied voltage V. The reflected irradiance remains constant up to the critical voltage V_c; it then falls with increasing voltage until it again becomes constant beyond V_{sat}.

4.10 Numeric displays

We turn now to the problem of combining our individual display elements into some pattern capable of conveying more information than just a simple 'on' or 'off' situation. There is no doubt that for medium area high resolution displays there is as yet no ready alternative to CRTs. These do, however, have the obvious disadvantages of a small display area to volume ratio, of requiring a high voltage supply and of being sensitive to adverse environmental conditions (e.g. stray magnetic fields).

Displays that require only a small number of basic elements are easily catered for by using LED, liquid crystal or plasma display elements. One of the simplest display formats commonly used to form the numbers 0 to 9 consists of seven bar segments and is illustrated in Fig. 4.28(a). Each bar might itself consist of several discrete display elements depending on its size. More complex characters may be obtained using a 7×5 matrix (Fig. 4.28(b)). A display of this latter type can be made either by assembling 35 discrete elements or, in the case of LEDs, a monolithic device can be constructed with all the elements grown onto a single substrate. This latter alternative is only feasible for smaller displays where the characters are less than 5 mm or so in size.

There are two basic methods of wiring up a matrix display. The simpler of these is known as the *common anode* (or *cathode*) and is shown in Fig. 4.29(a). The anodes (or cathodes) are all wired together whilst each cathode (or anode) has a separate connection made to it. Thus for N elements the number of external connections is $N + 1$. A technique requiring fewer connections is the *coordinate connected* (or *matrix connected*) method shown in Fig. 4.29(b). In this, all the anodes of the elements in each column

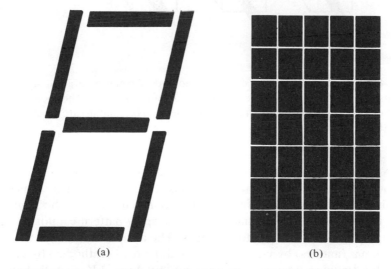

(a)　　　　　　　　　(b)

Fig. 4.28 Two common display formats: (a) illustrates the seven bar segment display used to form the numbers from 0 to 9; the 7×5 matrix display (b) may be used for more complex characters.

Individual cathode (or anode) connections

Display element

Common anode (or cathode) connection

(a)

y_2

y_1

x_1 x_2 x_3

(b)

Fig. 4.29 Common anode (or cathode) method of connecting an array of display elements (a). For N elements there will be $N + 1$ connections. (b) The coordinate connected method of wiring up an array of display elements. The columns are labeled x_1, x_2, x_3, etc. and the rows y_1, y_2, y_3 etc.

(row) are connected together as are all the cathodes of the elements in each row (column). For a square array of N elements we thus require $N^{\frac{1}{2}}$ external connections. If N is large, this number is considerably less than that required for the common anode method. It is obvious, however, that a general display pattern cannot be maintained merely by applying static voltages to the leads. The array must be *scanned*. Thus if we label the columns (anodes) by x_1, x_2, x_3, etc. and the rows (cathodes) by y_1, y_2, y_3, etc. then a possible scanning sequence is as follows: a voltage $+V$ is applied to column x_1 and zero volts to the rest. Voltages of either $-V$ or $+V$ are then applied simultaneously to the rows y_1, y_2, y_3 etc. the choice depending on whether we require the element at the intersection of column x_1 and the particular row to be on or off. After a time t_c the voltage $+V$ is switched to column x_2 and a new set of voltages applied to the rows.

If the number of columns is N_c, then the picture will be scanned in a time $N_c t_c$. To avoid flicker, we must have $N_c t_c \lesssim 1/45$ s and hence $t_c \lesssim 1/(45\,N_c)$ s. If N_c is very large, t_c may approach the response time of the device which will result in a reduced device output. Because each element in the display is on for at most a fraction $1/N_c$ of the total 'on' time of the display then the active display elements must be overdriven by a factor N_c to maintain the same average brightness levels possible from an equivalent common anode display. This places a limitation on the size of N_c unless a much reduced brightness is acceptable.

It will be noticed that even when an element is 'off' it still may have a voltage V across it. An 'ideal' element should thus show no output up to a threshold voltage V_{th} and then give its maximum saturated output at a voltage V_{sat}, where $V_{\mathrm{sat}} = 2 \times V_{\mathrm{th}}$ (see Fig. 4.30). Several devices do have characteristics that are not too far removed from this ideal, for example the LED and d.c. electroluminescent cells. Often, however, V_{sat} is greater than $2V_{\mathrm{th}}$ and so it is not possible to drive the device to saturation, thus reducing the maximum brightness available. In addition, V_{th} may vary with operating temperature and a 'safe' value somewhat below the actual value may have to be adopted, again leading to a reduced output.

Although liquid crystal displays have the attraction of requiring very little power, they have so far proved rather unsuitable for matrix displays. They have rather ill-defined threshold voltages which are quite strongly dependent on the operating temperature. In addition, their slow response times prevent them from responding fully to short voltage pulses. Nevertheless it is possible that new materials will be developed

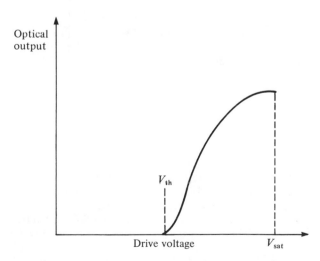

Fig. 4.30 Idealized characteristic for a device suitable for matrix addressed displays. The device shows no optical output up to a threshold drive voltage V_{th}. Thereafter the output increases to a maximum at a drive voltage of V_{sat}.

which will in part overcome some of these difficulties. (For example the positive cholesteric materials mentioned in the previous section have a faster response time than the
more usual nematic types.)

It is possible, when using a liquid crystal display, to hold each element on for a time
longer than the element address time. This may be achieved using the circuitry shown in
Fig. 4.31. Here each element is connected to the drain of an MOS transistor[†] whose
source and gate are connected to the appropriate column and row respectively. If the
element is to be 'on' then the voltage on the corresponding column is switched onto the
element by applying a voltage pulse to the appropriate row. Since the LCD element
consumes very little current, the drain capacitance of the MOST may enable the element to remain 'on' during the rest of the scan time when the MOST is switched 'off'.
There is, however, the obvious difficulty of reliably fabricating the required number of
MOS transistors if N is large.

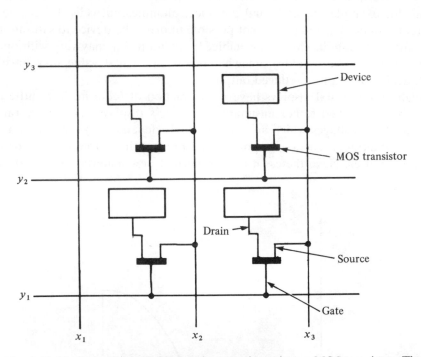

Fig. 4.31 Method used for LCD matrix operation using an MOS transistor. The element
may be held 'on' (or 'off') during the time between addressing by virtue of the charge
held on the drain terminal. (The LCD consumes minimal power.) To address an element
the gate of the attached MOST is 'opened' by applying a suitable potential to the appropriate y line. The voltage that is on the corresponding x line then appears across the device.
Note: only one connection to each device is shown here, there will be a further connection common to all elements.

[†] For those unfamiliar with the characteristics of an MOS transistor, further details may be found in any of
the many books available on solid state devices, for example refs. 2.1a and d.

Problems

4.1 Assuming an isotropic radiation source within a transparent medium of refractive index n_1, show that the fraction F of the radiation that can escape through a plane boundary into another medium of refractive index n_2 ($n_2 < n_1$) is given approximately by

$$F \simeq \frac{1}{4} \left(\frac{n_2}{n_1}\right)^2 \left[1 - \left(\frac{n_1 - n_2}{n_1 + n_2}\right)^2\right]$$

Hint: assume that when radiation is incident on the boundary at an angle θ (less than the critical angle), then the transmittance $T(\theta)$ is identical with that obtained at normal incidence, i.e. $T(\theta) = T(0) = 1 - [(n_1 - n_2)/(n_1 + n_2)]^2$.

4.2 A GaAs LED fabricated from fairly lightly doped materials has an effective recombination region of width 0.1 μm. If it is operated at a current density of 2×10^7 A m^{-2} estimate the modulation bandwidth that can be expected.

4.3 Design a simple power supply using a 9 V battery to drive an LED which has a maximum forward current of 20 mA at 1.9 V.

4.4 Estimate the average luminance of a CRT screen of area 0.05 m^2 when it is operated with a beam current of 1 μA and an accelerating potential of 15 kV. Assume the phosphor used has a power efficiency of 10% and peak emission at a wavelength of 500 nm.

4.5 A light source of area A has a luminance at normal incidence of $B(0)$ nits. Calculate the total flux from the source if (a) it is isotropic and (b) it is Lambertian.

4.6 An isotropic source is embedded in a transparent medium of refractive index n_1. Show that when the radiation passes into a second medium of refractive index n_2, then the source appears to be approximately Lambertian. You may make the same assumptions concerning the transmission of radiation through the interface as are made in Problem 4.1.

4.7 Derive a general expression for the free electrostatic energy ($= -\frac{1}{2}\vec{D} \cdot \vec{\mathscr{E}}$) of a liquid crystal molecule when it is in an electric field $\mathscr{E}$ in terms of the dielectric constants $\varepsilon_\perp$ and $\varepsilon_\parallel$, and hence show that if $\varepsilon_\parallel > \varepsilon_\perp$ then the free energy is minimized when the electric field is parallel to the molecular axis.

4.8 Unpolarized light is incident normally onto an LCD cell; if there is no applied voltage what is the maximum fraction of the incident light that can be reflected back?

References

4.1 G. K. Woodgate, *Elementary Atomic Structure*, McGraw-Hill, London, 1970, Chapter 3.

4.2 D. Curie, *Luminescence in Crystals*, Methuen and Co. Ltd., London, 1960, Chapter VI.

4.3 R. C. Alig and S. Bloom, 'Electron-hole pair creation energies in semiconductors', *Phys. Rev. Letters*, **35**, 1975, 1522.

4.4 T. E. Everhart and P. H. Hoff, 'Determination of kilovolt electron energy dissipation vs. distance in solid materials', *J. Appl. Phys.*, **42**, 1971, 5837.

4.5 S. Sherr, *Electronic Displays*, John Wiley, New York, 1979, Chapter 2.

4.6 H. K. Henisch, *Electroluminescence*, Pergamon Press, Oxford, 1962, section 1.3.1.

4.7 V. P. Varshi, 'Band-to-band radiative recombination in semiconductors, *Phys. Stat. Solidi*, **19**, 1967, 459.

4.8   (a) K. Gillessen and W. Schairer, *Light-Emitting Diodes, an Introduction*, Prentice Hall International, London, 1987, Chapter 3.

(b) A. A. Bergh and P. J. Dean, *Light-Emitting Diodes*, Oxford University Press, Oxford, 1976, Chapter 5.

4.9   *Ibid*, Chapter 1.

4.10   S. Chandrasekhar, *Liquid Crystals*, Cambridge University Press, Cambridge, 1980.

Chapter 5

Lasers I

The word 'laser' is an acronym for 'Light Amplification by Stimulated Emission of Radiation'. Albert Einstein in 1917 showed that the process of stimulated emission must exist but it was not until 1960 that T. H. Maiman (ref. 5.1) first achieved laser action at optical frequencies in ruby. The basic principles and construction of a laser are relatively straightforward and it is somewhat surprising that the invention of the laser was so long delayed. A rigorous analysis of the physics of the laser, on the other hand, is quite difficult and the treatment we give below is somewhat simplified. The development of lasers since 1960 has been extremely rapid and although applications for lasers had a very slow start during their first decade, new applications for laser radiation are being found now almost every day (see section 6.5). In view of the increased use of lasers and as laser radiation is potentialy hazardous, some comments on laser safety have been included as Appendix 7.

5.1 Emission and absorption of radiation

It was seen in Chapter 2 that when an electron in an atom undergoes transitions between two energy states or levels it either emits or absorbs a photon, which can be described in terms of a wave of frequency ν where $\nu = \Delta E/h$, ΔE being the energy difference between the two levels concerned. Let us consider the electron transitions which may occur between the two energy levels of the hypothetical atomic system shown in Fig. 5.1. If the electron is in the lower level E_1 then in the presence of photons of energy $(E_2 - E_1)$ it may be excited to the upper level E_2 by absorbing a photon. Alternatively if the electron is in the level E_2 it may return to the ground state with the emission of a photon. The emission process may occur in two distinct ways. These are: (a) the *spontaneous emission* process in which the electron drops to the lower level in an entirely random way and (b) the *stimulated emission* process in which the electron is 'triggered' to undergo the transition by the presence of photons of energy $(E_2 - E_1)$. There is nothing mystical in this, as the electron would undergo this process sooner or later spontaneously: the transition is simply initiated by the presence of the stimulating photon.

The absorption and emission processes are illustrated in Figs. 5.2(a), (b) and (c).

Fig. 5.1 Two energy level system.

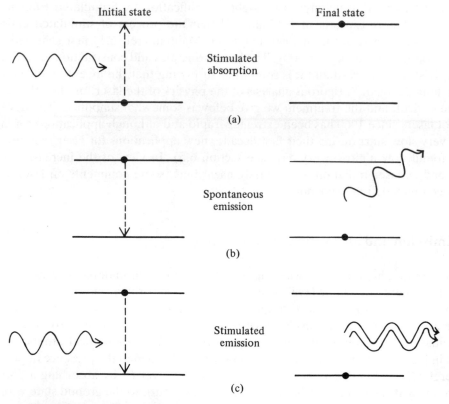

Fig. 5.2 Energy level diagram illustrating (a) absorption, (b) spontaneous emission and (c) stimulated emission. The black dot indicates the state of the atom before and after the transition.

Under normal circumstances we do not observe the stimulated process because the probability of the spontaneous process occurring is much higher. The average time the electron exists in the excited state before making a spontaneous transition is called the lifetime τ_{21} of the excited state. The '21' here indicates the energy levels involved. The probability that a particular atom will undergo spontaneous emission within a time interval dt is given by $A_{21}dt = dt/\tau_{21}$, where A_{21} is the spontaneous transition rate. Because the spontaneous radiation from any atom is emitted at random, the radiation emitted by a large number of atoms will clearly be incoherent. In contrast to this, the stimulated emission process results in coherent radiation since the waves associated with the stimulating and stimulated photons have identical frequencies (but see section 5.7), are in phase, have the same state of polarization and travel in the same direction. This means that with stimulated emission the amplitude of an incident wave can grow as it passes through a collection of excited atoms in what is clearly an amplification process. As the absorption transition, in common with stimulated emission, can only occur in the presence of photons of appropriate energy, it is often referred to as stimulated absorption. These two processes may be regarded as the inverse of one another.

5.2 Einstein relations

Einstein (ref. 5.2) showed that the parameters describing the above three processes are related through the requirement that for a system in thermal equilibrium the rate of upward transitions (from E_1 to E_2) must equal the rate of the downward transition processes (from E_2 to E_1).

If there are N_1 atoms per unit volume in the collection with energy E_1, then the upward transition or absorption rate will be proportional to both N_1 and to the number of photons available at the correct frequency. Now ϱ_ν, the energy density at frequency ν, is given by $\varrho_\nu = \mathcal{N}h\nu$ where $\mathcal{N}$ is the number of photons per unit volume having frequency ν. Therefore we may write the upward transition rate as $N_1\varrho_\nu B_{12}$, where B_{12} is a constant. Similarly if there are N_2 atoms per unit volume in the collection with energy E_2 then the induced transition rate from level 2 to level 1 is $N_2\varrho_\nu B_{21}$, where again B_{21} is a constant. The spontaneous transition rate from level 2 to level 1 is simply N_2A_{21}. The total downward transition rate is the sum of the induced and spontaneous contributions, i.e.

$$N_2\varrho_\nu B_{21} + N_2A_{21}$$

A_{21}, B_{21} and B_{12} are called the *Einstein coefficients*; the relationships between them can be established as follows.

For a system in equilibrium, the upward and downward transition rates must be equal and hence we have

$$N_1\varrho_\nu B_{12} = N_2\varrho_\nu B_{21} + N_2A_{21} \tag{5.1}$$

thus

$$\varrho_\nu = \frac{N_2A_{21}}{N_1B_{12} - N_2B_{21}}$$

or

$$\varrho_\nu = \frac{A_{21}/B_{21}}{\dfrac{B_{12}}{B_{21}}\dfrac{N_1}{N_2} - 1} \tag{5.2}$$

Now, the populations of the various energy levels of a system in thermal equilibrium are give by Boltzmann statistics to be:

$$N_j = \frac{g_j N_0 \exp\left(-E_j/kT\right)}{\Sigma\, g_i \exp\left(-E_i/kT\right)} \tag{5.3}$$

where N_j is the population density of the energy level E_j, N_0 is the total population density and g_j is the degeneracy of the jth level.[†] Hence

$$\frac{N_1}{N_2} = \frac{g_1}{g_2} \exp\left((E_2 - E_1)/kT\right) = \frac{g_1}{g_2} \exp\left(h\nu/kT\right) \tag{5.4}$$

Therefore, substituting eq. (5.4) into eq. (5.2) gives

$$\varrho_\nu = \frac{A_{21}/B_{21}}{\left[\dfrac{g_1}{g_2}\dfrac{B_{12}}{B_{21}} \exp\left(h\nu/kT\right)\right] - 1} \tag{5.5}$$

Since the collection of atoms in the system we are considering is in thermal equilibrium it must give rise to radiation which is identical with blackbody radiation, the radiation density of which can be described by (see Ref. 1.1c)

$$\varrho_\nu = \frac{8\pi h\nu^3}{c^3}\left(\frac{1}{\exp\left(h\nu/kT\right) - 1}\right) \tag{5.6}$$

Comparing eqs. (5.5) and (5.6) for ϱ_ν, we see that

$$g_1 B_{12} = g_2 B_{21} \tag{5.7}$$

and

$$\frac{A_{21}}{B_{21}} = \frac{8\pi h\nu^3}{c^3} \tag{5.8}$$

Equations (5.7) and (5.8) are referred to as the *Einstein relations*. The second relation enables us to evaluate the ratio of the rate of spontaneous emission to the rate of stimulated emission for a given pair of energy levels. We see that this ratio is given by

$$R = \frac{A_{21}}{\varrho_\nu B_{21}} \tag{5.9}$$

It is left as a problem for the reader to show that

$$R = \exp\left(h\nu/kT\right) - 1$$

[†] In general, the probability of occupation of all the available states or energy levels is not the same. The states have a probability g_j of occupation (g is often called the degeneracy). Frequently, the ratio g_1/g_2 in eqn. (5.5), for example, is of the order of unity and hence is often omitted.

Example 5.1 Ratio of rates of spontaneous and stimulated emission

Let us calculate this ratio for a tungsten filament lamp operating at a temperature of 2000 K. Taking the average frequency to be 5×10^{14} Hz we see that the ratio is

$$R = \exp \left(\frac{6.6 \times 10^{-34} \times 5 \times 10^{14}}{1.38 \times 10^{-23} \times 2000} \right) + 1 = e^{12} + 1 \simeq 1.5 \times 10^5$$

This confirms that under conditions of thermal equilibrium, stimulated emission is not an important process. For sources operating at lower temperatures and higher frequencies, stimulated emission is even less likely.

The above discussion indicates that the process of stimulated emission competes with the processes of spontaneous emission and absorption. Clearly, if we wish to amplify a beam of light by stimulated emission then we must increase the rate of this process in relation to the other two processes. Consideration of eq. (5.1) indicates that to achieve this for a given pair of energy levels we must increase both the radiation density and the population density N_2 of the upper level in relation to the population density N_1 of the lower level. Indeed, we shall show that to produce laser action we must create a condition in which $N_2 > (g_2/g_1)N_1$ even though $E_2 > E_1$, i.e. we must create a so-called *population inversion*. Before describing this situation in detail it will be instructive to look more closely at the process of absorption.

5.3 Absorption of radiation

Let us consider a collimated beam of perfectly monochromatic radiation of unit cross-sectional area passing through an absorbing medium. We assume for simplicity that there is only one relevant electron transition, which occurs between the energy levels E_1 and E_2. Then the change in irradiance of the beam as a function of distance is given by

$$\Delta I(x) = I(x + \Delta x) - I(x)$$

For a homogeneous medium $\Delta I(x)$ is proportional both to the distance traveled Δx and to $I(x)$. That is $\Delta I(x) = -\alpha I(x)\Delta x$, where the constant of proportionality, α, is the *absorption coefficient*. The negative sign indicates the reduction in beam irradiance due to absorption as α is a positive quantity. Writing this expression as a differential equation we have:

$$\frac{\mathrm{d}I(x)}{\mathrm{d}x} = -\alpha I(x)$$

Integrating this equation gives

$$I = I_0 \exp(-\alpha x) \tag{5.10}$$

where I_0 is the incident irradiance.

Let us consider the absorption coefficient in more detail. Clearly, the degree of absorption of the beam will depend on how many atoms N_1 there are with electrons in the lower energy state E_1 and on how many atoms N_2 there are in energy state E_2. If N_2 were zero, then absorption would be a maximum, while if all of the atoms were in the upper state the absorption would be zero and the probability of stimulated emission would be large.

From the discussion of induced or stimulated transitions given in section 5.2, we can write an expression for the net rate of loss of photons per unit volume, $-\mathrm{d}\mathcal{N}/\mathrm{d}t$, from the beam as it travels through a volume element of medium of thickness Δx and unit cross-sectional area (Fig. 5.3) as:

$$-\frac{\mathrm{d}\mathcal{N}}{\mathrm{d}t} = N_1 \varrho_v B_{12} - N_2 \varrho_v B_{21}$$

or substituting from eq. (5.7)

$$-\frac{\mathrm{d}\mathcal{N}}{\mathrm{d}t} = \left(\frac{g_2}{g_1} N_1 - N_2\right) \varrho_v B_{21} \tag{5.11}$$

In this discussion, we have deliberately ignored photons generated by spontaneous emission as these are emitted randomly in all directions and do not therefore contribute to the collimated beam. Similarly we have ignored scattering losses.

We can now link eq. (5.11), which contains the difference in populations of the two energy levels, to the absorption coefficient α. We recall that the irradiance of the beam is the energy crossing unit area in unit time and therefore is given by the energy density times the speed of light in the medium, that is $I = \varrho\, c/n$, or for photons of frequency v, $I_v = \varrho_v\, c/n = \mathcal{N}hv\, c/n$, where c is the speed of light *in vacuo* and n is the refractive index of the medium. Hence the change in photon density within the beam between the boundaries x and $x + \Delta x$ of the volume element can be written as

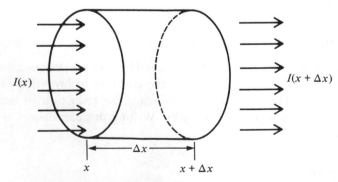

Fig. 5.3 Radiation passing through a volume element of length Δx and unit cross-sectional area.

$$-d\mathcal{N}(x) = [I(x) - I(x + \Delta x)] \frac{n}{hv_{21}c}$$

If Δx is sufficiently small we can rewrite this equation as

$$-d\mathcal{N}(x) = -\frac{dI(x)}{dx} \cdot \frac{\Delta x n}{hv_{21}c}$$

Thus the rate of decay of photon density in a time interval dt $(= \Delta x/(c/n))$ is

$$\frac{d\mathcal{N}}{dt} = \frac{dI(x)}{dx} \cdot \frac{1}{hv_{21}}$$

and substituting for dI/dx from eq. (5.10) gives

$$\frac{d\mathcal{N}}{dt} = -\alpha I(x) \cdot \frac{1}{hv_{21}} = -\alpha \varrho_v \frac{c}{n} \cdot \frac{1}{hv_{21}} \tag{5.12}$$

Hence comparing eqs. (5.11) and (5.12) we have

$$\alpha \varrho_v \frac{c}{n} \frac{1}{hv_{21}} = \left(\frac{g_2}{g_1} N_1 - N_2\right) \varrho_v B_{21}$$

Therefore the absorption coefficient α is given by

$$\alpha = \left(\frac{g_2}{g_1} N_1 - N_2\right) \frac{B_{21} hv_{21} n}{c} \tag{5.13}$$

We see from eq. (5.13) that α, as we supposed earlier, depends on the difference in the populations of the two energy levels E_1 and E_2. For a collection of atoms in thermal equilibrium, since $E_2 > E_1$, $(g_2/g_1)N_1$ will always be greater than N_2 (cq. (5.4)) and hence α is positive. If, however, we can create a situation, referred to in the last section, in which N_2 becomes greater than $(g_2/g_1)N_1$ then α is negative and the quantity $(-\alpha x)$ in the exponent of eq. (5.10) becomes positive. Thus the irradiance of the beam *grows* as it propagates through the medium in accordance with the equation:

$$I = I_0 \exp (kx) \tag{5.14}$$

where k is referred to as the *small signal gain coefficient* and is given by

$$k = \left(N_2 - \frac{g_2}{g_1} N_1\right) B_{21} \frac{hv_{21} n}{c} \tag{5.15}$$

5.4 Population Inversion

The population inversion condition required for light amplification is a non-equilibrium distribution of atoms among the various energy levels of the atomic system. The Boltzmann distribution which applies to a system in thermal equilibrium is given by eq. (5.3) and is illustrated in Fig. 5.4(a); N_j is the population density of the jth energy level and clearly as E_j increases N_j decreases for a constant temperature. We note that if the

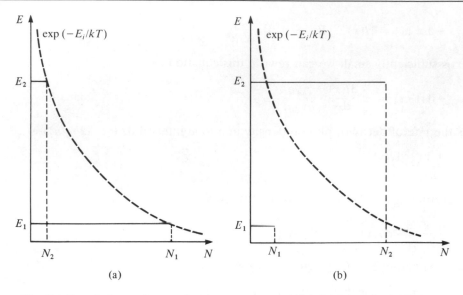

Fig. 5.4 Populations of a two level energy system: (a) in thermal equilibrium and (b) after a population inversion has been produced.

energy difference between E_1 and E_2 were nearly equal to kT ($\simeq 0.025$ eV at room temperature) then the population of the upper level would be $1/e$ or 0.37 of that of the lower level. For an energy difference large enough to give visible radiation ($\simeq 2.0$ eV), however, the population of the upper level is almost negligible as Example 5.2 shows.

Example 5.2 Relative populations of energy levels

We may estimate the relative populations of two energy levels such that a transition from the higher to the lower will give visible radiation.

Let us take the average wavelength of visible radiation as 550 nm, then $E_2 - E_1$ = $hc/(550 \times 10^{-9})$ = 3.6×10^{-19} J = 2.25 eV.

Assuming room temperature ($T \simeq 300$ K) and that $g_1 = g_2$ we have from eq. (5.4)

$$\frac{N_2}{N_1} = \exp \frac{-3.6 \times 10^{-19}}{1.38 \times 10^{-23} \times 300} \simeq e^{-87} \simeq 10^{-37}$$

Clearly then, if we are to create a population inversion, illustrated in Fig. 5.4(b), we must supply a large amount of energy to excite atoms into the upper level E_2. This excitation process is called *pumping* and much of the technology of lasers is concerned with how the pumping energy can be supplied to a given laser system. Pumping produces a nonthermal equilibrium situation; we shall now consider in general terms how pumping enables a population inversion to be achieved.

5.4.1 Attainment of a population inversion

One of the methods used for pumping is stimulated absorption, i.e. the energy levels which one hopes to use for laser action are pumped by intense irradiation of the system. Now as B_{12} and B_{21} are equal (assuming $g_1 = g_2$), once atoms are excited into the upper level the probabilities of further stimulated absorption or emission are equal so that even with very intense pumping the best that can be achieved with the two level system, considered hitherto, is equality of the populations of the two levels.

As a consequence we must look for materials with either three or four energy level systems; this is not really a disadvantage as atomic systems generally have a large number of energy levels.

The three level system first proposed by Bloembergen, ref. 5.3, is illustrated in Fig. 5.5. Initially the distribution obeys the Boltzmann law. If the collection of atoms is intensely illuminated the electrons can be excited (i.e. pumped) into the level E_2 from the ground state E_0. From E_2 the electrons decay by non-radiative processes to the level E_1 and a population inversion may be created between E_1 and E_0. Ideally, the transition from level E_2 to E_1 should be very rapid, thereby ensuring that there are always vacant states at E_2, while that from E_1 to E_0 should be very slow, i.e. E_1 should be a *metastable* state. This allows a large build-up in the number of atoms in level E_1, as the probability of spontaneous emission is relatively small. Eventually N_1 may become greater than N_0 and then population inversion will have been achieved.

The level E_2 should preferably consist of a large number of closely spaced levels so that pumping uses as wide a part of the spectral range of the pumping radiation as possible, thereby increasing the pumping efficiency. Even so, three level lasers, for

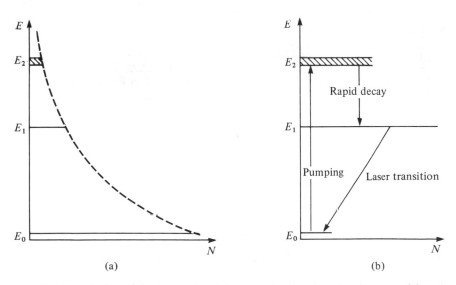

Fig. 5.5 Population of the energy levels by pumping in a three level system: (a) Boltzmann distribution before pumping and (b) distribution after pumping and the transitions involved.

example ruby, require very high pump powers because the terminal level of the laser transition is the ground state. This means that rather more than half of the ground state atoms (this number is usually very nearly equal to the total number of atoms in the collection) have to be pumped to the upper state to achieve a population inversion.

The four level system shown in Fig. 5.6 has much lower pumping requirements. If $(E_1 - E_0)$ is rather large compared to kT (the thermal energy at the temperature of operation), then the populations of the levels E_1, E_2 and E_3 are all very small in conditions of thermal equilibrium. Thus, if atoms are pumped from the ground state to the level E_3 from which they decay very rapidly to the metastable level E_2, a population inversion is quickly created between levels E_2 and E_1.

Again the upper level E_3 should preferably consist of a large number of levels for greatest pumping efficiency. If the lifetimes of the transitions $E_3 \rightarrow E_2$ and $E_1 \rightarrow E_0$ are short, the population inversion between E_2 and E_1 can be maintained with moderate pumping and continuous laser action can be achieved more readily. In the Nd:YAG laser, for example, $\tau_{21} \simeq 0.5$ ms while $\tau_{10} \simeq 30$ ns and, although there are many upper levels used for pumping, each has a lifetime of about 10^{-8} s (i.e. $\tau_{32} \simeq 10^{-8}$ s). The details of the mechanisms used for pumping lasers can be quite complicated and, in addition to optical pumping, pumping can occur in an electrical discharge or by electron bombardment, the release of chemical energy, the passage of a current, etc. The energy level schemes of the media used in lasers are often complex, but they can usually be approximated by either three or four level schemes.

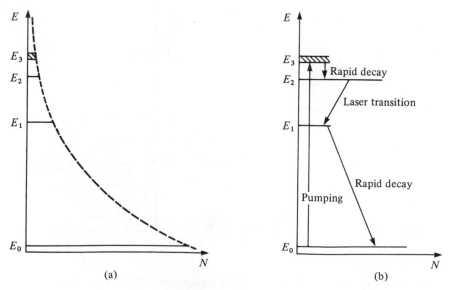

(a)   (b)

Fig. 5.6 Population of the energy levels in a four level system: (a) before pumping and (b) after pumping.

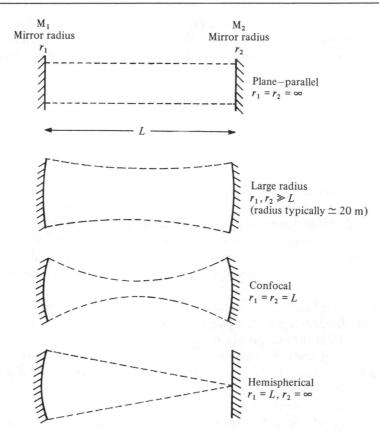

Fig. 5.7 Some commonly used laser cavity mirror configurations (the dashed lines show the extent of the mode volume in each case).

5.5 Optical feedback

The laser, despite its name, is more analogous to an oscillator than an amplifier. In an electronic oscillator, an amplifier tuned to a particular frequency is provided with positive feedback and, when switched on, any electrical noise signal of the appropriate frequency appearing at the input will be amplified. The amplified output is fed back to the input and amplified yet again and so on. A stable output is quickly reached, however, since the amplifier saturates at high input voltages, as it cannot produce a larger output than the supply voltage.

In the laser, positive feedback may be obtained by placing the gain medium between a pair of mirrors which, in fact, form an optical cavity (a Fabry–Perot resonator). The initial stimulus is provided by any spontaneous transitions between appropriate energy levels in which the emitted photon travels along the axis of the system. The signal is amplified as it passes through the medium and 'fed back' by the mirrors. Saturation is reached when the gain provided by the medium exactly matches the losses incurred during a complete round trip.

The gain per unit length of most active media is so small that very little amplification of a beam of light results from a single pass through the medium. In the multiple passes which a beam undergoes when the medium is placed within a cavity, however, the amplification may be substantial.

We have tacitly assumed that the radiation within the cavity propagates to and fro between two plane–parallel mirrors in a well-collimated beam. Because of diffraction effects, however, this cannot be the case as a perfectly collimated beam cannot be maintained with mirrors of finite extent; some radiation will spread out beyond the edges of the mirrors. Diffraction losses of this nature can be reduced by using concave mirrors. In practice a number of different mirror curvatures and configurations are used depending on the applications envisaged and the type of laser being used.

A detailed analysis of the effects of different mirror systems requires a rigorous application of diffraction theory and is beyond the scope of this book. Using simple ray tracing techniques, however, it is quite easy to anticipate the results of such an analysis in that mirror configurations which retain a ray of light, initially inclined at a small angle to the axis, within the optical cavity after several reflections are likely to be useful (see ref. 5.4). Such cavities are said to be stable.

The commonly used mirror configurations are shown in Fig. 5.7; they all have various advantages and disadvantages. The plane–parallel configuration, for example, is very difficult to align, for if the mirrors are not strictly parallel (to within about 1 sec of arc) the optical beam will 'walk off' the mirrors after a few reflections. On the other hand, the radiation beam makes maximum use of the laser medium (we say that it has a large *mode volume* – see also section 5.9) as there is no focusing of the beam within the cavity. In addition, the mirrors need to be flat to within $\lambda/100$. In contrast to the plane parallel case, the confocal arrangement is relatively easy to align (an accuracy of 1.5 minutes of arc is sufficient) but the use of the active medium is restricted (i.e. the mode volume is small). In gas lasers, if maximum power output is required we use a large radius resonator, while if uniphase operation (i.e. maximum beam coherence) is required we use the hemispherical system.

Sometimes mirror configurations are used which give rise to unstable cavities. In these a ray which is initially traveling at a small angle of incidence to the cavity axis will diverge away from the axis after a number of reflections. Such resonators are characterized by high losses, but even so they have some useful properties. In particular they can make efficient use of the mode volume. As unstable resonators have large losses they can be used effectively only with high-gain media such as carbon dioxide. As we mentioned earlier, the gain is usually very small so it is essential to minimize all losses in the laser (see section 5.6). One source of loss is absorption in the mirrors. To reduce this, high reflectance multilayer dielectric coatings on the mirrors are used rather than metallic coatings. In these so-called multilayer stacks there is a sequence of quarter-wave (i.e. $\lambda/4$) layers of alternate high and low refractive index dielectric materials on a glass substrate. Because of the phase changes occurring at alternate interfaces, all the reflected waves are in phase and add constructively. More than twenty such layers may be needed to give reflectances in excess of 99.9% – lower reflectances require fewer layers. Clearly, the mirrors will only be effective over a narrow wavelength range. A familiar example of this sort of process is the blooming of camera lenses to reduce unwanted reflections.

We can now derive the minimum pump power required (that is, the threshold condition) in terms of the parameters of the whole system for laser oscillations to occur.

5.6 Threshold conditions – laser losses

It was explained above that a steady state level of oscillation is reached when the rate of amplification is balanced by the rate of loss. This is the situation in continuous output (CW) lasers; it is a little different in pulse lasers. Thus, while a population inversion is a necessary condition for laser action, it is not a sufficient one because the minimum (i.e. threshold value) of the gain coefficient must be large enough to overcome the losses and sustain oscillations. The threshold gain, in turn, through eq. (5.15) specifies the minimum population inversion required.

The total loss of the system is due to a number of different processes, the most important ones include:

1. Transmission at the mirrors – the transmission from one of the mirrors usually provides the useful output, the other mirror is made as reflective as possible to minimize losses.
2. Absorption and scattering at the mirrors.
3. Absorption in the laser medium due to transitions other than the desired transitions (as mentioned earlier most laser media have many energy levels, not all of which will be involved in the laser action).
4. Scattering at optical inhomogeneities in the laser medium – this applies particularly to solid state lasers.
5. Diffraction losses at the mirrors.

To simplify matters, let us include all the losses except those due to transmission at the mirrors in a single effective loss coefficient γ which reduces the effective gain coefficient to $(k - \gamma)$. We can determine the threshold gain by considering the change in irradiance of a beam of light undergoing a round trip within the laser cavity. We assume that the laser medium fills the space between the mirrors M_1 and M_2 which have reflectances R_1 and R_2 and a separation L. Then in traveling from M_1 to M_2 the beam irradiance increases from I_0 to I where, from eq. (5.14),

$$I = I_0 \exp (k - \gamma)L$$

After reflection at M_2, the beam irradiance will be $R_2I_0 \exp (k - \gamma)L$ and after a complete round trip the final irradiance will be such that the round trip gain G is

$$G = \frac{\text{Final irradiance}}{\text{Initial irradiance}} = R_1R_2 \exp \{2(k - \gamma)L\}$$

If G is greater than unity a disturbance at the laser resonant frequency will undergo a net amplification and the oscillations will grow; if G is less than unity the oscillations will die out. Therefore we can write the *threshold condition* as:

$$G = R_1R_2 \exp \{2(k_{th} - \gamma)L\} = 1 \tag{5.16}$$

where k_{th} is the threshold gain. It is important to realize that the threshold gain is equal to the steady state gain in continuous output lasers, i.e. $k_{th} = k_{ss}$. This equality is due to a phenomenon known as *gain saturation*, which can be explained as follows. Initially, when laser action commences the gain may be well above the threshold value. The effect of stimulated emission, however, will be to reduce the population of the upper level of the laser transition so that the degree of population inversion and consequently the gain will decrease. Thus the net round trip gain may vary and be greater than or less than unity so that the cavity energy density will correspondingly increase or decrease. It is only when G has been equal to unity for a period of time that the cavity energy (and laser output power) settles down to a steady state value, that is when the gain just balances the losses in the medium. In terms of the population inversion there will be a threshold value $N_{th} = [N_2 - (g_2/g_1)N_1]_{th}$ corresponding to k_{th}. In steady state situations $[N_2 - (g_2/g_1)N_1]$ remains equal to N_{th} regardless of the amount by which the threshold pumping rate is exceeded (see section 5.8). The small signal gain required to support steady state operation depends on the laser medium through k and γ, and on the laser construction through R_1, R_2 and L. From eq. (5.16) we can see that:

$$k_{th} = \gamma + \frac{1}{2L} \ln \left(\frac{1}{R_1R_2} \right) \tag{5.17}$$

where the first term represents the volume losses and the second the loss in the form of the useful output. Equation (5.15) shows that k can have a wide range of values, depending not only on $[N_2 - (g_2/g_1)N_1]$ but also on the intrinsic properties of the active medium. If k is high then it is relatively easy to achieve laser action, mirror alignment is not critical and dust can be tolerated on the mirrors. With low gain media, on the other hand, such losses are unacceptable and the mirrors must have high reflectances, and be scrupulously clean and carefully aligned.

It should be noted that a laser with a high gain medium will not necessarily have a high efficiency. The efficiency is the ratio of the output light power to the input pumping power. It therefore depends on how effectively the pump power is converted into producing a population inversion, on the probabilities of different kinds of transitions from the upper level and on the losses in the system. With reference to Fig. 5.6(b), and confining our attention to optical pumping, we can easily see that the efficiency cannot exceed $(E_2 - E_1)/(E_3 - E_0) = \nu_{21}/\nu_{30}$ for the four level system and that it will be considerably less than this for the three level system, where over half the atoms have to be pumped out of the ground state before population inversion is produced. The actual efficiencies, as defined above, are usually very much less than this because of the energy loss in converting electrical energy into optical energy at the pump frequency and the fact that not all the atoms pumped into level 3 will necessarily make a transition to level 2. Certain lasers (for example CO_2) are characterized by having a high efficiency and a high small-signal gain. Other lasers such as argon although having a high gain have a very low efficiency.

5.7 Lineshape function

In deriving the expression for the small signal gain we assumed that all the atoms in either the upper or lower levels would be able to interact with the (perfectly) mono-chromatic beam. In fact this is not so; spectral lines have a finite wavelength (or fre-quency) spread, i.e. they have a spectral width. This can be seen in both emission and absorption and if, for example, we were to measure the transmission as a function of frequency for the transition between the two energy states E_1 and E_2 we would obtain the bell-shape curve shown in Fig. 5.8(a).

The emission curve would be the inverse of this (see Fig. 5.8(b)). The shape of these curves is described by the *lineshape function* $g(v)$, which can also be used to describe a frequency probability curve. Thus we may define $g(v)dv$ as the probability that a given transition between the two energy levels will result in the emission (or absorption) of a photon whose frequency lies between v and $v + dv$. $g(v)$ is normalized such that $\int_{-\infty}^{\infty} g(v)dv = 1$. Therefore we see that a photon of energy hv may not necessarily stimulate another photon of energy hv. We then take $g(v)dv$ as the probability that the stimulated photon will have an energy between hv and $h(v + dv)$.

It is shown in Appendix 4 that, when a monochromatic beam of frequency v_s interacts with a group of atoms with a lineshape function $g(v)$, the small signal gain coefficient may be written as:

$$k(v_s) = \left(N_2 - \frac{g_2}{g_1} N_1 \right) \frac{B_{21} h v_s n g(v_s)}{c} \tag{5.18}$$

The form of the lineshape function $g(v)$ depends on the particular mechanism respon-sible for the *spectral broadening* in a given transition. The three most important mechanisms are Doppler broadening, collision (or pressure) broadening and natural

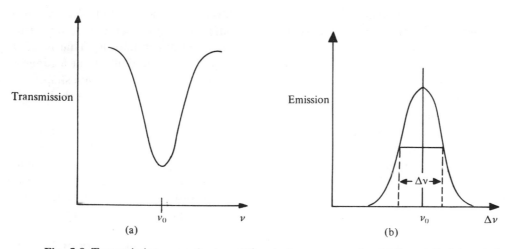

(a)   (b)

Fig. 5.8 Transmission curve for transitions between energy levels E_1 and E_2 (a) and the emission curve for transitions between E_2 and E_1 (b). The precise form of these curves (the lineshape) depends on the spectral broadening mechanisms.

(or lifetime) damping, which are described briefly below; the interested reader is referred to ref. 5.5 for further details.

Example 5.3 Small-signal gain coefficient

It may be instructive to calculate the value of population inversion required to give a gain coefficient of 1 m^{-1} in a given laser. We take Nd:YAG for which we have the following data: spontaneous lifetime, τ_{21} = 230 μs; wavelength, λ = 1.06 μm; refractive index, n = 1.82, and linewidth, Δv = 3 × 10^{12} Hz (see eq. (5.21)).

From eq. (5.8) we have $B_{21} = \lambda^3/8\pi h\tau_{21}$ = 3.1 × 10^{17} m^3 W^{-1} s^{-3}. Therefore from eq. (5.18) we have (with k = 1)

$$\left(N_2 - \frac{g_2}{g_1} N_1\right) = \frac{k\lambda\Delta v}{B_{21}hn} \simeq 8 \times 10^{22} \text{ m}^{-3}$$

Doppler broadening. We are familiar with the Doppler effect, which occurs because of the relative motion of a source and observer. The frequency as measured by the observer increases if the source and observer approach one another and decreases as they recede. This effect applies to a collection of atoms emitting at an optical frequency v_{12}, so that the observed frequency is given by

$$v'_{12} = v_{12}\left(1 \pm \frac{v_x}{c}\right)$$

where v_x is the component of the velocity of the atom along the direction of observation (we assume $v_x \ll c$). Since the atoms are in random motion, an observer would measure a range of frequencies depending on the magnitude and direction of v_x. That is, as far as the observer is concerned the collection of atoms would be emitting at a range of different resonant frequencies resulting in a broadening of the emission lineshape. The individual Doppler shifted resonant frequencies contribute to the smooth Doppler broadened lineshape.

The mean squared velocity components v_x depend on the temperature according to $\frac{1}{2}Mv_x^2 = \frac{1}{2}kT$, where M is the atomic mass, so that the halfwidth (full width of the curve at half the maximum intensity of emission) of the curve is proportional to the square root of T. Doppler broadening is the predominant mechanism in most gas lasers emitting in the visible region. For example, the halfwidth of the 632.8 nm transition of the He–Ne laser is about 2 × 10^{-3} nm, assuming a temperature of operation of about 400 K (see Problem 5.5).

Halfwidths are often expressed in terms of frequency, thus a halfwidth of 2 × 10^{-3} nm corresponds to a frequency halfwidth of 1500 MHz. (As $c = v\lambda$, we may write dv = $-(c/\lambda^2)$ dλ and hence dv = (3 × 10^8 × 2 × 10^{-12})/(632.8 × 10^{-9})^2 = 1500 MHz.)

Collision broadening. The Doppler linewidth of molecular lasers such as the CO_2 laser is relatively small because of their low resonant frequencies (in the infrared) and comparatively large molecular masses. In such lasers, collision broadening becomes important. Collision broadening also occurs in doped insulator lasers. In these lasers the ions of the active medium may suffer collisions with phonons, i.e. quantized lattice vibrations.

If an atom which is emitting a photon suffers a collision, then the phase of the wave train associated with the photon is suddenly altered. This in effect shortens the emitted wave trains, which can be shown by Fourier techniques, ref. 5.6, to be equivalent to a broadening of the spectral line. Clearly, the higher the pressure (and temperature) of the gas the more frequently will the atoms suffer collisions and the greater will be the spectral broadening.

Natural damping. It can be shown that the very act of an atomic electron emitting energy in the form of a photon leads to an exponential damping of the amplitude of the wave train. The effect of this is similar to collision broadening in that it effectively shortens the wave trains and produces a broadened spectral line.

Broadening mechanisms can be classified into *homogeneous* and *inhomogeneous* broadening. If all of the atoms of the collection have the same transition center frequency and the same resonance lineshape then the broadening is termed homogeneous; such is the case for collision broadening. On the other hand, in some situations each atom has a slightly different resonance frequency or lineshape for the same transition. The observed lineshape is then the average of the individual ones, such as in Doppler broadening, and the mechanism is termed inhomogeneous. Local variations of temperature, pressure, applied magnetic field as well as local variations due to crystal imperfections also lead to inhomogeneous broadening of the emission or absorption lineshapes. The nature of the broadening is important in several aspect of laser theory, for example in the discussion of gain saturation mentioned earlier.

Homogeneous broadening mechanisms lead to a Lorentzian lineshape which may be written as:

$$g(v)_L = \frac{\Delta v}{2\pi} \left[(v - v_0)^2 + \left(\frac{\Delta v}{2} \right)^2 \right]^{-1}$$

where Δv is the linewidth, that is the separation between the two points on the (frequency) curve where the function falls to half of its peak value which occurs at frequency v_0. Putting $v = v_0$ gives

$$g(v_0)_L = \frac{2}{\pi \Delta v} \tag{5.19}$$

Inhomogeneous broadening mechanisms, on the other hand, lead for a gas to a Gaussian frequency distribution, given by:

$$g(v)_G = \frac{2}{\Delta v} \left(\frac{\ln 2}{\pi} \right)^{\frac{1}{2}} \exp \left[-(\ln 2) \left(\frac{v - v_0}{\Delta v / 2} \right)^2 \right]$$

and putting $\nu = \nu_0$ gives

$$g(\nu_0)_G = \frac{2}{\Delta\nu}\left(\frac{\ln 2}{\pi}\right)^{\frac{1}{2}} \tag{5.20}$$

For the purpose of later calculations we may, in fact, approximate both eqs. (5.19) and (5.20) by

$$g(\nu_0) \simeq \frac{1}{\Delta\nu} \tag{5.21}$$

Because of these various broadening mechanisms, we can no longer treat a group of atoms as though they all radiate at the same frequency. Instead, we must consider a small spread of frequencies about some central value. It might then be expected that the output of the laser would contain the same distribution of frequencies as the broadened transitions of the atoms in the medium. This is, in fact, not the case as the spectral character of the laser output is different from that of spontaneous emission in the same medium. Two factors account for this difference: namely, the effects of the optical resonator (discussed below) and the effect of the amplification process on the irradiance. As light travels through an amplifying medium the irradiance varies as

$$I_\nu(\nu, x) = I(\nu, 0) \exp \{k(\nu)x\}$$

Equation (5.18) shows that $k(\nu)$ depends on the lineshape function $g(\nu)$; hence $I_\nu(\nu, x)$ is related exponentially to $g(\nu)$. Consequently the function $I_\nu(\nu, x)$ is much greater at the center and smaller in the 'wings' than the atomic lineshape. $I_\nu(\nu, x)$ is therefore narrower than the atomic lineshape – this effect is known as spectral narrowing. In fact, as we shall see below, laser light has an even narrower spectral range than suggested by this argument.

5.8 Population inversion and pumping threshold conditions

We may now use eq. (5.18) to calculate the population inversion required to reach the lasing threshold. From eq. (5.18) we have

$$\left(N_2 - \frac{g_2}{g_1}N_1\right) = \frac{k(\nu_s)c}{B_{21}h\nu_s ng(\nu_s)}$$

At threshold, the small-signal gain coefficient is given by eq. (5.17), i.e.

$$k(\nu_s) = k_{\text{th}} = \gamma + \frac{1}{2L}\ln\left(\frac{1}{R_1 R_2}\right)$$

and therefore

$$\left(N_2 - \frac{g_2}{g_1}N_1\right)_{\text{th}} = \frac{k_{\text{th}}c}{B_{21}h\nu_s ng(\nu_s)}$$

From eq. (5.8) we have $B_{21} = c^3 A_{21}/(8\pi h\nu^3 n^3)$. The quantity A_{21} may be determined experimentally by noting that it is the reciprocal of the spontaneous radiation lifetime τ_{21} from level 2 to level 1.

Thus combining the above equations we can write the threshold population inversion as

$$N_{th} = \left(N_2 - \frac{g_2}{g_1} N_1\right)_{th} = \frac{8\pi v_s^2 k_{th} \tau_{21} n^2}{c^2 g(v_s)} \tag{5.22}$$

We note that the lasing threshold will be achieved most readily when $g(v_s)$ is a maximum, i.e. when v_s has the value v_0 corresponding to the center of the natural linewidth. We may therefore replace $g(v_s)$ by $1/\Delta v$ (see eq. (5.21)) to give:

$$N_{th} = \frac{8\pi v_0^2 k_{th} \tau_{21} \Delta v n^2}{c^2} \tag{5.23}$$

We now proceed to calculate the pumping power required to reach threshold. To do this, we must solve the rate equations for the particular system. The rate equations describe the rate of change of the populations of the laser medium energy levels in terms of the emission and absorption processes and pump rate. We shall consider the ideal four level system shown in Fig. 5.9. We assume that $E_1 \gg kT$ so that the thermal population of level 1 is negligible; we also assume that the threshold population density N_{th} is very small compared to the ground state population so that during lasing the latter is hardly affected. If we let $\mathcal{R}_2$ and $\mathcal{R}_1$ be the rates at which atoms are pumped into levels 2 and 1 respectively, we can write the rate equations for these levels (assuming $g_1 = g_2$ for simplicity, and hence $B_{21} = B_{12}$) as

$$\frac{dN_2}{dt} = \mathcal{R}_2 - N_2 A_{21} - \varrho_v B_{21}(N_2 - N_1) \tag{5.24}$$

and

$$\frac{dN_1}{dt} = \mathcal{R}_1 + \varrho_v B_{21}(N_2 - N_1) + N_2 A_{21} - N_1 A_{10} \tag{5.25}$$

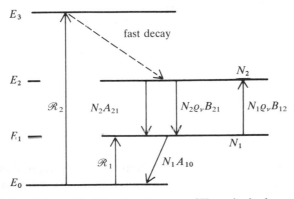

Fig. 5.9 Transitions within an ideal four level system. (The principal pumping mechanism excites atoms to the level E_3, from where they rapidly decay to the level of E_2. As E_2 is the upper level of the laser transition, we have designated the corresponding pumping rate $\mathcal{R}_2$ rather then $\mathcal{R}_3$.)

Process $\mathscr{R}_1$, which populates the lower laser level 1, is detrimental to laser action as it clearly reduces the population inversion. Although such pumping is unavoidable in many lasers, for example gas lasers pumped via an electrical discharge (section 5.10.3), we shall henceforth ignore $\mathscr{R}_1$. If we assume that the system is being pumped at a steady rate then we have $dN_2/dt = dN_1/dt = 0$. Hence we may solve eqs. (5.24) and (5.25) for N_1 and N_2. We leave it as an exercise for the reader to show that:

$$N_1 = \mathscr{R}_2/A_{10}$$

$$N_2 = \mathscr{R}_2 \left[1 + \frac{\varrho_v B_{21}}{A_{10}} \right] (A_{21} + \varrho_v B_{21})^{-1}$$

and hence

$$N_2 - N_1 = \mathscr{R}_2 \left(\frac{1 - A_{21}/A_{10}}{A_{21} + \varrho_v B_{21}} \right) \tag{5.26}$$

We can see that unless $A_{21} < A_{10}$, the numerator will be negative and no population inversion can take place. As the Einstein A coefficients are the reciprocals of the spontaneous lifetimes, the condition $A_{21} < A_{10}$ is equivalent to the condition $\tau_{10} < \tau_{21}$, i.e. that the upper lasing level has a longer spontaneous emission lifetime than the lower level. In most lasers $\tau_{21} \gg \tau_{10}$ and $(1 - A_{21}/A_{10}) \simeq 1$.

Now, below threshold we may neglect ϱ_v since lasing has not yet commenced and most of the pump power appears as spontaneous emission; thus eq. (5.26) can be written as

$$N_2 - N_1 = \mathscr{R}_2 \left(\frac{1 - A_{21}/A_{10}}{A_{21}} \right)$$

i.e. there is a linear increase in population inversion with pumping rate but insufficient inversion to maintain amplification.

At threshold, ϱ_v is still small and assuming $g_1 = g_2$, we can express the threshold population inversion in terms of the threshold pump rate, i.e.

$$(N_2 - N_1)_{th} = N_{th} = \mathscr{R}_{th} \left(\frac{1 - A_{21}/A_{10}}{A_{21}} \right) \tag{5.27}$$

or inserting the above approximation that $(1 - A_{21}/A_{10}) \simeq 1$

$$\mathscr{R}_{th} = N_{th} A_{21}$$

or

$$\mathscr{R}_{th} = \frac{N_{th}}{\tau_{21}}$$

Each atom raised into level 2 requires an amount of energy E_3 so that the total pumping power per unit volume P_{th} required at threshold may be written as:

$$P_{th} = \frac{E_3 N_{th}}{\tau_{21}}$$

We may substitute for N_{th} from eq. (5.23) to give

$$P_{\text{th}} = \frac{E_3 8\pi v_0^2 k_{\text{th}} \tau_{21} \Delta v n^2}{\tau_{21} c^2}$$

or

$$P_{\text{th}} = \frac{E_3 8\pi v_0^2 k_{\text{th}} \Delta v n^2}{c^2} \tag{5.28}$$

This is the point at which the gain due to the population inversion exactly equals the cavity losses. Further increase of the population inversion with pumping is impossible in a steady state situation, since this would result in a rate of induced energy emission which exceeds the losses. Thus the total energy stored in the cavity would increase with time in violation of the steady state assumption (this is the phenomenon of gain saturation described earlier).

This argument suggests that $[N_2 - (g_2/g_1)N_1]$ must remain equal to N_{th} regardless of the amount by which the threshold pump rate is exceeded. Equation (5.26) shows that this is possible providing $\varrho_v B_{21}$ is able to increase (once $\mathscr{R}_2$ exceeds its threshold value given by eq. (5.27)) so that the equality

$$N_{\text{th}} = \mathscr{R}_2 \left[\frac{1 - A_{21}/A_{10}}{A_{21} + \varrho_v B_{21}} \right]$$

is satisfied. Now combining this equation with eq. (5.27) we have

$$\frac{\mathscr{R}_{\text{th}}}{A_{21}} = \frac{\mathscr{R}_2}{A_{21} + \varrho_v B_{21}}$$

and hence

$$\varrho_v = \frac{A_{21}}{B_{21}} \left(\frac{\mathscr{R}_2}{\mathscr{R}_{\text{th}}} - 1 \right) \tag{5.29}$$

Since the power output W of the laser will be directly proportional to the optical power density within the laser cavity and the pump rate into level 2 (i.e. $\mathscr{R}_2$) will be proportional to the pump power P delivered to the laser, we may rewrite eq. (5.29) as

$$W = W_0 \left(\frac{P}{P_{\text{th}}} - 1 \right) \tag{5.30}$$

where W_0 is a constant.

Thus if the pump rate is increased above the value P_{th} the beam irradiance is expected to increase linearly with pump rate. This is born out in practice and plots of population inversion and laser output as a function of pump rate are of the form shown in Fig. 5.10. The additional power above threshold is channeled into a single (or a few) cavity mode(s) (see section 5.9). Spontaneous emission still appears above threshold but it is extremely weak in relation to the laser output as it is emitted in all directions and has a much greater frequency spread.

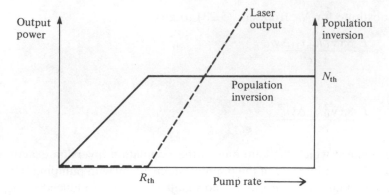

Fig. 5.10 Population inversion and laser power output as a function of pump rate.

5.9 Laser modes

Examination of a laser output with a spectrometer of very high resolving power, such as the scanning Fabry–Perot interferometer, reveals that it consists of a number of discrete frequency components (or very narrow spectral lines). To appreciate how these discrete lines arise and how they are related to the laser transition lineshape we need to examine the effects of the mirrors on the light within the laser cavity (see ref. 5.7).

5.9.1 Axial modes

The two mirrors of the laser form a resonant cavity and standing wave patterns are set up between the mirrors in exactly the same way that standing waves develop on a string or within an organ pipe. The standing waves satisfy the condition

$$p\,\frac{\lambda}{2} = L$$

or

$$\nu = \frac{pc}{2L} \tag{5.31}$$

where, strictly speaking, L is the *optical* pathlength between the mirrors, in which case λ would be the vacuum wavelength and p is an integer, which may be very large (for example if $L = 0.5$ m and $\lambda \simeq 500$ nm then $p \simeq 2 \times 10^6$). As p has such a large value, many different values of p are possible for only a small change in wavelength. Each value of p satisfying eq. (5.31) defines an *axial* (or *longitudinal*) *mode* of the cavity.

From eq. (5.31), the frequency separation $\delta\nu$ between adjacent modes ($\delta p = 1$) is given by

$$\delta\nu = \frac{c}{2L} \tag{5.32}$$

and therefore for $L = 0.5$ m, $\delta v = 300$ MHz. As eq. (5.32) is independent of p, the frequency separation of adjacent modes is the same irrespective of their actual frequencies. The modes of oscillation of the laser cavity will consist, therefore, of a large number of frequencies, each given by eq. (5.31) and separated by $c/2L$, as illustrated in Fig. 5.11(b).

It should be appreciated, however, that while all the integers p give *possible* axial

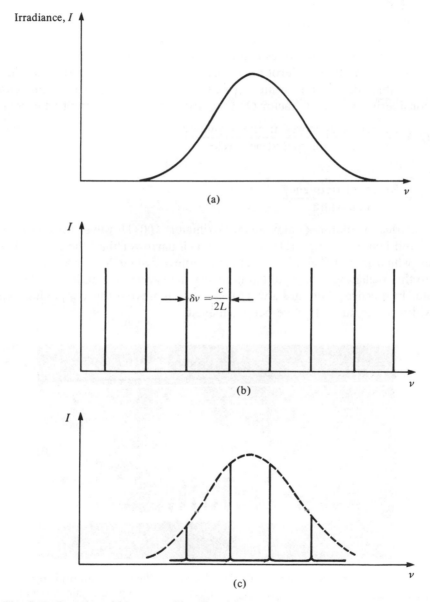

Fig. 5.11 Broadened laser transition line (a); cavity modes (b); and axial modes in the laser output (c).

cavity modes only those which lie within the gain curve or laser transition line will actually oscillate. Thus the broadened laser transition (Fig. 5.11 (a)) for the 632.8 nm wavelength emitted by neon is about 1.5×10^9 GHz wide so that with the 0.5 m long cavity in the above example we would expect four or five modes to be present as illustrated in Fig. 5.11(c). Figure 5.12 shows the axial modes of an He–Ne laser about 1 m in length which are displayed with the aid of an optical frequency analyzer and silicon photodiode.

The axial modes all contribute to a single 'spot' of light in the laser output, whereas the transverse modes discussed below may give rise to a pattern of spots in the output. If the linewidth of the axial modes is measured it will be found to be much narrower than the width of the Fabry–Perot resonances to be expected from treating the cavity simply as a Fabry–Perot interferometer (see ref. 5.8). We can appreciate the reason for this by considering the quality factor Q of the resonator. Q can be defined in general by

$$Q = \frac{2\pi \times \text{Energy stored in the resonator}}{\text{Energy dissipated per cycle}}$$

or

$$Q = \frac{\text{Resonant frequency}}{\text{Linewidth}} = \frac{\nu}{\Delta \nu}$$

For an electrical oscillator, Q may be approximately 100. However for a laser, Q may be $\simeq 10^8$ and hence $\Delta \nu \simeq 1$ MHz, which is much narrower than the Fabry–Perot resonances, which are $\simeq 10^9$ Hz. Indeed, in lasers the active medium is actually supplying energy to the oscillating modes so that in theory the energy dissipation can be zero and Q infinite. In practice, there are always losses which prevent this happening; but even so linewidths of about 1 Hz have been achieved.

Fig. 5.12 Axial modes formed in the He–Ne laser – the mode pattern is repeated (here, three times) as the optical frequency analyzer scans through the gain curve of the laser. [Photograph courtesy of Dr I. D. Latimer, Dept of Physics, Newcastle upon Tyne Polytechnic.]

5.9.2 Transverse modes

Axial modes are formed by plane waves traveling axially along the laser cavity on a line joining the centers of the mirrors. For any real laser cavity there will probably be waves traveling just off-axis that are able to replicate themselves after covering a closed path such as that shown in Fig. 5.13(a). These will also give rise to resonant modes, but because they have components of their electromagnetic fields which are transverse to the direction of propagation they are termed transverse electromagnetic (or TEM) modes. A complete analysis of TEM modes is quite complicated and will not be attempted here (but see Chapter 8 for a further discussion). They are characterized by two integers q and r so that, as Fig. 5.13(b) shows, we have TEM_{00}, TEM_{01}, TEM_{11}, etc, modes (q gives the number of minima as the beam is scanned horizontally and r the number of minima as it is scanned vertically).

In a TEM_{00} mode, the irradiance distribution across the beam is in fact Gaussian, and so we may write the electric field variation as

$$\mathscr{E}(x, y) = \mathscr{E}_0 \exp\left(-\frac{x^2 + y^2}{w^2}\right) \tag{5.33}$$

where x and y are measured in directions perpendicular to the laser axis which is taken to be along the z direction. The sideways spread of the beam is determined by the value of the parameter w, which is a function of the distance z. When $x^2 + y^2 > w^2$, the field falls off rapidly with distance away from the laser axis. The value of w is determined by the locus of points where the field amplitude has fallen to $1/e$ of its maximum value (that is, where $x^2 + y^2 = w^2$). Figure 5.14 shows the typical variation of w, with position, within a cavity formed by two concave mirrors of radius of curvature r_1 and r_2 separated by L. Such a cavity can be shown to be stable when $L \leqslant r_1 + r_2$. The surfaces of constant phase are not in general plane, but are perpendicular to the contours of constant field strength. It can be seen that the mirrors themselves are surfaces of constant phase. This is no accident, but merely a direct consequence of the requirement that the mode be self-replicating as the light energy flows backwards and forwards between the mirrors. At one position within the cavity, the wavefronts become plane and, in fact, a plane mirror placed at this point would give rise to a hemispherical cavity. At this point also w has its smallest value, w_0. The variation of w with z is given by (ref. 5.9)

$$w(z) = w_0\left[1 + \left(\frac{z\lambda}{\pi w_0^2}\right)^2\right]^{\frac{1}{2}} \tag{5.34}$$

where z is measured from the position of minimum beam diameter. The precise value of w_0 depends on the type of cavity. In the case of a nearly confocal cavity where $r_1 = r_2 = r \geqslant L$ it can be shown that (ref. 5.5)

$$w_0^2 \simeq \frac{\lambda}{2\pi}(L(2r - L))^{\frac{1}{2}} \quad \text{or} \quad w_0^2 \simeq \frac{\lambda r}{2\pi} \tag{5.35}$$

Combining eq. (5.34) and (5.35) yields

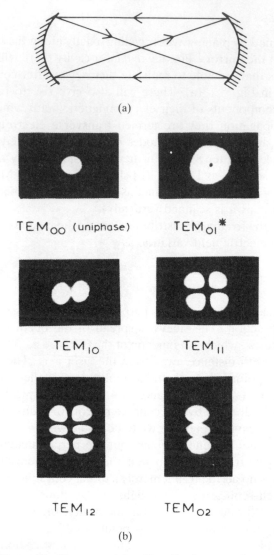

(a)

TEM$_{00}$ (uniphase)   TEM$_{01}$*

TEM$_{10}$ TEM$_{11}$

TEM$_{12}$ TEM$_{02}$

(b)

Fig. 5.13 (a) Example of a nonaxial self-replicating ray that gives rise to transverse modes. (b) Some low order transverse modes of a laser. The modes are designated TEM$_{qr}$, where q and r are integers referring to the number of minima as the laser beam is scanned horizontally and vertically. The TEM$_{01}^*$ mode is a combination of TEM$_{01}$ and TEM$_{10}$ modes. [From M. J. Beesley, *Lasers and their Applications*, 1972; courtesy of Taylor and Francis Ltd.]

$$w(z) = w_0 \left[1 + \left(\frac{2z}{r} \right)^2 \right]^{\frac{1}{2}}$$

(5.36)

Equation (5.36) applies outside as well as inside the cavity, and at large distances from the laser, that is when $z \gg r$, we have

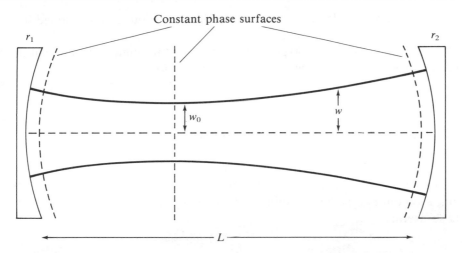

Constant phase surfaces

Fig. 5.14 TEM_{00} mode within a laser cavity. The mode 'adjusts' itself so that the mirror surfaces are surfaces of constant phase. The value of w is determined by the locus of points where the field amplitude has fallen to $1/e$ of its maximum value.

$$w(z) = \frac{z\lambda}{\pi w_0} \tag{5.37}$$

As the TEM_{01} and higher order modes extend further from the cavity axis than the TEM_{00} mode, they will only oscillate if the aperture of the cavity is large enough. These and higher order modes can, therefore, be eliminated by narrowing the laser cavity leaving just the TEM_{00} mode oscillating. The TEM_{00} mode is often called the uniphase mode, since all parts of the propagating wavefront are in phase; this is not the case with higher order modes and in fact phase reversals account for the higher order transverse mode patterns. Consequently, a laser operating only in the TEM_{00} mode has the greatest spectral purity and degree of coherence, while operation in multimode form provides considerably more power.

It should be noted that each transverse mode will have the axial modes discussed above associated with it, so that the total spread in the laser spectrum may be (relatively) large.

5.10 Classes of laser

In the thirty years since Maiman reported the first observation of successful laser action in ruby, there has been an extremely rapid increase in the types of lasers and in the range of materials in which lasing has been shown to occur. It is not possible to describe all of these developments, so in this section we have concentrated on a description of the construction and mode of operation of some of the more commonly available and important lasers. These are classified into four groups: doped insulator, semiconductor, gas and dye lasers.

Before discussing these laser types it might be useful to remind ourselves of the basic requirements which must be satisfied for laser operation.

Firstly, there must be an active medium which emits radiation in the required region of the electromagnetic spectrum.

Secondly, a population inversion must be created within the medium; this, in turn, requires the existence of suitable energy levels associated with the lasing transition for pumping.

Thirdly, for true laser oscillation there must be optical feedback at the ends of the medium to form a resonant cavity (satisfying the first two conditions can provide light amplification but not the highly collimated, monochromatic beam of light which makes lasers so useful).

5.10.1 Doped insulator lasers

The term doped insulator laser is used to describe a laser whose active medium consists of an array of atoms, usually in crystalline form, with impurity atoms intentionally introduced (doped) into the array during growth. Such lasers are rugged, easy to maintain and capable of generating high peak powers; typical examples are the ruby and the Nd:YAG lasers. Although the ruby laser is interesting in that it was the first successful laser, the Nd:YAG laser is now much more widely used and we describe it in some detail, giving only passing reference to the ruby laser.

Nd:YAG laser. The active medium for this laser is yttrium aluminum garnet ($Y_3Al_5O_{12}$) with the rare earth metal ion neodymium Nd^{3+} present as an impurity. The Nd^{3+} ions, which are randomly distributed as substitutional impurities on lattice sites normally occupied by the yttrium ions, provide the energy levels for both the lasing transitions and pumping. Though the YAG host itself does not participate directly in the lasing action, it does have two important roles.

When an Nd^{3+} ion is placed in a host crystal lattice it is subject to the electrostatic field of the surrounding ions, the so-called *crystal field*. The crystal field of the host interacts with the electron energy levels in a variety of ways depending on such factors as its strength and symmetry, and on the electron configuration of the impurity.

A neodymium ion which is free to move in a gaseous discharge, for example, has many of its energy levels with the same energy; these are said to be degenerate. When the ion is placed in the host the crystal field splits some of the energy levels, thereby partly removing the degeneracy. A rather more important effect in the case of the Nd:YAG laser is that the crystal field modifies the transition probabilities between the various energy levels of the Nd^{3+} ion so that some transitions, which are forbidden in the free ion, become allowed.

The net result is that the ground and first excited state energy levels of the Nd^{3+} ion split into the groups of levels shown in Fig. 5.15. The symbols used to describe the energy levels in this and succeeding diagrams depend on the exact nature of the ions and atoms

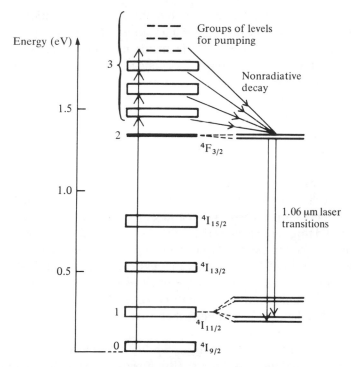

Fig. 5.15 Simplified energy level diagram for the neodymium ion in YAG showing the principal laser transitions. Laser emission also results from transitions between the $^4F_{3/2}$ levels and the $^4I_{15/2}$ and $^4I_{13/2}$ levels, but at only one tenth of the intensity of the transition shown.

involved. In the case under discussion, the symbols arise from Russel–Saunders or LS coupling (ref. 5.10). The symbol for an energy level is written $^{2S+1}X_J$. Here S is the vector sum of the electron spins of the ion. X gives the vector sum L of the orbital angular momentum quantum numbers, where values of $L = 0, 1, 2, 3, 4 \ldots$ are designated by S, P, D, F, G ... Finally J is the vector sum of S and L (see ref. 5.10). Thus in the $^4F_{3/2}$ level, $S = 3/2$ (there are three unpaired electrons in the 4f subshell of the Nd^{3+} ion), $L = 3$ and $J = 3/2$. Whilst familiarity with this and similar notations for electron energy levels is not essential to appreciate the basics of a given laser, it is a prerequisite for a detailed understanding of the mechanisms involved.

As we can see from Fig. 5.15 the Nd:YAG laser is essentially a four level system, i.e. the terminal laser level $^4I_{11/2}$ is sufficiently far removed from the ground state $^4I_{9/2}$ that its room temperature population is very small. Whilst a number of laser transitions may occur between some of the pairs of levels shown to the right of the figure, the most intense line at 1.064 µm arises from the superposition of the two transitions shown.

Pumping is normally achieved by using an intense flash of white light from a xenon flashtube. This excites the Nd^{3+} ions from the ground state to the various energy states above the $^4F_{3/2}$ state; there are, in fact, many more states at higher energies than are shown in Fig. 5.15. The presence of several possible pumping transitions contributes to the efficiency of the laser when using a pumping source with a broad spectral output. To ensure that as much radiation as possible from the flashtube is absorbed in the laser medium, close optical coupling is required. The usual arrangement is shown in Fig. 5.16; a linear flashtube and the lasing medium in the form of a rod are placed inside a highly reflecting elliptical cavity. If the flashtube is along one focal axis and the laser rod along the other, then the properties of the ellipse ensure that most of the radiation from the flashtube passes through the laser. The flashtube is fired by discharging a capacitor bank through the tube; the discharge is often initiated by using a secondary high voltage (≈ 20 kV) trigger pulse.

As the pumping flash lasts for only a short time (≈ 1 ms) the laser output is in the form of a pulse, which starts about 0.5 ms after the pumping flash starts. This represents the time for the population inversion to build up. Once started, stimulated emission builds up rapidly and thus depopulates the upper lasing level 2 – much faster in fact than the pumping can replace the excited atoms so that laser action momentarily stops until population inversion achieved again. This process then repeats itself so that the output consists of a large number of spikes of about 1 μs duration with about 1 μs separation. As the spikes are unrelated, the coherence of the laser pulse, (which lasts a total of about 0.5 ms) while being much greater than that of classical light sources, is not as great as might be expected.

The optical cavity may be formed by grinding the ends of the Nd:YAG rod flat and parallel and then silvering them. More usually, however, external mirrors are used as shown in Fig. 5.16. One mirror is made totally reflecting while the other is about 10% transmitting to give an output.

Fig. 5.16 Typical construction of a doped insulator laser showing the ellipsoidal reflector used to maximize optical coupling between the flashtube and laser rod.

A large amount of heat is dissipated by the flashtube and consequently the laser rod quickly becomes very hot. To avoid damage resulting from this, and to allow a reasonable pulse repetition rate, cooling has to be provided by forcing air over the crystal using the reflector as a container. For higher power lasers it is necessary to use water cooling. Provided sufficient cooling is available it is possible to replace the xenon flashtube with quartz–halogen lamps and to operate the laser continuously.

A glance at the energy level diagram. Fig. 5.15, shows that the maximum possible power efficiency of the laser, v_{21}/v_{03}, is about 80%. In practice, because of losses in the system (which include the loss in converting electrical to optical energy in the pumping source, the poor coupling of the pumping source output to the laser rod and the small fraction of the pumping radiation which is actually absorbed), the actual power efficiency is typically 0.1%. Thus, a laser pumped by a flash lamp operated by the discharge of a 1000 μF capacitor charged to 4–5 kV (that is an input energy of about 10 kJ) may produce an output pulse of about 10 J. As the pulse lasts for only about 0.5 ms, the average power, however, is then about 2×10^4 W. This may be greatly increased by a technique known as Q-switching, which is described in Chapter 6.

Nd:Glass laser. Glass, with its high optical homogeneity, provides an excellent host material for neodymium. Local electric fields within the glass modify the Nd^{3+} ion energy levels in much the same way as the crystal field in YAG. The Nd^{3+} ions, however, may be in a variety of slightly different environments causing the linewidth to be much broader than in YAG and thereby raising the threshold pumping power required for laser action (eq. (5.28) showed that the threshold pump power is proportional to Δv). In consequence, Nd:glass lasers are operated in the pulsed mode and the output spectral line width is greater than in Nd:YAG. On the other hand, glass can be doped more heavily than YAG (6% as opposed to 1.5%) and up to three times as much energy can therefore be produced by Nd:glass lasers; it is also much easier and cheaper to prepare glass rods than to grow YAG crystals.

Ruby laser. The basic principle of operation of the ruby laser is the same as that of the Nd:YAG laser. The active medium is a synthetically grown crystal of ruby, i.e. aluminium oxide, with about 0.05% by weight of chromium as an impurity. Chromium ions, Cr^{3+}, replace aluminium ions in the lattice and the crystal field partially removes the degeneracy of the isolated ions to provide levels for pumping and for the laser transitions. In this case, some of the energy levels of the Cr^{3+} ions are almost independent of the crystal field and they remain sharp. Other levels, however, are strongly dependent on the crystal field so that lattice vibrations, which cause fluctuations in the crystal field, broaden these levels quite substantially. The ^2E and $^4A_2$ levels remain sharp while the $^4T_1$ and $^4T_2$ levels become broad as shown in Fig. 5.17 (the symbols in this case are derived from crystal field theory ref. 5.11). Thus the pump transitions are spectrally broad while the transitions R_1 and R_2 are narrow. The energy level diagram shows that ruby is basically a three level system. As explained previously, rather more than half of the total number of ions have to be pumped to level 2 via level 3 to create a population inversion. Thus the laser has a very low efficiency compared to a four level

Fig. 5.17 Three level ruby laser system. Pumping is due to the Cr^{3+} ions absorping blue (excitation to $^4T_1$ levels) and green light (excitation to $^4T_2$ levels). The wavelengths of the R_1 and R_2 laser lines are temperature dependent; the values given are typical.

system such as Nd:YAG. Pumping is achieved through the absorption of the green and blue spectral regions of a white light discharge; this absorption, of course, accounts for the color of ruby.

Diode pumped lasers. A recent development has been the introduction of the diode pumped Nd:YAG laser. Here, a semiconductor laser is used as the source of pumping radiation in place of the more conventional flashtube. Although we have not yet discussed the semiconductor laser, for the present purpose we need only regard it as a source of nearly monochromatic radiation whose wavelength can be adjusted (for example by changing the temperature of the diode) so that it coincides with a strong absorption peak in Nd:YAG at about 807 nm. In this way, the diode radiation is very efficiently utilized; in fact some 25% of the diode radiation can be converted into laser output. Since the diode itself is an efficient converter of electrical energy into light, overall efficiencies can be as high as 5%. Other advantages accruing are small overall size, high beam quality and good frequency stability (see section 6.1).

Pumping is usually carried out longitudinally, i.e. the pumping radiation enters through one end of the laser rod. This end has a dielectric coating so that it reflects 1.06 µm radiation (and hence acts as one of the cavity mirrors) whilst being transparent at the pumping wavelength. The cavity output coupling mirror is then situated some distance beyond the other end of the rod. Such lasers are usually CW operated with output powers of some tens of milliwatts, this figure being limited at present by the maximum power available from diode lasers.

Glass fiber laser. An interesting extension of the idea of an end pumped solid rod laser is the fiber laser (ref. 5.12). These use optical fiber instead of solid rods. Optical fibers will be dealt with in detail in Chapter 8; however, here we may simply regard them as being small diameter 'flexible' rods consisting of a narrow diameter 'core' region surrounded by a somewhat thicker 'cladding'. Light may readily propagate down the core region of such fibers with minimal sideways loss, even when the fiber is bent. The fibers are usually made from silica (SiO_2) which may be doped with Nd to create a lasing medium. A laser cavity may be formed by butting dielectric mirrors up against the ends of the fibers. Pumping may be carried out longitudinally using diode lasers. Other dopants, such as erbium (Er), have also been used to give different lasing wavelengths. Although power outputs are relatively modest ($\sim$1 mW), such lasers are potentially interesting for use in fiber based communication and sensing systems (see Chapters 9 and 10).

5.10.2 Semiconductor lasers

Semiconductor lasers are not very different in principle from the light emitting diodes discussed in Chapter 4. A $p-n$ junction provides the active medium; thus, to obtain laser action we need only meet the other necessary requirements of population inversion and optical feedback. To obtain stimulated emission, there must be a region of the device where there are many excited electrons and vacant states (i.e. holes) present together. This is achieved by forward biasing a junction formed from very heavily doped n and p materials. In such n^+-type material, the Fermi level lies within the conduction band. Similarly, for the p^+-type material the Fermi level lies in the valence band. The equilibrium and forward biased energy band diagrams for a junction formed from such so-called degenerate materials are shown in Fig. 5.18. When the junction is forward biased with a voltage that is nearly equal to the energy gap voltage E_g/e, elcctrons and holes are injected across the junction in sufficient numbers to create a population inversion in a narrow zone called the *active region* (Fig. 5.19).

The thickness t of the active region can be approximated by the diffusion length L_e of the electrons injected into the p region, assuming that the doping level of the p region is less than that of the n region so that the junction current is carried substantially by electrons (sections 2.7.1 and 2.8.2). For heavily doped GaAs at room temperature L_e is 1–3 μm.

In the case of those materials such as GaAs which have a direct bandgap (section 2.4) the electrons and holes have a high probability of recombining radiatively. The recombination radiation produced may interact with valence electrons and be absorbed, or interact with electrons in the conduction band thereby stimulating the production of further photons of the same frequency ($\nu = E_g/h$). If the injected carrier concentration becomes large enough, the stimulated emission can exceed the absorption so that optical

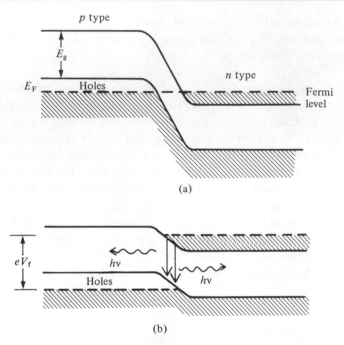

Fig. 5.18 Heavily doped $p-n$ junction: (a) in equilibrium and (b) with forward bias (the broken lines represent the Fermi level in equilibrium (a) and with forward bias (b)).

gain can be achieved in the active region. Laser oscillations occur, as usual, when the round trip gain exceeds the total losses over the same distance. In semiconductors, the principal losses are due to scattering at optical inhomogeneities in the semiconductor material and free carrier absorption. The latter results when electrons and holes absorb a photon and move to higher energy states in the conduction band or valence band respectively. The carriers then return to lower energy states by nonradiative processes.

Fig. 5.19 Diagram showing the active region and mode volume of a semiconducting laser.

In the case of diode lasers, it is not necessary to use external mirrors to provide positive feedback. The high refractive index of the semiconductor material ensures that the reflectance at the material/air interface is sufficiently large even though it is only about 0.32.

Example 5.4 Reflectance at a GaAs/air interface

We may confirm that the reflectance at a GaAs surface is quite high by virtue of the high refractive index (= 3.6) of GaAs. From the Fresnel equations (section 8.1) we have

$$R = \left(\frac{n_2 - n_1}{n_2 + n_1}\right)^2 = \left(\frac{3.6 - 1}{3.6 + 1}\right)^2 \simeq 0.32$$

The diode is cleaved along natural crystal planes normal to the plane of the junction so that the end faces are parallel; no further treatment of the cleaved faces is necessary, although occasionally optical coatings are added for various purposes. For GaAs, the junction plane is (100) and the cleaved faces are (110) planes.

The radiation generated within the active region spreads out into the surrounding lossy GaAs, although there is, in fact, some confinement of the radiation within a region called the *mode volume* (Fig. 5.19). The additional carriers present in the active region increase its refractive index above that of the surrounding material, thereby forming a dielectric waveguide (see Chapter 8). As the difference in refractive index between the center waveguiding layer and the neighboring regions is only about 0.02, the waveguiding is very inefficient and the radiation extends some way beyond the active region, thereby forming the mode volume. The waveguiding achieved in simple homojunction laser diodes of the form shown in Fig. 5.20 only works just well enough to allow laser action to occur as a result of very vigorous pumping. Indeed homojunction lasers can usually only be operated in the pulsed mode at room temperature because the threshold pumping current density (see below) required is so high, being typically of the order of 400 A mm^{-2}.

The onset of laser action at the threshold current density is detected by an abrupt increase in the radiance of the emitting region, as shown in Fig. 5.21. Energy is channeled into one of the modes shown in Fig. 5.22(b) so that a marked decrease in spectral width is noted above threshold, as illustrated in Fig. 5.22(c).

5.10.2.1 THRESHOLD CURRENT DENSITY FOR SEMICONDUCTOR LASERS

An exact calculation of the threshold current for a semiconductor laser is complicated by the difficulty of defining what is meant by a population inversion between two *bands* of energy levels. To simplify the problem, however, and to gain some insight into the important factors, we use the idealized structure shown in Fig. 5.19. We let the active volume, where population inversion is maintained, have thickness t and the

(a)

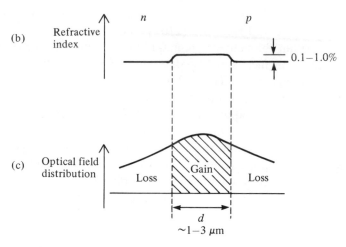

Fig. 5.20 Schematic construction of a GaAs homojunction semiconductor diode laser having side lengths 200–400 μm (a). The emission is confined to the junction region. The narrow thickness d of this region causes a large beam divergence. The very small change in refractive index in the junction region is shown in (b) and (c) shows the resulting poor confinement of the optical radiation to the gain region.

mode volume, where the generated electromagnetic mode is confined, be of thickness d ($d > t$). In other lasers, the mode volume is usually smaller than the volume within which population inversion is maintained.

A consequence of the situation in semiconductor lasers is that the portions of the mode propagating outside the active region may be absorbed. This offsets to some extent the gain resulting from those parts of the mode propagating within the active

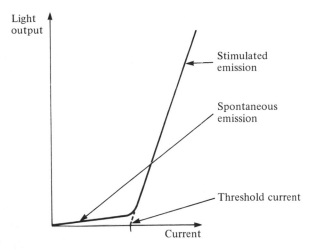

Fig. 5.21 Light output–current characteristic of an ideal semiconductor laser.

region. We allow for this by assuming that the effective population inversion within the mode volume ($d \times l \times w$) is given by reducing the actual population inversion in the active region by the factor t/d.

Referring to eq. (5.23), the threshold condition will thus be reached when

$$N_{th} = \left(N_2 - \frac{g_2}{g_1} N_1\right)_{th} = \frac{d}{t} \left(\frac{8\pi v_0^2 k_{th} \tau_{21} \Delta v n^2}{c^2}\right)$$

We next assume that within the active region we can ignore N_1, that is, there is a large number of holes in the valence band; hence,

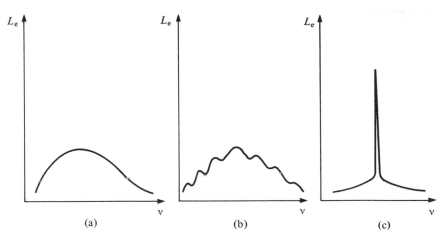

Fig. 5.22 Radiant output as a function of frequency for a p–n junction laser: (a) below threshold (spontaneous emission); (b) with the laser modes at threshold; and (c) with the dominant laser mode above threshold (the radiance scales are compressed (a) → (c)).

$$(N_2)_{th} = \frac{d}{t}\left(\frac{8\pi v_0^2 k_{th}\tau_{21}\Delta v n^2}{c^2}\right) \tag{5.38}$$

If the current density flowing through the laser diode is J A m^{-2}, then the number of electrons per second being injected into a volume t (i.e. a region of thickness t and of unit cross-sectional area) of the active region is J/e. Thus the number density of electrons being injected per second is J/et electrons s^{-1} m^{-3}. The equilibrium number density of electrons in the conduction band required to give a recombination rate equal to this injection rate is N_2/τ_e, where τ_e is the electron lifetime (τ_e is not necessarily equal to τ_{21}, the spontaneous lifetime, since some nonradiative recombination mechanisms are likely to be present).

The threshold current density is then given by

$$\frac{(J)_{th}}{et} = \frac{(N_2)_{th}}{\tau_e}$$

and substituting from eq. (5.38) we have

$$(J)_{th} = \frac{et\,d}{\tau_e t}\frac{(8\pi v_0^2 k_{th}\tau_{21}\Delta v n^2)}{c^2}$$

Example 5.5 Threshold current density in a GaAs laser

We may use the following data to estimate the threshold current density of a GaAs junction laser: wavelength, $\lambda = 0.84$ μm; transition linewidth, $\Delta v = 1.45 \times 10^{13}$ Hz; loss coefficient, $\gamma = 3.5 \times 10^3$ m^{-1}; refractive index, $n = 3.6$; dimensions, $l = 300$ μm, $d = 2$ μm; and internal quantum efficiency, $\eta_i \simeq 1$. Now, $n = 3.6$ gives $R = 0.32$.

The threshold gain is given by eq. (5.17) as

$$(k)_{th} = \gamma + \frac{1}{2l}\ln\frac{1}{R_1 R_2}$$

Therefore

$$(k)_{th} = 3500 + \frac{1}{8 \times 10^{-4}}\ln\left(\frac{1}{0.32}\right)^2$$

i.e.

$$(k)_{th} = 7298 \text{ m}^{-1}$$

Hence, from eq. (5.39)

$$(J)_{th} = 15.5 \times 10^6 \text{ A m}^{-2} = 15.5 \text{ A mm}^{-2}$$

This value is in reasonable agreement with those measured at low temperature in GaAs lasers.

Substituting for k_{th} from eq. (5.17) then gives

$$(J)_{th} = \frac{8\pi v_0^2 e d\tau_{21} \Delta v n^2}{\tau_e c^2} \left[\gamma + \frac{1}{2l} \ln \left(\frac{1}{R_1 R_2} \right) \right] \tag{5.39}$$

The ratio τ_e/τ_{21} in this equation is often written as η_i, the internal quantum efficiency, which is the fraction of the injected electrons (or holes) which recombine radiatively.

5.10.2.2 POWER OUTPUT OF SEMICONDUCTOR LASERS

A discussion of power output and saturation in semiconducting lasers is basically the same as that for other lasers given in section 5.8. As the injection current increases above threshold, laser oscillations build up and the resulting stimulated emission reduces the population inversion until it is clamped at the threshold value. We can then express the power emitted by stimulated emission as

$$P = A[J - (J)_{th}] \frac{\eta_i h v}{e}$$

where A is the junction area.

Part of this power is dissipated inside the laser cavity and the rest is coupled out via the end crystal faces. These two components are proportional to γ and $(1/2l) \ln (1/R_1 R_2)$ respectively. Hence we can write the output power as

$$P_0 = \frac{A[J - (J)_{th}] \eta_i h v}{e} \frac{\left[\frac{1}{2l} \ln \left(\frac{1}{R_1 R_2} \right) \right]}{\left[\gamma + \frac{1}{2l} \ln \left(\frac{1}{R_1 R_2} \right) \right]} \tag{5.40}$$

The external differential quantum efficiency η_{ex} is defined as the ratio of the increase in photon output rate resulting from an increase in the injection rate (i.e. carriers per second), that is

$$\eta_{ex} = \frac{d(P_0/hv)}{d\left(\frac{A}{e} [J - (J)_{th}] \right)}$$

From eq. (5.40) we can write η_{ex} as

$$\eta_{ex} = \eta_i \left[\frac{\ln \left(\frac{1}{R_1} \right)}{\gamma l + \ln \left(\frac{1}{R_1} \right)} \right] \tag{5.41}$$

assuming that $R_1 = R_2$. Equation (5.41) enables us to determine the internal quantum efficiency from the experimentally measured dependence of η_{ex} on l; η_i in GaAs is approximately equal to 0.7–1.0.

Now if the forward bias voltage applied to the laser is V_f, then the power input is $V_f A J$ and the efficiency of the laser in converting electrical input to laser output is

$$\eta = \frac{P_0}{V_{\mathrm{f}}AJ} = \eta_{\mathrm{i}} \left[\frac{J - (J)_{\mathrm{th}}}{J} \right] \left(\frac{h\nu}{eV_{\mathrm{f}}} \right) \frac{\ln\left(\frac{1}{R_1}\right)}{\gamma l + \ln\left(\frac{1}{R_1}\right)} \tag{5.42}$$

From Fig. 5.18, $eV_{\mathrm{f}} \simeq h\nu$ and therefore, well above threshold ($J \gg (J)_{\mathrm{th}}$) where optimum coupling ensures that $(1/l)\ln(1/R_1) \gg \gamma$, η approaches η_{i}. As noted above, η_{i} is high (~ 0.7) and thus semiconductor lasers have a very high power efficiency.

5.10.2.3 HETEROJUNCTION LASERS

As we noted above, the threshold current density for homojunction lasers is very large due to poor optical and carrier confinement, which results in the parameters d and γ in eq. (5.39) being large. Dramatic reductions in the threshold current density to values of the order of 10 A mm^{-2} at room temperature coupled with higher efficiency can be achieved using lasers containing heterojunctions (see section 2.8.6 and ref. 5.13). The properties of heterostructure lasers which permit a low threshold current density and CW operation at room temperature can be illustrated with the double heterostructure (DH) laser illustrated in Fig. 5.23(b). In this structure, a layer of GaAs, for example, is sandwiched between two layers of the ternary compound $Ga_{1-x}Al_xAs$ which has a wider energy gap than GaAs and also a lower refractive index. Both $N-n-P$ and $N-p-P$ structures show the same behavior (where N and P represent the wider bandgap semiconductor, according to carrier type).

Figure 5.23(b) also shows that carrier and optical confinement may be achieved simultaneously. The bandgap differences form potential barriers in both the conduction and valence bands which prevent electrons and holes injected into the GaAs layer from diffusing away. The GaAs layer thus becomes the active region, and it can be made very narrow so that d is very small, typically about 0.2 μm. Similarly, the step change in refractive index provides a very much more efficient waveguide structure than was the case in homojunction lasers. The radiation is therefore confined mainly to the active region. In addition, the fraction of the propagating mode which lies outside the active region is in a wider bandgap semiconductor and is therefore not absorbed, so that γ is much smaller than in homojunction lasers.

Further reductions in threshold current can be obtained by restricting the current along the junction plane into a narrow 'stripe' which may only be a few microns wide. Such stripe geometry lasers have been prepared in a variety of different ways; typical examples are shown in Fig. 5.24. In Fig. 5.24(a), the stripe has been defined by proton bombardment of the adjacent regions to form highly resistive material, whereas in Fig. 5.24(b) a mesa structure has been formed by etching; an oxide mask prevents shorting of the junction during metallization to form contacts. With stripe geometry structures, operating currents of less than 50 mA can produce output powers of about 10 mW.

Stripe geometry devices have further advantages including the facts that (a) the radiation is emitted from a small area which simplifies the coupling of the radiation into optical fibers (see Chapter 9) and (b) the output is more stable than in other lasers. A

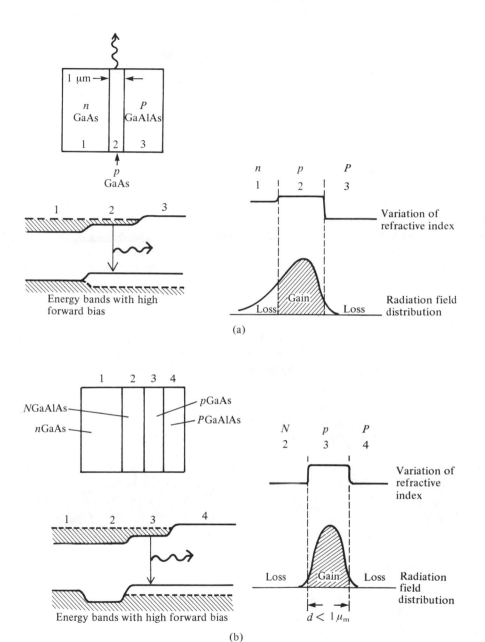

Fig. 5.23 Diagram illustrating the action of single (a) and double (b) heterojunction structures in confining the carriers and radiation to the gain region (as before, in the diagrams of the energy bands, the broken lines represent the Fermi levels after forward bias has been applied).

Fig. 5.24 Schematic cross section (end view) of two typical stripe geometry laser diodes: (a) the stripe is defined by proton bombardment of selected regions to form high resistivity material and (b) the stripe is formed by etching a mesa structure and then GaAlAs is grown into the previously etched outsides of the active region to form a 'buried stripe' structure.

close examination of typical light output–current characteristics reveals the presence of 'kinks' as shown in Fig. 5.25(a). These 'kinks' are associated with a sideways displacement of the radiating filament within the active region (the radiation is usually produced from narrow filaments within the active region rather than uniformly from the whole active region). This lateral instability is caused by interaction between the optical and carrier distributions which arises because the refractive index profile, and hence the waveguiding characteristic, is determined, to a certain extent, by the carrier distribu-

Fig. 5.25 Light output–current characteristics of (a) a laser showing a lateral instability or 'kink' and (b) stripe lasers, in which the 'kinks' have disappeared.

tion within the active region. The use of very narrow stripe regions limits the possible movement of the radiating filament and eliminates the 'kinks' in the light output–current characteristic as shown in Fig. 5.25(b). Although the application of diode lasers to fiber optical communications will be discussed in detail in Chapter 9 it is perhaps appropriate to mention at this stage one or two points of relevance. These include the temperature dependence of the threshold current, output beam spread, the use of materials other than GaAlAs and degradation.

The threshold current density J_{th} increases with temperature in all types of semi-conductor laser but, as many factors contribute to the temperature variation, no single expression is valid for all devices and temperature ranges. Above room temperature, which is usually the region of practical interest, it is found that the ratio of J_{th} at 70 °C to J_{th} at 22 °C for GaAlAs lasers is about 1.3–1.5 with the lowest temperature dependence occurring for an aluminum concentration such that the bandgap energy difference is 0.4 eV. Typical light output–current characteristics for a GaAlAs DH laser are shown in Fig. 5.26.

The angular spread of the output beam depends on the dimensions of the active region and the number of oscillating modes (which, in turn, depends on the dimensions of the active region, the refractive index and the pump power). For wide active regions, we find that the beam divergence both parallel to ($\theta_{\parallel}$) and perpendicular to ($\theta_{\perp}$) the plane of the junction is given approximately by simple diffraction theory. Thus, normal to the junction plane we have $\theta_{\perp} \simeq 1.22\lambda/d$. For DH lasers, where the active region is much narrower, $\theta_{\perp}$ is given approximately by $\theta_{\perp} \simeq 1.1 \times 10^3 x(d/\lambda)$, where x is the mole fraction of aluminum. Thus for a DH laser with $d = 0.1$ μm, $x = 0.3$ and $\lambda = 0.9$ μm, we find $\theta_{\perp} = 37°$ (in good agreement with experimental observations).

Fig. 5.26 Light output–current characteristics of a 20 μm stripe laser as a function of temperature.

Until recently, the system $Ga_{1-x}Al_xAs/GaAs$ was the most widely investigated and used for the production of DH lasers. There are many reasons for this, including the facts that (a) GaAs is a direct bandgap semiconductor which can easily be doped n or p type, (b) the ternary compound $Ga_{1-x}Al_xAs$ can be grown over a wide range of compositions and has a very close lattice match to GaAs ($\simeq 0.1\%$) for all values of x (thus there is low interfacial strain between adjacent layers and consequently very few strain induced defects at which nonradiative recombination may occur) and (c) the relative refractive indices and bandgaps of GaAs and $Ga_{1-x}Al_xAs$ provide for optical and carrier confinement.

In optical fiber communications, however, it is desirable to have a laser emitting at wavelengths in the region 1.1 to 1.6 μm where present optical fibers have minimum attenuation and dispersion. Wavelengths in this range can be obtained from lasers fabricated from quaternary compounds such as $Ga_xIn_{1-x}As_{1-y}P_y$ because of the wide range of bandgaps and lattice constants spanned by this alloy. Figure 5.27 shows the

Fig. 5.27 Lattice constant versus energy gap for various III−V compounds. The solid lines correspond to direct bandgap materials, the dashed lines to indirect bandgap materials. The binary compounds which may be used for lattice-matched growth are indicated on the right. The shaded area shows the quaternary compound $Ga_xIn_{1-x}As_{1-y}P_y$; by adjusting x and y, lattice matching to both GaAs and InP is possible; the latter match, for example, occurs for $x = 0.8$, $y = 0.35$.

lattice constant variation with bandgap (and emission wavelength) for this alloy. By suitable choice of x and y, exact lattice matching to an InP substrate can be achieved and strain-free heterojunction devices can be produced. The GaInAsP layers may be grown on InP substrates by liquid phase, vapor phase or molecular beam epitaxial methods (see ref. 5.14). A typical DH stripe contact laser diode of GaInAsP/InP emitting at $1.1-1.3\ \mu m$ is shown schematically in Fig. 5.28.

The question of laser reliability is also important in relation to applications such as telecommunications. Laser life may be limited by 'catastrophic' or 'gradual' degradation. Catastrophic failure results from mechanical damage to the laser facets due to too great an optical flux density. The damage threshold is reduced by the presence of flaws on the facets; however, it may be increased by the application of half-wave coatings of materials such as Al_2O_3. While facet damage is more likely in lasers operating in the pulse mode, it can also occur in CW operated lasers. This is so especially in the central portion of the active region of stripe lasers where the optical flux density is greatest. Uncoated lasers with stripes about $20\ \mu m$ wide tend to fail catastrophically when the optical flux exceeds about $10^9\ W\ m^{-2}$.

Gradual degradation depends principally on the current density, but also on the duty cycle and fabrication process. It has been observed that as time elapses the threshold current density increases, 'dark' lines develop in the emission and then the CW output falls off drastically.

The development of dark lines is apparently related to the formation, in the vicinity of the active region, of so-called *dark line defects*, which act as nonradiative recombination centers. Dark line defects are attributed to defects such as dislocations, which may have a number of sources. These include (a) edge dislocations formed to relieve stress caused by interfacial lattice mismatch, (b) bonding of the laser to the heat sink and (c) impurities introduced during substrate preparation and heteroepitaxial wafer growth.

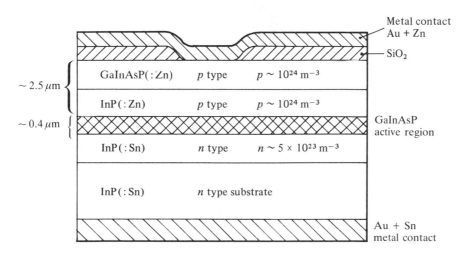

Fig. 5.28 Schematic diagram of a double heterojunction stripe contact laser diode of the quaternary compound $Ga_xIn_{1-x}As_{1-y}P_y$ on an InP substrate with (100) orientation.

Furthermore, the energy released by nonradiative electron–hole recombination may result in the creation and migration of defects.

Defects may be formed in the active region during device fabrication or penetrate into it during subsequent operation. Dark line defects may grow due to a process called dislocation climb (i.e. the movement of dislocations involving atomic transport to or away from the dislocation) and extend throughout the device structure.

Dislocation growth may be stimulated by carrier injection and recombination. GaAs lasers initially containing dislocations are found to degrade at a much higher rate than those that are initially dislocation free. Furthermore, devices with exposed edges that contain edge defects also degrade more rapidly than those in which recombination is restricted to internal regions of the crystal.

Thus, to produce lasers with long lifetimes great care must be taken with substrate selection, and wafer processing and crystal growth must be carried out under ultraclean conditions to fabricate a laser with a strain-free structure. Despite these problems, lasers with lifetimes in excess of 40 000 hours are now available corresponding to continuous operation over a five-year period.

5.10.2.4 QUANTUM WELL LASERS

In normal double heterostructure lasers, the active layer thicknesses are typically in the range 0.1–0.3 μm. Recently, however, considerable interest has been shown in structures where the active layer is much thinner, that is, about 10 nm or less (see ref. 5.15). Interest in such devices stems mainly from the very low threshold currents that can result. To understand why this should be so, we must examine the electron energy levels. We suppose that the active layer has thickness L_z with length and breadth L_x and L_y both being much greater than L_z. Now, according to the theory leading up to eq. (2.13), we may write the total energy of an electron in terms of three quantum numbers n_1, n_2 and n_3 as

$$E(n_1, n_2, n_3) = \frac{h^2}{8m_e}\left(\frac{n_1^2}{L_x^2} + \frac{n_2^2}{L_y^2}\right) + \frac{h^2}{8m_e}\left(\frac{n_3^2}{L_z^2}\right)$$

Because of the relative magnitudes of L_x, L_y and L_z, the lowest energy states will all have $n_3 = 1$ and will constitute effectively a 'two-dimensional' group of states with varying n_1 and n_2 values. Now, the number of states per unit energy interval is given by the density of states function, and for a two-dimensional potential well this is given by (see Problem 2.13) $Z(E) = 4m_e/h^2$, that is the two-dimensional density of states is independent of E. We may contrast this with the corresponding three-dimensional expression (eq. (2.29)) where the density of states varies as $E^{1/2}$. These expressions for the densities of states should apply to electrons in the conduction band (and also to holes in the valence band with the appropriate modifications) provided the electron (hole) energy is referenced to the bottom of the conduction band (top of the valence band) and the electron (hole) mass replaced by an appropriate effective mass. Now, the probability of occupation of these levels is given by the Fermi–Dirac distribution function (eq. (2.30)), which decreases rapidly as E increases. Thus, at the bottom of the conduction band the electrons are spread over a relatively wide energy range with quite small

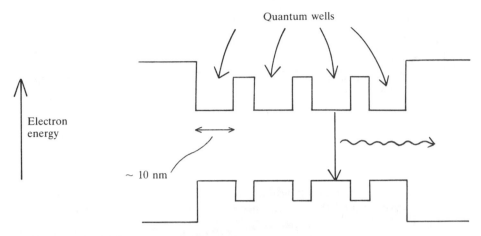

Fig. 5.29 Multiple quantum well structure.

densities at the band edges (see Fig. 2.14). By contrast, in a quantum well structure where we have effectively a two-dimensional density of states, the electrons are spread over a smaller energy range with relatively high densities at the band edge. This situation implies that population inversion is easier to achieve in a quantum well laser than in a conventional diode laser.

One of the problems with the single quantum well structure described above is that, because of the extreme narrowness of the gain region, optical confinement is very poor. This causes higher laser losses and tends to negate the potential advantage of a very low threshold current. One way out of this difficulty is to have a multiple quantum well structure (Fig. 5.29) which allows a much higher optical confinement factor. Quantum well lasers have been constructed with threshold currents as low as a few milliamps.

5.10.3 Gas lasers

Gas lasers are the most widely used type of laser; they range from the low power helium–neon (He–Ne) laser commonly found in teaching laboratories to very high power carbon dioxide lasers, which have many industrial applications. Basically, there are three different classes of gas laser, according to whether the transitions are between the electronic energy levels of atoms or ions, or between the vibrational/rotational levels of molecules. In general, the energy levels involved in the lasing process are well defined and the absence of broad bands effectively eliminates the possibility of optical pumping. Though other methods can be used, most gas lasers are excited by electron collisions in a gas discharge. We shall now consider a typical example from each of those classes mentioned above.

5.10.3.1 ATOMIC LASERS – THE HE–NE LASER
In the He–Ne laser the active medium is a mixture of about ten parts of helium to one part of neon. The neon provides the energy levels for the laser transitions (about 150 different laser transitions have been observed, although only the four shown in Fig. 5.31

Fig. 5.30 Construction of a typical low power laser such as the He−Ne laser. The load resistor serves to limit the current once the discharge has been initiated.

Fig. 5.31 Energy levels relevant to the operation of the He−Ne laser. M indicates a metastable state.

are reasonably strong), while the helium atoms, though not directly involved in the laser transitions, have an important role in providing an efficient excitation mechanism for the neon atoms. Excitation usually takes place in a d.c. discharge created by applying a high voltage ($\simeq$ 2 to 4 kV) across the gas contained in a narrow diameter glass tube at a pressure of about 10 torr, as illustrated in Fig. 5.30. As the discharge tube exhibits a negative dynamic resistance when a discharge has been initiated, it is necessary to include a load resistor to limit the current and protect the power supply.

The pumping process can be described as follows. The first step is the electron impact excitation of helium atoms to one of two metastable states designated 2^1S and 2^3S; this is represented by

$$e_1 + He = He^* + e_2$$

where e_1 and e_2 are the electron energies before and after the collision. While in one of their excited states (He^*), helium atoms can transfer their energy to neon atoms, with which they may collide. The probability of this *resonant transfer* of energy is proportional to exp $(-\Delta E/kT)$ where ΔE is the energy difference between the excited states of the two atoms involved. The energy level diagram for helium and neon is shown in Fig. 5.31. Here, the neon states are labeled by the so called Paschen notation (the numerical subscripts have no spectroscopic significance and are used only as labels) while the helium states are labeled according to the L-S representation mentioned on page (183). Fig. 5.31 shows that there is a group of four neon levels at almost the same energies as each of the two excited helium states, and resonant transfer thus occurs quite readily. The energy transfer is represented by

$$He^* + Ne = Ne^* + He$$

A population inversion is thus created between the 3s and (3p, 2p) groups of levels and also between the 2s and 2p levels. Transitions between the 3s and 2s levels and between the 3p and 2p levels are forbidden by quantum mechanical selection rules.

The He–Ne laser is another example of a four level system, and as such we require that the population of the lasing transition terminal level be kept as low as possible. This implies that electrons in the terminal level should decay as rapidly as possible back to the ground state. In neon this is a two-step process; the first, 2p to 1s, is a rapid transition, but the second, 1s to the ground state, is not so rapid. The latter transition rate is, however, enhanced by collisions with the walls of the discharge tube, and indeed the gain of the laser is found to be inversely proportional to the tube radius. For this reason, the discharge tube diameter should be kept as narrow as possible.

The transition 2p to 1s is also of interest since it gives rise to the familiar color of 'neon lights'. Thus the 2p level itself must be populated by the discharge. This is unfortunate since an increase in the population of the 2p level implies a decrease in population inversion (at least as far as the 1.15 μm, 632.8 nm and 543.5 nm wavelengths are concerned), and in fact this effect is to a large extent responsible for lasing ceasing at high tube currents. We cannot therefore increase the output simply by increasing the current indefinitely. Thus the He–Ne laser is destined to remain a relatively low-power device.

Example 5.6 Efficiency of an He–Ne laser

We may estimate the efficiency of a low power He–Ne laser from the following.

A typical laser operates with a current of 10 mA at a d.c. voltage of 2500 V and gives an optical output of 5 mW. Its overall power efficiency is then

$$\frac{5 \times 10^{-3}}{2500 \times 1 \times 10^{-2}} = 0.02\%$$

Each of the four main laser transitions (at 3.39 μm, 1.15 μm, 632.8 nm and 543.5 nm) shares with one of the others either a common starting or terminating level. Thus the transitions are always competing with each other and precautions must be taken to prevent the two unwanted wavelengths from lasing. This can be achieved quite easily by using the multilayer coated mirrors discussed in section 5.5, which have a wavelength-dependent reflectance. The very low absorption loss of such mirrors is an essential feature as the gain in the He–Ne medium is rather small; indeed, the use of such mirrors is quite general in gas lasers.

The basic structure of the He–Ne laser is relatively simple. The essential elements are shown in Fig. 5.30. The discharge is usually initiated by a high voltage 'trigger' pulse and then maintained at a current of 10 to 20 mA. The mirrors forming the resonant cavity are sometimes cemented to the ends of the discharge tube, thereby forming a gas-tight seal. Alternatively, the mirrors can be external to the tube which is then sealed with glass windows orientated at the Brewster angle to the axis of the tube. This arrangement allows 100% transmission for the radiation with its electric vector vibrating parallel to the plane of incidence, thereby ensuring the maximum possible gain (minimum losses) in each round trip. The Brewster windows also result in the output being plane polarized. Although this arrangement is slightly more complicated than the former one, it enables us to insert frequency stabilizing, mode selecting and other devices into the cavity. The mirrors can also be changed to allow operation with other output characteristics and at other wavelengths.

The power output from He–Ne lasers is rather small (up to about 100 mW maximum, though more typically a few milliwatts); however, the radiation is extremely useful in a wide range of applications because it is highly collimated, coherent and has an extremely narrow linewidth.

5.10.3.2 ION LASERS–THE ARGON LASER

The most powerful CW lasers operating in the visible region are the inert gas ion lasers such as the argon and krypton ion lasers. CW outputs of several watts can readily be obtained while, if the laser is pulsed, powers up to a kilowatt in microsecond pulses can be generated.

The gas atoms are ionized by electron collision in a high current discharge ($\simeq$ 15 to 50 A). The ions are excited by further electron collisions up to a group of energy levels

(4p) some 35 eV above the atomic ground state. As the electron energies are only a few electron volts, the excitation must be the result of multiple collisions. A population inversion forms between the 4p levels and the 4s level, which is about 33.5 eV (Fig. 5.32) above the ground state so that a series of stimulated lines is emitted. These range in wavelength from 351 nm to 520 nm, although most of the energy is concentrated in the 488 nm and 514.5 nm lines.

The tube design of the argon laser is much more complicated than that of the He–Ne laser, principally because of the much higher energy required to pump the ionic levels and the need to dissipate the heat energy released. The current density can be increased by concentrating the discharge with a magnetic field applied along the axis of the tube (the ions spiral about the magnetic lines of force). This has the added advantage of reducing the number of ions which collide with and damage the walls of the tube. The tube is made of a refractory material such as graphite or beryllium oxide. To dissipate the large amount of heat generated, most ion laser tubes are water cooled and often include a series of metal disks to act as heat exchangers, as shown in Fig. 5.33. Holes in the center of these disks define the active region of the laser. Because of the high current involved, the cathode must be an excellent electron emitter and a getter is often incorporated to 'clean up' any impurities which might otherwise poison the cathode.

Again, the discharge is initiated by a high voltage pulse and then maintained by a d.c.

Fig. 5.32 Simplified energy level diagram for the argon ion laser. Ten or more laser lines are produced, but the two shown are by far the most intense.

Fig. 5.33 Construction of a typical argon ion laser.

voltage of about 200 V. During operation, the positive ions tend to collect at the cathode and may eventually cause the discharge to be extinguished. To prevent this, a gas return path is provided between the cathode and anode to equalize the pressure. Pulsed ion lasers tend to be simpler and, with a low duty cycle, the heat generated is small enough to be dissipated by convective cooling.

To select any desired wavelength, a small prism is inserted into the cavity and the position of the end mirror is changed by rotating it to be normal to the path of the radiation with the desired wavelength. This ensures that radiation of this particular wavelength will be reflected to and fro, while that of other wavelengths will be lost from the cavity after only a few round trips.

Krypton ion lasers are becoming increasingly used as excitation sources for dye lasers (see below) and in physical and chemical investigations. They produce a wealth of spectral lines ranging across the entire visible spectrum from about 340 nm to 800 nm, with the most intense line at 647 nm.

5.10.3.3 MOLECULAR LASERS – THE CARBON DIOXIDE LASER

The carbon dioxide laser is the most important molecular laser and indeed it is arguably the most important of all lasers from the standpoint of technological applications. In molecular lasers, the energy levels are provided by the quantization of the energy of vibration and rotation of the constituent gas molecules. The CO_2 molecule is basically an in-line arrangement of the two oxygen atoms and the carbon atom, which can undergo three fundamental modes of vibration as shown in Fig. 5.34. At any one time, the molecule can be vibrating in any linear combination of these fundamental modes. The modes of vibration are denoted by a set of three quantum numbers (v_1, v_2, v_3) which represent the amount of energy or number of energy quanta associated with each mode. The set (100), for example, means that a molecule in this state is vibrating in a pure symmetric mode with one quantum of vibrational energy; it has no energy associated with the asymmetric or bending modes.

In addition to these vibrational modes, the molecule can also rotate, and thus it has closely spaced rotational energy levels associated with each vibrational energy level.

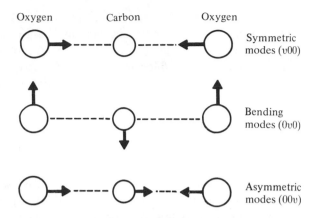

Fig. 5.34 Vibrational modes of the CO_2 molecule.

The rotational levels are designated by an integer $\mathscr{I}$. The energy separation between these molecular levels is small and the laser output is therefore in the infrared. The important parts of the CO_2 energy level arrangement are shown in Fig. 5.35, which also shows the ground state and first excited state of the vibrational modes of nitrogen.

Many CO_2 lasers contain a mixture of CO_2, nitrogen and helium in the ratio $1:4:5$. Nitrogen plays a similar role to that of helium in the He–Ne laser. Excited nitrogen molecules transfer energy to the CO_2 molecules in resonant collisions, exciting them to the (001) levels. The (100) CO_2 levels have a lower energy and cannot be populated in

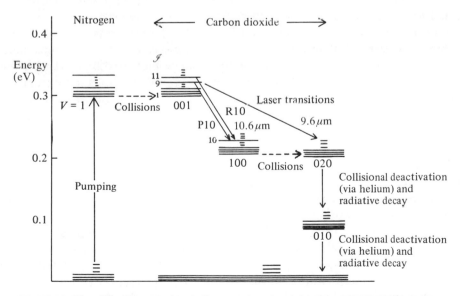

Fig. 5.35 Simplified energy level diagram for the CO_2 laser. Each vibrational level has many rotational levels associated with it, $\mathscr{I} = 1, 2, \ldots$ The 10.6 µm line is the strongest.

this way, so that population inversion is created between the (001) and (100) levels giving stimulated emission at about 10.6 μm. The helium has a dual role. Firstly, it increases the thermal conductivity to the walls of the tube, thereby decreasing the temperature and Doppler broadening, which in turn increases the gain. Secondly, it increases the laser efficiency by indirectly depleting the population of the (100) level, which is linked by resonant collisions to the (020) and (010) levels, the latter being depleted via collisions with the helium atoms.

While other gas lasers have efficiencies of 0.1% or less, the CO_2 laser may have an efficiency up to about 30%. This is essentially due to the ease with which electrons in the discharge can cause excitation and because of the strong coupling of the various levels involved. Because of this high efficiency, it is relatively easy to obtain CW outputs of 100 W for a laser 1 m long. Powers of this magnitude and greater (see below) mean that the mirrors should have very low absorption, while operation in the infrared also means that special materials must be used for windows, mirrors and other laser components. Materials which have been used successfully include germanium, gallium arsenide, zinc sulfide, zinc selenide and various alkali halides, although these suffer from being relatively soft and hygroscopic. In some cases, a diffraction grating mounted on a piezoelectric transducer is used instead of the high reflectance mirror. The grating permits tuning of the laser output over the large range of distinct lines which Fig. 5.35 shows are possible. In the transition from the (001) to the (100) group of levels, the selection rule $\Delta \mathcal{J} = \pm 1$ operates. Thus for example, for $\mathcal{J} = 10$, in the upper level, $\mathcal{J}$ can be 11 or 9 in the lower level; the corresponding transitions give rise to the P10 ($\Delta \mathcal{J} = +1$) and R10 ($\Delta \mathcal{J} = -1$) branches respectively. The strongest lines are given by the P18, P20 and P22 transitions.

The power output of the CO_2 laser is approximately proportional to the tube length and, in attempts to obtain greater powers, lasers tens of meters long have been built, which give power outputs of several tens of kilowatts. Two other techniques have been used to produce very high powers. The first is the transverse excitation atmospheric (or TEA) laser. In the CO_2 laser the power output can be increased by increasing the gas pressure. The problem is that it then becomes increasingly difficult to create a discharge. At atmospheric pressure, for example, the breakdown voltage is 1.2 kV mm^{-1} and thus even for a laser 1 m long we would need an unacceptably high voltage. To overcome this, the discharge is struck transversely across the tube so that the discharge path length is about a centimeter. The high voltage is applied to a set of electrodes placed along the tube as shown in Fig. 5.36. With this arrangement, peak powers in the gigawatt range are obtained in very short pulses with about 20 pulses occurring per second. Cool gas flowing through the lasing region further increases population inversion and hence power output.

The second approach to very high power devices is to create population inversion through the application of thermodynamic principles rather than via a discharge – this gives the *gas dynamic laser*. A nitrogen–carbon dioxide mixture is heated and compressed and then allowed to expand into a low pressure region. During heating and compression the population of the energy states reaches the Boltzmann distribution appropriate to the higher temperature. At high temperatures most of the energy is

Pin cathodes and ballast resistors

M_1

M_2

Output

High voltage

Anode bar

Fig. 5.36 Schematic diagram of the TEA CO_2 laser. The discharge is perpendicular to the axis of the laser cavity.

stored in the vibrational modes of the nitrogen molecule. At lower temperatures, after expansion into the low pressure region, resonant collisions of the nitrogen molecules with the carbon dioxide molecules populate the (001) state of CO_2 and create a population inversion. With very active pumping of this type, CW output powers in excess of 100 kW have been achieved. Gas dynamic lasers suffer from the disadvantage of large size and the rocket-like roar as the gas expands.

Other molecular lasers. As the number of molecular systems with energy levels which may be suitable for laser action is very large, it is not surprising that there are other molecular lasers and it is likely that many more will be discovered. Two such molecular lasers emitting in the ultraviolet–visible region have been developed, namely nitrogen and excimer lasers.

The nitrogen laser differs markedly from the CO_2 laser. In the latter the transitions are between molecular rotational/vibrational energy levels, while in nitrogen the laser transitions are between electronic energy states. A requirement for CW operation is that the upper level of the lasing transition should have a long lifetime while the lower level should be rapidly depopulated. In nitrogen, however, the converse is true; the upper level lifetime is exceedingly short (of the order of nanoseconds) while the lower level lifetime is of the order of microseconds. Hence population inversion and lasing action can only be maintained for a few nanoseconds. Very fast rise time pumping mechanisms similar to those used for TEA lasers must be used.

The gain in nitrogen is so large that it can be used as a simple amplifier; i.e. in many applications it is not necessary to provide feedback (such high gain is termed *superradiant*). Commercial nitrogen lasers are capable of producing 100 kW peak power pulses. They are often used in photochemical investigations and for pumping other lasers, for example dye lasers (section 5.10.4).

In contrast to nitrogen, excimers provide a metastable excited state. An excimer (or *exci*ted di*mer*) refers to a molecule formed by the association of one excited atom (or molecule) with another atom (or molecule) which is in the ground state. If both constituents are in the ground state, then at interatomic distances characteristic of molecules they repel each other. Consequently, the excimer readily dissociates thereby effectively reducing the population of the lower lasing level and increasing the ease with which population inversion is established. If one or both of the constituents of the excimer are rare gas atoms, the excitation energy is extremely large and the metastable excimer state is an important system for storing energy.

Since 1972, a large number of lasers based on excimers have been developed; these cover the wavelength range from about 120 nm to 500 nm. Rare gas halide excimer lasers are especially efficient with XeF and KrF giving the highest efficiencies (some 10–15%).

Excimer lasers are usually pumped by an intense electron beam source or by a fast discharge. The electrons in the electron beam are accelerated until they have energies of 1 MeV and then transmitted to the laser chamber in pulses giving beam currents of about 100 kA. Electron beam generators with this sort of capability are rather large and alternative pumping mechanisms are being investigated.

5.10.4 Liquid dye lasers

Liquids have useful advantages in relation to both solid and gas laser media. Solids are very difficult to prepare with the requisite degree of optical homogeneity and they may suffer permanent damage if overheated. Gases do not suffer from these difficulties but have a much smaller density of active atoms. Several different liquid lasers have been developed, but the most important is the dye laser. It has the advantage that it can be tuned over a *significant* wavelength range. This is extremely useful in many applications such as spectroscopy and the study of chemical reactions.

The active medium is an organic dye dissolved in a solvent. When the dye is excited by short wavelength light it emits radiation at a longer wavelength, i.e. it *fluoresces*. The energy difference between the absorbed and emitted photons ultimately appears as heat – typical absorption and emission spectra are shown in Fig. 5.37. The broad fluore-

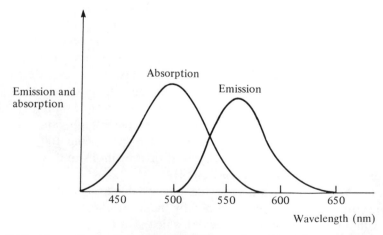

Fig. 5.37 Absorption and emission (fluorescence) spectra of a typical dye laser.

scence spectrum can be explained by the energy level diagram of a typical dye molecule. As Fig. 5.38 shows, the molecule has two groups of closely spaced electronic energy levels; the singlet states (S_0, S_1 and S_2) and the triplet states (T_1 and T_2). The singlet states occur when the total spin of the excited electrons in each molecule is zero (the value of $2S + 1$ is thus unity). The triplet states occur when the total spin is unity ($2S + 1 = 3$). Each electronic energy level is broadened into a near continuum of levels by the effects of the vibration and rotation of the dye molecule and also by the effects of the solvent molecules. Pumping results in the excitation of the molecule from the ground state S_0 to the first excited state S_1. This is followed by very rapid nonradiative decay processes to the lower of the energy levels in S_1. The laser transition is then from these levels to a level in S_0. Since there are many such rotational/vibrational levels within S_0 and S_1, there are many transitions resulting in an emission line which is very broad. As the termination of the laser transition in S_0 is at an energy much larger than kT above the bottom of S_0, the dye laser is a four level system and threshold is reached with a very small population inversion.

Although the triplet states are not directly involved in the laser action, they have a profound effect as there is a small probability of a transition $S_1 \rightarrow T_1$, even though this is forbidden by quantum mechanical selection rules. Since the transition $T_1 \rightarrow S_0$ is also forbidden, molecules 'pile up' in T_1. The transition $T_1 \rightarrow T_2$ is allowed, however, and unfortunately the range of frequencies required for this transition is almost exactly the same as the laser transition frequencies. Thus, once a significant number of molecules have made the $S_1 \rightarrow T_1$ transition, $T_1 \rightarrow T_2$ absorption reduces the gain and may stop the lasing action. For this reason, most dye lasers operate in short pulses – shorter, in fact, than the time taken for T_1 to acquire a significant population, which is typically

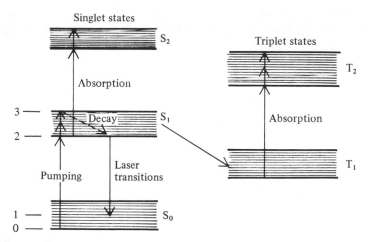

Fig. 5.38 Energy level scheme for a dye molecule. The laser transitions terminate above the lowest energy in the S_0 singlet state, so the laser is a four level system. $S_1 \rightarrow T_1$ transitions lead to strong absorption at the laser wavelengths, thereby quenching laser action. $S_1 \rightarrow S_2$ transitions may also quench laser action in some dyes.

1 µs. For long pulse or CW operation, the population in T_1 will build up to equilibrium values, in which case the absorption is high and becomes the ultimate limitation on the efficiency of the laser.

Many dyes have been used as laser media, and Fig. 5.39 shows that, by tuning, laser wavelengths covering the whole of the visible spectrum can be obtained. The dye called rhodamine 6G with methanol as a solvent is one of the most successful, having an efficiency of about 20% and a broad tuning range (570 to 660 nm).

All dye lasers are optically pumped, the pumping source having a wavelength slightly less than that of the laser output. Commercial pumping methods include flashtubes, nitrogen lasers, solid state lasers and ion (A^+ or Kr^+) lasers. As the pumping radiation is in the visible or ultraviolet part of the spectrum, we must use harmonics of the output from neodymium lasers. The choice of the pump source depends on the absorption spectrum of the dye being used and the type of output desired. For CW output, the usual commercial pump source is an ion laser; all the other pumping sources operate in the pulsed mode and produce a pulsed output from the dye laser. As the population of T_1 builds up in about 1 µs, the flashtubes must discharge in about 1 µs (in contrast to the millisecond time of tubes used with solid lasers). For this reason, specially strengthened flashtubes must be used. To avoid absorption losses due to a buildup of molecules in triplet states, the dye is passed in a liquid jet through the pumping radiation. The dye emerges from a specially shaped nozzle in the form of a smooth laminar sheet; the small-signal gain is so high that this very small thickness of the active medium is sufficient to give laser action.

The laser output can be tuned using, for example, a prism, wedge filter or a diffraction grating, which can serve as a combined end mirror and dispersing element. A typical dye laser arrangement is shown in Fig. 5.40.

We have by no means covered the entire range of lasers or fully discussed the various modifications and refinements of the lasers which have been described. It is hoped, however, that the basic laser physics covered, together with the descriptions of some

Fig. 5.39 Relative outputs of some common laser dyes pumped by ion lasers. Rhodamine 6G (R6G), for example, is here pumped by 5 W of power from all the argon lines. Coumarin dyes are labeled C; C490 is pumped by 2.3 W at 488 nm.

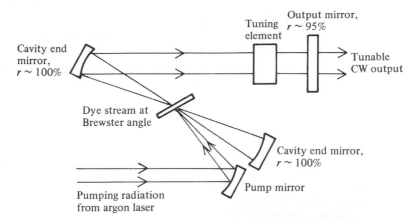

Fig. 5.40 Schematic diagram of a tunable CW laminar flow dye laser. The dye stream flow is perpendicular to the page.

laser types, will enable the reader to understand the mode of operation of other lasers which might be encountered or which might be developed in the future.

Problems

5.1 At what temperature are the rates of spontaneous and stimulated emission equal (take $\lambda = 500$ nm)? At what wavelength are they equal at room temperature ($T = 300$ K)?

5.2 If 1% of the light incident into a medium is absorbed per millimeter, what fraction is transmitted if the medium is 0.1 m long? Calculate the absorption coefficient α.

5.3 If the irradiance of light doubles after passing once through a laser amplifier 0.5 m long, calculate the small-signal gain coefficient k assuming no losses. If the increase in irradiance were only 5%, what would k be?

5.4 An atom has two energy levels with a transition wavelength of 694.3 nm. Assuming that all the atoms in an assembly are in one or other of these levels, calculate the percentage of the atoms in the upper level at room temperature ($T = 300$ K) and at $T = 500$ K.

5.5 Calculate the degree of population inversion required to give a small-signal gain coefficient for a CO_2 laser ($\lambda = 10.6$ μm) of 0.5 m^{-1}. Take the Einstein A coefficient for the upper laser level to be 200 s^{-1}.

5.6 Calculate the Doppler broadened linewidth for the CO_2 laser transition ($\lambda = 10.6$ μm) and the He−Ne laser transition ($\lambda = 632.8$ nm) assuming a gas discharge temperature of about 400 K. Take the relative atomic masses of carbon, oxygen and neon to be 12, 16 and 20.2 respectively.

5.7 Calculate the mirror reflectances required to sustain laser oscillations in a laser which is 0.1 m long, given that the small-signal gain coefficient is 1 m^{-1} (assume the mirrors have the same values of reflectance).

5.8   Calculate the threshold small-signal gain coefficient for ruby given the following data: threshold population inversion $= 5 \times 10^{22}$ m^{-3}, refractive index $= 1.5$, linewidth $= 2 \times 10^{11}$ Hz, Einstein A coefficient $= 300$ s^{-1} and wavelength $= 694.3$ nm.

5.9   What is the mode number nearest the line center of the 632.8 nm transition of the He–Ne laser; what is the mode separation; how many longitudinal modes could possibly oscillate given that the width of the gain curve is 1.5×10^9 Hz? Take the mirror separation to be 0.5 m.

5.10   By considering the rate equations for a three level laser, derive an expression for the threshold population inversion in terms of the laser parameters. Hence compare the pump power required to bring a three level and a four level laser to threshold.

5.11   Calculate the threshold pumping power for an Nd:glass laser, given that the critical population inversion is 9×10^{21} m^{-3}, the spontaneous lifetime is 300 μs and that the upper laser level is at an energy of 1.4 eV.

5.12   Assuming that the exciting lamp is 10% efficient, that 10% of the light produced actually falls on the crystal, which has a diameter of 2 mm and is 0.1 m long and that 5% of the exciting light energy falls within useful absorption bands, estimate the power to be supplied to the lamp used to pump the laser in Problem 5.11.

5.13   Estimate the efficiency of a GaAs laser operating well above threshold, given that $n = 3.6$ and that the length of the laser cavity is 200 μm. Take the loss coefficient to be 800 m^{-1} and the internal quantum efficiency to be 0.8.

References

5.1   T. H. Maiman, 'Stimulated optical radiation in ruby masers', *Nature,* **187**, 1960, 493.

5.2   A. Einstein, 'Zur Quantentheorie de Strahlung', *Phy. Z.,* **18**, 1917, 121.

5.3   N. Bloembergen, 'Proposal for a new type of solid state maser', *Phys. Rev.,* **105**, 1957, 762.

5.4   (a) A. E. Siegman, *Introduction to Masers and Lasers*, McGraw-Hill, New York, 1971, Chapter 8.
(b) T. Li and H. Kogelnik, 'Resonator stability curves', *Appl. Opt.,* **5**, 1966, 1550.

5.5   (a) A. Yariv, *Introduction to Optical Electronics*, Holt Rinehart & Winston, New York, 1971, Chapter 5.
(b) R. W. Ditchburn, *Light* (2nd edn), Blackie, Glasgow, 1962, pp. 91–7.

5.6   *Ibid*, Chapter 4, pp. 85–91 and 106–17.

5.7   (a) A. E. Siegman, *Introduction to Masers and Lasers*, McGraw-Hill, New York, 1971.
(b) 'Resonant modes in a maser interferometer', *Bell System Tech. J.,* **40**, 1961, 453.

5.8   G. R. Fowles, *Introduction to Modern Optics* (2nd edn), Holt Rinehart & Winston, New York, 1975, Chapter 4.

5.9   J. T. Verdeyen, *Laser Electronics*, Prentice-Hall, Englewood Cliffs, NJ, 1981, section 5.1.

5.10   (a) H. Semat and J. R. Albright, *Introduction to Atomic and Nuclear Physics* (5th edn), Holt Rinehart & Winston, New York, 1972, pp. 259–64.
(b) D. Eastham, *Atomic Physics of Lasers*, Taylor & Francis, London & Philadelphia, 1986.

5.11   J. S. Griffith, *The Theory of Transition-metal Ions*, Cambridge University Press, Cambridge, 1964, Chapter 9.

5.12   R. J. Mears and L. Reekie, 'Neodymium–doped Silica Single-Mode Fibre Laser', *Electron Letters*, **21**, 1985, 738.

5.13   (a) H. C. Casey and M. B. Panish, *Heterojunction Lasers*, Academic Press, New York, 1978.

      (b) H. Kressel and J. K. Butler, *Semiconductor Lasers and Heterojunction LEDs*, Academic Press, New York, 1977.

5.14   (a) H. C. Casey and M. B. Panish, *op. cit.*, Part B, Chapters 5–7.

      (b) H. Kressel and J. K. Butler, *op. cit.*, Chapter 9.

5.15   E. E. Basch (ed.), *Optical-fiber Transmission*, Howard W. Sams, Indianapolis, 1987, Chapter 9, section 4.

Chapter 6

Lasers II

In the previous chapter, we saw that the output of lasers does not always consist of a beam of very coherent, almost monochromatic radiation. The output, for example, may be continuous or in the form of mutually incoherent spikes within a pulse and consist of several longitudinal and transverse modes of slightly different wavelength. In considering the applications of lasers we often find that such characteristics are quite acceptable, but equally it is often desirable to modify the laser output to suit a particular application. Some modifications to the output can be achieved quite simply; for instance, we can select one of the many wavelengths produced by the argon ion laser by introducing a prism or grating into the optical cavity. The prism or grating disperses the light so that after transmission only one wavelength falls normally onto the end mirror and is reflected back into the cavity. Other modifications of the output, though often quite readily achieved in practice, require a clear understanding of the concepts of modes, population inversion, threshold gain and the like. Before discussing the applications of lasers, we now consider some of the ways in which the laser output may be modified.

6.1 Single mode operation

In many applications including chemical and physical investigations it is desirable to have the greatest possible spectral purity. We can achieve this by operating a CW laser in a single longitudinal, single transverse mode. Since an inhomogeneously broadened laser (see section 5.9 and below) can support several longitudinal and transverse modes simultaneously, single mode operation can be achieved only by arranging for one mode to have a higher gain than all the others. We can ensure that the cavity will support a single transverse mode only, the TEM_{00} mode, by placing an aperture within the cavity. As the higher order TEM modes spread out further than the TEM_{00} mode, an aperture of suitable diameter will transmit the TEM_{00} mode while eliminating the others. All but one of the longitudinal modes can then be rejected by reducing the length L of the laser cavity until the frequency separation between adjacent modes, that is $\delta v = c/2L$ (see section 5.9), is greater than the linewidth of the laser transition. Fig. 5.11 then shows that the single mode which falls within the transition linewidth is the only one that can oscillate.

The disadvantage of this system is that the active length of the laser cavity may become so small as to severely limit the power output. This may be overcome using other techniques involving, for example, a Fabry–Perot resonator either inside or outside the laser cavity, third mirror techniques or absorbers within the cavity (ref. 6.1). It should be stressed that to maintain the wavelength of the single mode output at a constant value, we must stabilize the cavity dimensions by rigid construction and temperature control or by introducing compensating systems. If this not done, L will change and the frequency and power of the laser will change as a consequence. In passing, we note that it is possible to stabilize the operating frequency of a laser to better than 1 MHz, or about 1 part in 10^9. Ways of doing this are described in the next section.

6.2 Frequency stabilization

In lasers with homogeneously broadened transitions, an increase in pumping cannot increase the population inversion beyond the threshold value where the gain per pass equals the losses. This is because the spectral lineshape function $g(v)$ describes the response of each individual atom, which are all considered to behave identically. Thus, as the pumping is increased from below the threshold value the laser will begin to oscillate at the center frequency v_0. The gain at other frequencies will remain below threshold, however, so that an ideal homogeneously broadened laser will oscillate only at a single frequency.

In inhomogeneously broadened lasers, on the other hand, where individual atoms are considered to behave differently from one another, the population inversion and gain profile can increase above the threshold values at frequencies other than v_0. The gain at v_0, however, remains clamped at the threshold value due to gain saturation (section 5.6). Further pumping may increase the gain at other frequencies until oscillations commence at those frequencies also. This results in decreases in both the population inversion and the gain to their threshold values. The gain curve therefore acquires depressions or 'holes' in it at these oscillating frequencies – this is referred to as *hole burning*. The gain curves for homogeneous and inhomogeneous atomic systems are illustrated in Fig. 6.1, where the curves labeled A, B and C correspond to pumping levels respectively below threshold, at threshold and above threshold.

Let us now consider gas lasers, which are usually inhomogeneously broadened due primarily to the Doppler effect as explained in section 5.7. We suppose that a single mode is oscillating at frequency v_m which is greater than the natural emission frequency of the atoms, v_0. The oscillation, being a standing wave within the cavity, consists of two sets of waves traveling in opposite directions, say the positive and negative x directions respectively. Both of these waves have the frequency v_m.

The interaction of the waves traveling in the positive x direction with the atoms in the laser medium will be greatest for those atoms that have a velocity component in the x direction of $+v_x$ such that

Fig. 6.1 Gain curves for (a) a homogeneously broadened atomic system and (b) an inhomogeneously broadened atomic system (A – below threshold; B – at threshold; C – above threshold). In (b), different groups of atoms respond to the stimulating radiation at different cavity mode frequencies. The gain is saturated at each of these frequencies independently, creating 'holes' in the gain curve.

$$\nu_m = \nu_0 \left(1 + \frac{\upsilon_x}{c}\right) \tag{6.1}$$

For this group of atoms, the apparent frequency of the waves is ν_m and the atoms are stimulated to emit. The argument also holds for a second group of atoms and waves moving in the negative x direction. There are therefore two groups of atoms whose stimulated emission contributes to the laser output intensity; the population inversion is reduced for these atoms and gain saturation occurs.

We have plotted the population inversion N as a function of the x component of velocity in Fig. 6.2(a) where we see that stimulated emission produces a saturation in the excited state atomic velocity distribution similar to the hole burning in the gain curve. Two 'holes are burned'; these are symmetrically placed about $\upsilon_x = 0$ and correspond to atoms with velocities of plus and minus υ_x.

Suppose that now the frequency of the oscillating mode is changed until it equals the peak frequency of the laser line, that is $\nu_m = \nu_0$. This may be accomplished, for example, by varying the temperature to change the cavity length slightly. Under these cirumstances, only a single group of atoms can contribute to the lasing process, namely those with zero x component of velocity, and there is a single 'hole' in the population inversion–velocity curve as shown in Fig. 6.2(b). When this happens, the laser output power drops as the available inverted population is smaller than before. A plot of output power as a function of frequency ν_m as in Fig. 6.3 then shows a dip, the *Lamb dip*, at the center frequency $\nu_m = \nu_0$.

The increase in power resulting from any slight deviation from the center of the laser line can be used, as the basis of a feedback system, to stabilize the frequency of the laser at the line center by minimizing the output. Such techniques enable the frequency to be stabilized to better than 1 part in 10^9. The long coherence length makes the output of lasers stabilized in this way useful in applications such as long path difference interference measurements (section 6.6.1).

Another technique for stabilising gas lasers relies on the fact that the gain profile

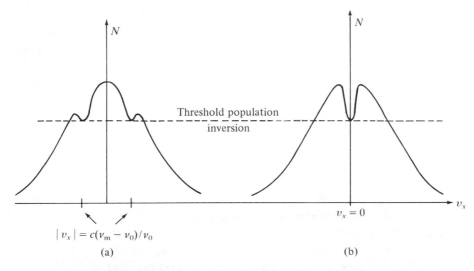

$$|v_x| = c(\nu_m - \nu_0)/\nu_0$$

(a)   (b)

Fig. 6.2 Distribution of the population inversion N as a function of the x component of velocity of the atoms v_x. In (a) the cavity frequency ν_m is different from the natural emission frequency of the atoms ν_0 and two 'holes' are created in the distribution. In (b) the cavity frequency equals the atomic resonant frequency, i.e. $\nu_m = \nu_0$ and only one hole is created, corresponding to $v_x = 0$.

is symmetrical about its midpoint. Consider the situation when only two modes are operating with equal irradiances. They must be equally spaced in frequency either side of the gain profile maximum. Any drift in the mode frequency will cause one mode to increase in irradiance and the other to decrease. Thus, if we are able to monitor the two-mode irradiance and then use the difference to operate a feedback loop that

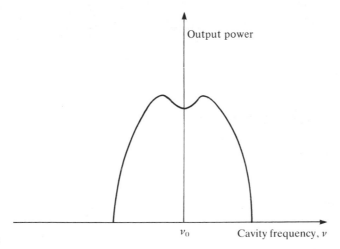

Fig. 6.3 Output power as a function of frequency of a single mode laser. The minimum at the center of the curve ($\nu = \nu_0$) is known as the Lamb dip.

controls the cavity length, we should be able to stabilize the operating frequency. At first sight, the measurement of the mode irradiances might seem difficult. Fortunately it is usually found that adjacent cavity modes are plane polarized with their planes of polarization at right angles to each other. Thus we need only split the laser output into two, insert suitably orientated pieces of polaroid into the two beams, and then allow them to fall onto two detectors. The output of the detectors is then proportional to the irradiances of the modes. A simple way of applying feedback is to let any difference in out-put of the detectors modulate the current passing through a heating coil wrapped round the laser tube. Any changes in relative mode irradiance will then alter the tube temperature and hence the effective length of the cavity. This in turn will vary the mode frequencies.

As might be expected, the techniques used to frequency stabilize semiconductor lasers are somewhat different from those used in gas lasers. Obviously, the first step is to ensure that only a single mode can oscillate. Obtaining a single transverse mode is relatively easy; the gain region may be narrowed using the techniques discussed in section 5.10.2.3. The gain curve is relatively wide, however, and several longitudinal modes are usually present. Even if we insure that only one of these can oscillate, there are problems with temperature stability. These arise because the position of the gain profile depends on the bandgap, and this in turn is temperature dependent. Thus, a change in temperature could cause the position of the gain profile to alter sufficiently to cause the laser to 'hop' to another longitudinal mode more favorably placed with regard to the gain profile.

One way to improve this situation is to provide some wavelength sensitive feedback that is relatively insensitive to temperature. A very effective way of implementing this is to use a 'distributed feedback' structure (ref. 6.2). The way in which this operates may be appreciated by reference to Fig. 6.4. This shows a laser diode structure with a region

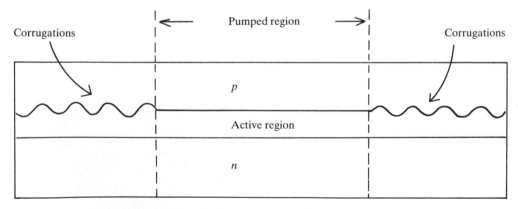

Fig. 6.4 Schematic diagram of a heterostructure laser incorporating regions at either end which have corrugations etched along one side. Such regions act as frequency selective mirrors.

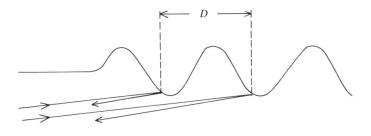

Fig. 6.5 Two beams of light being reflected from corresponding adjacent regions of the corrugation shown in Fig. 6.4. The two reflected beams will remain in phase provided that the path differences are multiples of the wavelength. Assuming that the beams make small angles with the laser axis, this path difference is very nearly equal to $2D$.

at either end containing corrugations. These act as frequency selective mirrors. To see this, we may consider two beams emerging from the pumped region at a small angle to the axis and being reflected from different parts of the corrugations as shown in Fig. 6.5. There will be constructive interference between the two beams when $2D = m \lambda_0/n$, where D is the 'wavelength' of the corrugations, λ_0 the vacuum wavelength, n the refractive index of the laser medium, and m an integer. The corrugated structure will only act as a mirror for a particular mode if its wavelength satisfies the above equation. Thus, by choosing an appropriate value of D we can insure that only one particular longitudinal mode can oscillate.

Although the structure of Fig. 6.4 undoubtedly works, it has the disadvantage that light enters the unpumped end regions, where it may suffer a considerable amount of absorption. For this reason a more efficient structure is given by extending the corrugations throughout the pumped region, thus giving rise to the *distributed feedback* (DFB) laser shown in Fig. 6.6. Here, it will be noticed that the corrugated region is in a layer

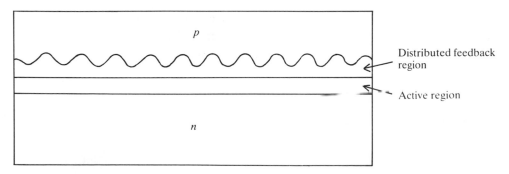

Fig. 6.6 Distributed feedback laser structure.

adjacent to the active layer, there being sufficient coupling between the two for the mirror action of the corrugations to be effective. This is because it is difficult to form the corrugations directly on the active layer without introducing into the active region defects that degrade the performance of the laser.

6.3 Mode locking

Mode locking is a technique for producing periodic, high power, short duration laser pulses. As we saw in the previous section, a typical inhomogeneously broadened laser cavity may support oscillations in many modes simultaneously. The output of such a laser as a function of time depends on the relative phases, frequencies and amplitudes of the modes. The total electric field as a function of time can be written as

$$\mathcal{E}(t) = \sum_{n=0}^{N-1} (\mathcal{E}_0)_n \exp\{i(\omega_n t + \delta_n)\} \tag{6.2}$$

where $(\mathcal{E}_0)_n$, ω_n and δ_n are the amplitude, angular frequency and phase of the nth mode. Usually these parameters are all time varying, so that the modes are incoherent and the total irradiance is simply the sum of the irradiances of the individual modes as we saw in section 1.2.2. Hence, for this situation, which is illustrated in Fig. 6.7(a),

$$I = N\mathcal{E}_0^2$$

where we have assumed for simplicity that all N modes have the same amplitude $\mathcal{E}_0$. The irradiance may exhibit small fluctuations if two or three of the modes happen to be in phase at any given time.

Suppose that we now force the various modes to maintain the same relative phase δ to one another, i.e. we *mode lock* the laser such that $\delta_n = \delta$. The total irradiance must now be found by adding the individual electric fields rather than the irradiances. Using eq. (6.2), the resultant electric field can now be written as

$$\mathcal{E}(t) = \mathcal{E}_0 \exp(i\delta) \sum_{n=0}^{N-1} \exp(i\omega_n t) \tag{6.3}$$

For convenience, let us write the angular frequency ω_n as $\omega_n = \omega - n\delta\omega$, where ω is the angular frequency of the highest frequency mode and $\delta\omega$ is the angular frequency separation between modes, which from eq. (5.32) we can write as

$$\delta\omega = \pi \frac{c}{L}$$

Equation (6.3) for $\mathcal{E}(t)$ can then be rewritten as

$$\mathcal{E}(t) = \mathcal{E}_0 \exp(i\delta) \sum_{n=0}^{N-1} \exp\{i(\omega - n\delta\omega)t\}$$

$$= \mathcal{E}_0 \exp\{i(\omega t + \delta)\} \sum_{n=0}^{N-1} \exp(-\pi inct/L)$$

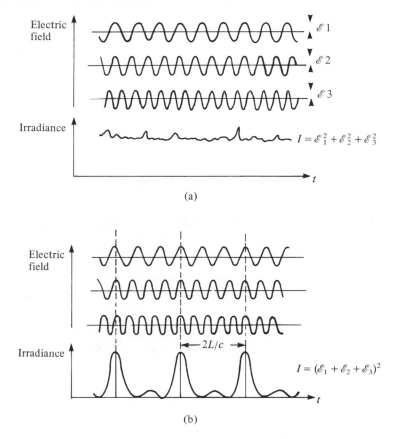

Fig. 6.7 Comparison of (a) non mode locked and (b) mode locked laser outputs. In (a) the irradiance shows random fluctuations while in (b) the phase relationships of the modes are kept constant and the output consists of a series of narrow intense pulses of time spacing $2L/c$ and duration $(1/N)(2L/c)$.

or

$$\mathcal{E}(t) = \mathcal{E}_0 \exp\{i(\omega t + \delta)\}(1 + \exp(-i\phi) + \exp(-2i\phi) + \ldots + \exp\{-(N-1)i\phi\}) \tag{6.4}$$

where $\phi = \pi ct/L$. The term in brackets in eq. (6.4) is a geometric progression and we can thus write

$$\mathcal{E}(t) = \mathcal{E}_0 \exp\{i(\omega t + \delta)\} \frac{\sin(N\phi/2)}{\sin(\phi/2)}$$

The irradiance I is then $I = \mathcal{E}(t) \cdot \mathcal{E}^*(t)$ or

$$I(t) = \mathcal{E}_0^2 \frac{\sin^2(N\phi/2)}{\sin^2(\phi/2)} \tag{6.5}$$

The form of this equation (for $N = 3$) is illustrated in Fig. 6.7(b).

We see that the irradiance $I(t)$ is periodic ($\Delta\phi = 2\pi$) in the time interval $t = 2L/c$, which equals the round trip transit time for light within the cavity. The maximum value of the irradiance is $N^2\mathscr{E}_0^2$. This occurs for values of $\phi = 0$ or $p\pi$, p being an integer, where the value of the function $[\sin^2(N\phi/2)/\sin^2(\phi/2)]$ equals N^2.

Similarly, the irradiance has minimum values of zero when $N\phi/2 = p\pi$, p being an integer which is not zero, i.e. when $\phi = 2p\pi/N$ or $t = (1/N)(2L/c)p$. Thus the time duration of the maxima, which is the time taken for the irradiance to fall from its maximum value to an adjacent zero ($p = 1$) is $(1/N)(2L/c)$. We can see, therefore, that the output of a mode locked laser consists of a sequence of short pulses, separated in time by $2L/c$, each of peak power equal to N times the average power (or N times the power of the same laser with the modes uncoupled). The ratio of the pulse spacing to the pulse width is approximately equal to the number of modes, i.e. $(2L/c)/[(2L/c)(1/N)] = N$. Thus, to obtain high power short duration pulses there should be a large number of modes; this requires a broad laser transition and a long laser cavity.

Example 6.1 Mode locked pulses

Let us compare the pulse separation and pulse duration in a mode locked Nd: YAG laser where the fluorescent linewidth is 1.1×10^{11} Hz and the laser rod is 0.1 m long.

The mode separation $c/2L = 1.5 \times 10^9$ Hz, thus the number of modes oscillating is $(1.1 \times 10^{11})/(1.5 \times 10^9)$, i.e. about 73. The pulse separation is $2L/c \simeq 0.7$ ns and the pulse duration is $(1/N)(2L/c) \simeq 10$ ps.

The situation can be visualized by considering a short wave packet bouncing back and forth between the cavity mirrors; the pulses emitted by the laser appear each time the wave packet is partially transmitted by the output mirror as indicated in Fig. 6.8. This physical picture is particularly useful when describing the active mode locking mechanism used with argon ion and Nd: glass lasers.

Fig. 6.8 Packet of energy resulting from the mode locking of N modes bouncing to and fro between the laser mirrors. Each time the packet is incident on the output mirror, a 'mode locked' pulse is emitted.

6.3.1 Active mode locking

Mode locking is achieved by forcing the longitudinal modes to maintain fixed phase relationships. This can be accomplished by modulating the loss (or gain) of the laser cavity at a frequency equal to the intermode frequency separation $\delta\nu = c/2L$ (or $\delta\omega = \pi c/L$). Let us imagine that the loss modulation is provided by a shutter placed near one of the mirrors. The shutter is closed (corresponding to very high losses) most of the time and is only opened very briefly every $2L/c$ seconds (corresponding to the cavity round trip time of the wave packet mentioned above). Now if the wave packet is exactly as long in time as the shutter stays open, and if it arrives exactly when the shutter is open, it will be unaffected by the presence of the shutter. Any parts of the wave packet, however, that arrive before the shutter opens or after it closes will be eliminated. Thus, the phase relationships of the oscillating modes are continuously restored by the periodic operation of the shutter.

The electro-optic or acousto-optic modulators discussed in Chapter 3 can be used as shutters, giving rise to mode locked pulses from a Nd:YAG laser, for example, of about 50 ps duration. In Nd:glass lasers, on the other hand, which generate a very large number of modes because of the broad laser transition line, the pulses can be less than 1 ps duration (see Problem 6.7).

6.3.2 Passive mode locking

Mode locking can also be accomplished by using certain dyes whose absorption decreases with increasing irradiance as shown in Fig. 6.9. Materials exhibiting this behavior are called *saturable absorbers*. A dye is chosen which has an absorption band at the lasing transition frequency. Initially, at low light levels, the dye is opaque owing to the large number of unexcited molecules which can absorb the light. As the irradiance increases, however, more and more of the excited states are populated until eventually

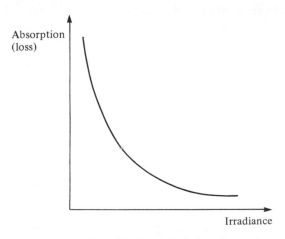

Fig. 6.9 Absorption as a function of incident light irradiance for a saturable absorber.

all of them are filled so that the dye becomes transparent. The dye is now said to be *bleached*.

The growth of the mode locked pulses can be envisaged as follows. Initially, the laser medium emits spontaneous radiation which gives rise to incoherent fluctuations in the energy density within the cavity. Some of these fluctuations, which can be of short duration, may be amplified by the laser medium and grow in irradiance to such an extent that the peak part of the fluctuation is transmitted by the saturable absorber with little attenuation. The low power parts of the fluctuation, however, are much more strongly attenuated and thus a high power pulse can grow within the cavity providing the dye can recover in a time short compared with the duration of the pulse. Because of the non-linear behavior of the dye, the shortest and most intense fluctuations grow at the expense of the weaker ones. With careful adjustment of the concentration of the dye within the cavity, an initial fluctuation may grow into a narrow pulse 'bouncing' to and fro within the cavity producing a periodic train of mode locked pulses.

Saturable absorbers provide a simple, inexpensive and rugged method of mode locking high power lasers such as Nd:glass and ruby; the so-called 9740 or 9860 dye solutions and cryptocyanine may be used as the saturable absorber for Nd:glass and ruby respectively. When a saturable absorber is used to mode lock a laser, the laser is simultaneously Q-switched (section 6.4). The result is the production of a series of narrow ($\simeq 10$ ps), mode locked pulses contained within an envelope which may be several hundred nanoseconds long. The peak power within the individual pulses may be enormous because of their very short duration.

6.4 Q-switching

Q-switching is another technique for obtaining short, intense bursts of oscillation from lasers. Single high power pulses can be obtained by introducing time- or irradiance-dependent losses into the cavity. The effects of such losses can be interpreted in terms of the 'spiking' oscillations discussed in section 5.10.1. If there is initially a very high loss in the laser cavity, the gain due to population inversion can reach a very high value without laser oscillations occurring. The high loss prevents laser action while energy is being pumped into the excited state of the medium. If, when a large population inversion has been achieved, the cavity loss is suddenly reduced (i.e., the cavity Q is switched to a high value), laser oscillations will suddenly commence. On Q-switching, the threshold gain decreases immediately (to the normal value associated with a cavity of high Q) while the actual gain remains high because of the large population inversion. Owing to the large difference between the actual and the threshold gain, laser oscillations within the cavity build up very rapidly and all of the available energy is emitted in a single large pulse. This quickly depopulates the upper lasing level to such an extent that the gain is reduced below threshold and the lasing action stops. The time variation of some of the laser parameters during Q-switching is shown schematically in Fig. 6.10. Q-switching dramatically increases the peak power obtainable from lasers.

In the ordinary pulsed mode, the output of an insulating crystal laser such as the

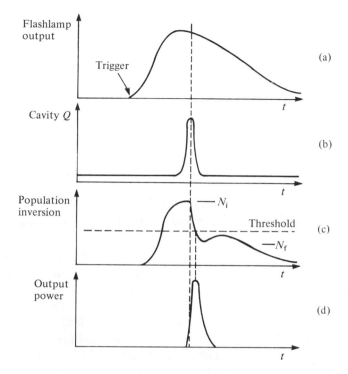

Fig. 6.10 Schematic representation of the variation of the parameters – (a) flashlamp output, (b) cavity *Q*, (c) population inversion and (d) output power – as a function of time during the formation of a *Q*-switched laser pulse.

Nd:YAG consists of many random 'spikes' of about 1 µs duration with a separation of about 1 µs; the length of the train of spikes depends principally on the duration of the exciting flashtube source, which may be about 1 ms. Peak powers within the 'spikes' are typically of the order of kilowatts. When the laser is *Q*-switched, however, the result is a single 'spike' of great power, typically in the megawatt range, with a duration of 10–100 ns. It should be noted that, although there is a vast increase in the peak power of a *Q*-switched laser, the total energy emitted is less than in non-*Q*-switched operation owing to losses associated with the *Q*-switching mechanism.

  Q-switching is carried out by placing a closed shutter (i.e. the *Q*-switch) within the cavity, thereby effectively isolating the cavity from the laser medium. After the laser has been pumped, the shutter is opened so restoring the *Q* of the cavity. A little thought reveals that there are two important requirements for effective *Q*-switching. These are:

(a) the rate of pumping must be faster than the spontaneous decay rate of the upper lasing level, otherwise the upper level will empty more quickly than it can be filled so that a sufficiently large population inversion will not be achieved; and

(b) the *Q*-switch must switch rapidly in comparison to the buildup of laser oscillations, otherwise the latter will build up gradually and a longer pulse will be obtained so reducing the peak power. In practice, the Q-switch should operate in a time less than 1 ns.

6.4.1 Methods of Q-switching

6.4.1.1 ROTATING MIRROR METHOD

This method, which was the first to be developed, involves rotating one of the mirrors at a very high angular velocity so that the optical losses are large except for the brief interval in each rotation cycle when the mirrors are very nearly parallel. Just before this point is reached a trigger mechanism initiates the flashlamp discharge to pump the laser (here assumed to be of the insulating crystal type). As the mirrors are not yet parallel, population inversion can build up without laser action starting. When the mirrors become parallel, Q-switching occurs allowing the Q-switched pulse to develop as illustrated in Fig. 6.10.

It should be pointed out that the repetition rate of laser firing is determined by control of the flashlamp and not by the speed of rotation of the mirror, which may be as high as 60 000 rev min^{-1}. If the laser fired every revolution, the repetition rate would be ≈ 1000 times per second, a rate which is prohibitive in insulating crystals due to the excessive heating of the laser rod which would occur. Although rotating mirror type Q-switches are cheap, reliable and rugged, the method suffers from the major disadvantage of being slow. This results in an inefficient production of Q-switched pulses with lower peak power than can be produced by other methods.

Example 6.2 Energy of Q-switched pulses

We may estimate the energy of the output pulses from a Q-switched laser as follows. We assume that the population inversion is N_i before the cavity is switched and that it falls to N_f at the end of the pulse (as shown in Fig. 6.10(c)).

The total energy emitted in the pulse is thus $E = \frac{1}{2}h\nu_{21}(N_i - N_f)V$, where V is the volume of the laser medium. The factor $\frac{1}{2}$ appears because population inversion changes by two units each time a photon is emitted (i.e. the population of the upper level decreases by one while that of the lower level increases by one).

In a typical laser, N_i may be approximately 10^{24} m^{-3} and assuming $N_f \ll N_i$ and that the laser frequency is 5×10^{14} Hz and its volume is 10^{-5} m^3 we have that the energy of the pulses is

$$E = \tfrac{1}{2}(6.63 \times 10^{-34}) \times (5 \times 10^{14}) \times 10^{24} \times 10^{-5} \simeq 1.7 \text{ J}$$

It can be shown (ref. 6.3) that the peak power, i.e. the greatest rate of change of population inversion and hence of photon emission, occurs when the population inversion drops to the threshold inversion N_{th}. To estimate the average power in the pulse, we need to evaluate the pulse duration. We imagine that the Q-switched pulse oscillates to and fro between the laser mirrors and that each time it strikes one of the mirrors a fraction $(1 - R)$ of its energy is lost by transmission. The pulse will then make $1/(1 - R)$ passes along the length of the cavity, which it accomplishes in a time $(1/(1 - R))(L/c)$. This is often referred to as the *cavity lifetime* t_c and may be taken as the duration of the pulse. The power of the pulse is then approximately $P = E/t_c$, which, as $E = \frac{1}{2}h\nu_{21}(N_i - N_f)V$ (see example 6.2), we can write as

$$P = \frac{(N_i - N_f)h\nu_{21}Vc(1 - R)}{2L} \tag{6.6}$$

Example 6.3 Power in *Q*-switched pulses

Using the data given in Example 6.2 we can estimate the power in the *Q*-switched pulses from a laser with a cavity length of 0.1 m and a mirror reflectance of 0.8.

The cavity lifetime $t_c = L/(1 - R)c = 1.7$ ns. Then the pulse power is given by $E/t_c = 1.7/1.7 \times 10^{-9} = 10^9$ W.

In practice, owing to losses associated with the *Q*-switch, the actual power in the pulse would probably be nearer 10^8 W.

6.4.1.2 ELECTRO-OPTIC *Q*-SWITCHING

The electro-optic, magneto-optic and acousto-optic modulators described in Chapter 3 can be used as fast *Q*-switches. If a Pockels cell, for example, is used and the laser output is not naturally polarized, then a polarizer must be placed in the cavity along with the electro-optic cell as shown in Fig. 6.11.

A voltage is applied to the cell to produce a quarter-wave plate which converts the linearly polarized light incident on it into circularly polarized light. The laser mirror reflects this light and in so doing reverses its direction of rotation so that on re-passing through the electro-optic cell it emerges as plane polarized light, but at 90° to its original direction of polarization. This light is therefore not transmitted by the polarizer and the cavity is 'switched off'. When the voltage is reduced to zero, there is no rotation of the plane of polarization and *Q*-switching occurs. The change of voltage, which is synchronized with the pumping mechanism, can be accomplished in less than 10 ns and very effective *Q*-switching occurs.

Alternative arrangements using Kerr cells and acousto-optic modulators are available. In the case of the acousto-optic modulator, application of an acoustical signal to the modulator deflects some of the beam out of the cavity (see Fig. 3.21), thereby creating a high loss. When the sound wave is shut off, *Q*-switching occurs as before. Acousto-optic devices are often used when the laser medium is continuously pumped and repetitively *Q*-switched, as is frequently the case with Nd:YAG and CO_2 lasers.

6.4.1.3 PASSIVE *Q*-SWITCHING

Passive *Q*-switching may be accomplished by placing a saturable absorber (bleachable dye) of the type mentioned in section 6.3.2 in the cavity. At the beginning of the excitation flash the dye is opaque, thereby preventing laser action and allowing a larger population inversion to be achieved than would otherwise be the case. As the light irradiance

Fig. 6.11 Electro-optic crystal used as a Q-switch. With the voltage V on, the electro-optic crystal acts as a quarter-wave plate and converts the vertically polarized light at b into circularly polarized light at c. The reflected light is converted to horizontally polarized light and eliminated by the polarizer so that the cavity Q is low. With V off, the crystal is ineffective and the cavity Q is high.

within the cavity increases, the dye can no longer absorb, i.e. it bleaches and Q-switching occurs. Passive Q-switching has the great advantage of being extremely simple to implement involving nothing more than a dye in a suitable solvent held in a transparent cell. Suitable dyes include cryptocyanine for ruby lasers and sulfur hexafluoride for CO_2 lasers.

As we mentioned in section 6.3.2, lasers that use a saturable absorber for Q-switching are also mode locked if the dye, once bleached, recovers in a time short compared with the duration of the mode locked pulses.

6.5 Laser applications

In the time which has elapsed since Maiman first demonstrated laser action in ruby in 1960, the applications of lasers have multiplied to such an extent that almost all aspects of our daily lives are touched upon, albeit indirectly, by lasers. They are used in many types of industrial processing, engineering, metrology, scientific research, communications, holography, medicine, and for military purposes. It is clearly impossible to give an exhaustive survey of all of these applications and the reader is referred to the selection of texts and journals given in ref. 6.4. Rather than attempt the impossible, we discuss the properties of laser radiation which make it so useful and reinforce this discussion by brief mention of some appropriate applications. In addition, a rather more detailed description of one or two selected applications is given in sections 6.6 and 6.7.

6.5.1 *Properties of laser light*

In considering the various properties of laser light we must always remember that not all of the different types of laser exhibit these properties to the same degree. This may often limit the choice of laser for a given application.

6.5.1.1 DIRECTIONALITY

Perhaps the most arresting property of laser light is its directionality. Apart from semiconductor junction lasers, lasers emit radiation in a highly directional, collimated beam with a low angle of divergence. This is important because it means that the energy carried by the laser beam can be collected easily and focused onto a small area. For conventional sources, where the radiation spreads out into a solid angle of 4π sr, efficient collection is almost impossible, while for lasers the beam divergence angle is so small that efficient collection is possible even at large distances from the laser.

The extent of beam divergence is set by diffraction (section 1.2.4). This is a fundamental physical phenomenon, rather than an engineering limit that can be improved by better optical design. The angle of divergence in radians at the diffraction limit is given by θ, where

$$\theta = \mathcal{K}\frac{\lambda}{D} \tag{6.7}$$

D is the diameter of the aperture through which the beam emerges, and $\mathcal{K}$ is a numerical factor of the order of unity. The precise value of $\mathcal{K}$ depends on the nature of the beam. For example, a TEM_{00} beam has a Gaussian profile (section 5.9.2), and Fig. 6.12 illustrates how the beam diverges outside the laser cavity. We see that the beam divergence angle θ tends asymptotically to the value $\sin^{-1}(w/z)$. Since from eq. (5.37) we also have that $w = z\lambda/\pi w_0$, it follows that at relatively large distances from the laser cavity $\theta = \sin^{-1}(\lambda/\pi w_0)$. Now $\lambda/\pi w_0$ is usually much less than unity, so that we finally obtain $\theta \simeq \lambda/\pi w_0$. Assuming that we may associate the minimum beam diameter $2w_0$ with an aperture diameter D, we see that this result agrees with the general divergence equation (6.7) with the parameter $\mathcal{K}$ taking the value $2/\pi$. The beam divergence tends to increase with increasing power output and mode content. Table 6.1 gives some typical beam divergence angles.

Fig. 6.12 Divergence of a laser beam outside the laser cavity.

Table 6.1 Typical laser beam divergence angles

Laser	He–Ne	Ar	CO_2	Ruby	Nd:glass	Dye	GaAs
Beam divergence (mrad)	0.5	0.8	2	5	5	2	20×200

The beam may be further collimated by passing it in the reverse direction through a telescope as illustrated in Fig. 6.13. The beam is enlarged by the factor f_2/f_1 and hence the divergence, which is inversely proportional to the beam diameter, is decreased by the factor f_1/f_2. The ratio of the beam diameters before and after the collimator is given by

$$\frac{D_1}{D_2} = \frac{f_1}{f_2} = \frac{\theta_2}{\theta_1}$$

The directional nature, in particular, of gas laser beams readily lends them to applications involving accurate alignment, including civil engineering projects such as drainage and tunnel boring, surveying and the assembly of large aircraft and ships, etc. (for additional references see ref. 6.4e, pp 245–6).

Example 6.4 Beam collimation

We may calculate the reduction in the divergence of a laser beam which is collimated by passing it through a telescope with an objective lens to eyepiece lens focal length ratio of 30:1. Let us consider an He–Ne laser with a plasma tube diameter of 3 mm.

The divergence θ of the beam from the laser is given approximately by λ/D. Therefore $\theta_1 \simeq 633 \times 10^{-9}/(3 \times 10^{-3}) \simeq 2.1 \times 10^{-4}$ rad (or 0.7′).

Hence, after collimation the angle of divergence will be reduced by a factor of 30 to $\theta_2 \simeq 7 \times 10^{-6}$ rad (or 1.4″).

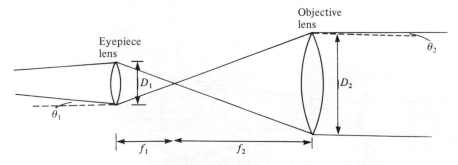

Fig. 6.13 Schematic diagram showing the collimation of a laser beam using a telescope having an eyepiece lens of diameter D_1 and focal length f_1 and an objective lens of diameter D_2 and focal length f_2. The beam width is enlarged by the factor $D_2/D_1 = f_2/f_1$ and the divergence angle is decreased by the factor f_1/f_2.

6.5.1.2 LINEWIDTH

Laser light is potentially extremely monochromatic but, as we saw in Chapter 5, the spectral content of the laser radiation may extend over almost as wide a range as the fluorescent linewidth of the laser medium. In other words, although the linewidth of an individual cavity mode may be extremely small there may be many modes present in the laser output. We saw in section 6.1 how single mode operation and frequency stabilization can be achieved. The high spectral purity of laser radiation leads directly to applications in basic scientific research including photochemistry, luminescence excitation spectroscopy, absorption and Raman spectroscopy and also in communications. Many other applications also depend, in part, on this property (ref. 6.4e).

6.5.1.3 BEAM COHERENCE

One of the characteristics of stimulated emission is that the stimulated wave is in phase with the stimulating wave; i.e. the spatial and temporal variation of the electric field of the two waves are the same. Thus in a 'perfect' laser we would expect the electric field to vary with time in an identical fashion for every point on the beam cross section. Such a beam would have perfect *spatial coherence*. Another related property is *temporal coherence*, which refers to the relative phase relationship of the electric field at the same place as a function of time. If the phase changes uniformly with time, then the beam is said to show perfect temporal coherence. These ideas as illustrated in Fig. 6.14.

Coherence is often specified in terms of the *mutual coherence function* $\gamma_{12}(\tau)$ (see ref. 6.5). This quantity, which is in fact a complex number, is a measure of the correlation between the light wave at two points P_1 and P_2 along the direction of propagation at different times t and $t + \tau$ (temporal coherence), and at points P_1 and P_3 on a plane perpendicular to the direction of propagation at the same time (spatial coherence) (see the representative points on Fig. 6.14a). It has an absolute value between 0 and 1. When it has the value zero the light is completely incoherent, while a value of unity implies complete coherence. Although these extreme values are never achieved in practice, the light from a gas laser operating in a single transverse mode has a value quite close to unity.

Two useful quantities that are related to temporal coherence are the coherence time and the coherence length. To understand these we consider what happens when we take a beam, split it into two equal parts, let the two components travel different distances and then recombine them to form interference fringes as in the Michelson interferometer described in section 6.6.1. Interference effects will only be observed if the path difference of the two beams is such that they are still coherent when they recombine. Light beams from 'real life' sources cannot be represented by infinitely long wave trains, so there is a limit to the path difference. Disregarding lasers for a moment, let us consider a group of atoms undergoing spontaneous emission. Each atom emits radiation independently of the other and does so for only a finite time. This is often because the emission process is perturbed in some way, for example during a collision with a neighboring atom. Thus each atom generates a finite-length wave train and for simplicity we regard the wave trains from all the atoms as having the same length (L_c). Since, however, the atoms are emitting spontaneously, the wave trains are not in phase with each

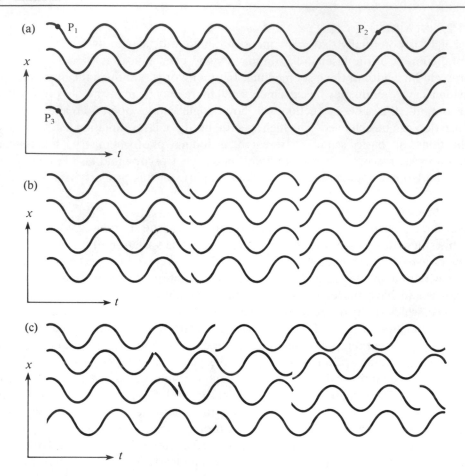

Fig. 6.14 Illustration of coherence. in (a) a perfectly coherent beam. All the constituent waves are in phase at all times. In (b) a beam which is spatially coherent but which exhibits only partial temporal coherence. This is because the waves simultaneously change their phase by an identical amount every few oscillations. In (c) an almost completely incoherent beam where the phases of each wave change randomly at random times. Note, however, that even in this case some small degree of temporal coherence remains, since over very short intervals the phases are to some extent predictable.

other. If now we pass such a beam into the Michelson interferometer, each individual wave train will be split into two, and the two split wave trains will be able to interfere *provided* the path difference does not exceed L_c. We refer to L_c as the *coherence length*. If the path difference does exceed L_c, then the two halves of each wave train cannot overlap in time when they are recombined, and hence they cannot interfere. In addition, when the path difference lies between zero and L_c, then only a part of each wave train can take part in interference (Fig. 6.15). This implies that as we increase the path difference from zero up to L_c, the irradiance fluctuations corresponding to constructive and destructive interferences will gradually diminish and die out altogether at L_c.

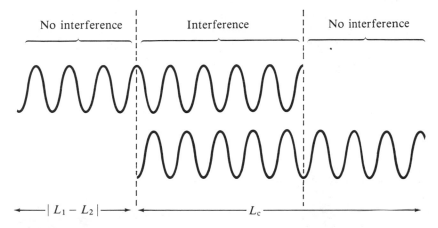

Fig. 6.15 When two identical wave trains of length L_c which have traveled different distances (L_1 and L_2) are recombined, they can only interfere over a length $L_c - |L_1 - L_2|$.

In reality of course, the wave trains are not all the same length but are distributed in some way about a mean value. The conclusion that the irradiance fluctuations will gradually reduce, however, still remains valid. Similarly the exact definition of coherence length is somewhat more involved, but we may still take it to be the path difference at which the irradiance fluctuations die out.

The *coherence time* t_c may then be defined as the time taken for the source to emit a wave train of length L_c. Thus:

$$t_c = L_c/c \tag{6.8}$$

where c is the velocity of light. In fact it can be shown [ref. 3.2b] that the coherence time is related to the linewidth of the emission (Δv) via the equation

$$t_c = \frac{1}{\Delta v} \tag{6.9}$$

Although we started this discussion of coherence by considering spontaneous emission, the ideas of coherence length and coherence time apply equally well to laser radiation.

Example 6.5 Coherence lengths of conventional and laser radiation sources

We consider first of all the light emitted from a low-pressure sodium lamp. A typical linewidth of the sodium D lines (both lines taken together) at $\lambda = 589$ μm is 5.1×10^{11} Hz. Equation (6.9) then gives $t_c = 2 \times 10^{-12}$ s and from eq. (6.8), $L_c = 0.6$ mm. We may contrast these values with those applicable to an He–Ne laser. If many modes are operating, then the linewidth is about 1500 MHz giving a coherence length of 0.2 m. However, if the laser is operating in a single mode stabilized to 1 MHz, then the coherence length is some 1500 times greater, i.e. 300 m.

Table 6.2. Summary of coherence lengths of some common lasers

Laser	Typical coherence length
He–Ne single tansverse, single longitudinal mode	up to 1000 m
He–Ne multimode	0.1 to 0.2 m
Argon multimode	0.02 m
Nd:YAG	10^{-2} m
Nd:glass	2×10^{-4} m
GaAs	1×10^{-3} m
Ruby:	
for whole output pulse	10^{-2} m
within a spike forming part of the pulse	$\leqslant c$ times spike length, i.e. $\leqslant 30$ m

Thus for there to be coherence at the points P_1 and P_2 in the direction of propagation, the separation of P_1 and P_2 should be less than L_c.

Consider now a perfectly collimated beam of monochromatic light, then the phase is the same at any point on a surface perpendicular to the beam, and the points P_1 and P_3 may be separated by quite large distances. On the other hand, for a nonmonochromatic, incompletely collimated beam the points P_1 and P_3 can only be a fraction of a millimeter apart if there is to be any coherence between them. We sometimes define a transverse coherence length, L_t, which represents the distance perpendicular to the main propagation direction over which the phases at two points remain correlated. If there is coherence between the points P_1 and P_3 then interference effects should result if the light from these points can be brought together. In principle, this can be accomplished using the Young's slits apparatus described in section 1.2.3.

Example 6.5 illustrates how the presence of several modes can dramatically reduce the temporal coherence of a laser, and the same is also true of the spatial coherence. A well-stabilized laser emitting a single transvere mode exhibits almost perfect spatial coherence across the entire beam cross section. It has been possible to expand the TEM$_{00}$ mode of gas lasers to give transverse coherence lengths of several meters for holographic purposes. On the other hand, if a number of transverse modes are present the spatial coherence is considerably reduced.

While the coherence of the output from CW gas lasers can be very high, that from pulsed lasers is usually somewhat smaller. This is because the temporal coherence may be limited by the duration of the spikes within the laser pulse or by shifts in frequency during emission. Thus, ruby lasers emit spikes that are typically 0.6 µs in duration and exhibit coherence times of about 0.1 µs. The coherence lengths of some common lasers are given in Table 6.2.

Coherence is important in any application where the laser beam will be split into parts. These include interferometric measurement of distance (section 6.6) and deformation, where the light is split into parts that traverse different distances, and holography (section 6.7), where the light beams traverse different paths which may be approximately equal but which may have spatially different distributions.

One of the most striking characteristics of laser light reflected from rough surfaces is its speckled or granular appearance. This is the result of a random interference pattern formed from the contributions of light reflected from neighboring portions of the surface. In some regions, these contributions will interfere constructively, while in others they interfere destructively. This behavior is a direct consequence of the high coherence of laser light.

In many applications, for example holography, the speckle pattern is a nuisance, though the phenomenon is finding application in a number of areas including metrology and vibration analysis (ref. 6.6).

6.5.1.4 BRIGHTNESS

The primary characteristic of laser radiation is that lasers have a higher brightness than any other light source. We define *brightness* as the power emitted per unit area per unit solid angle (sometimes the term *specific brightness*, that is brightness per unit wavelength range, $W\,m^{-2}\,sr^{-1}\,\Delta\lambda^{-1}$, is used). (In radiometry this unit – brightness – is called the radiance, see section 1.4, but in laser work the term brightness as defined here is used.) The relevant solid angle is that defined by the cone into which the beam spreads. Hence, as lasers can produce high levels of power in well-collimated beams, they represent sources of great brightness.

The brightness is also affected by the presence of additional modes, for often as the laser power is increased the number of modes increases but the brightness remains almost constant. Typical values of brightness are: for an He–Ne laser, $10^{10}\,W\,m^{-2}\,sr^{-1}$; for a Q-switched ruby laser, $10^{16}\,W\,m^{-2}\,sr^{-1}$; and for an Nd:glass laser followed by amplifiers, $10^{21}\,W\,m^{-2}\,sr^{-1}$ has been achieved. For comparison, the brightness of the sun is about $1.3 \times 10^6\,W\,m^{-2}\,sr^{-1}$!

High brightness is essential for the delivery of high power per unit area to a target; this in turn depends on the size of the spot to which the beam can be focused.

6.5.1.5 FOCUSING PROPERTIES OF LASER RADIATION

The minimum spot size to which a laser beam can be focused is determined by diffraction. A single mode beam can be focused into a spot which has dimensions of the order of the wavelength of the light, though imperfections in the optical system may mean that we cannot achieve this in practice. A useful 'rule of thumb' for estimating the spot size is that the radius r_s, at the focal plane of a lens of focal length f is given by

$$r_s = f\theta$$

where θ is the beam divergence angle in radians. Since θ is given approximately by λ/D, D being the limiting aperture diameter, we have (assuming the beam fills the lens)

$$r_s \simeq f\frac{\lambda}{D} \simeq \lambda F$$

where F is the F number of the lens. It is impracticable to work with F numbers much smaller than unity, so that r_s is of the order of λ.

Thus, for example, if we have a 10 mW He–Ne laser with a beam divergence angle of 10^{-4} rad, then an F:1 lens will produce a focused spot with an area of about 10^{-12} m^2 and the power per unit area near the center of the spot will be around 10^{10} W m^{-2}.

Once again, the presence of a complicated mode structure in the beam is deleterious in that, in this case, the focused spot size is much larger and the power density (irradiance) is correspondingly much smaller for a given laser power. Again, if the beam divergence is large the power density is reduced. Nevertheless, doped insulator lasers, because of the very high peak powers they generate, can easily produce very high irradiances. A focal area of 10^{-7} m^{-2} is typical for such lasers, giving rise to typical average irradiances of 10^5 W m^{-2} and peak irradiances of 10^8 W m^{-2}.

Such high irradiances lead to the use of lasers in the drilling, cutting, welding and heat treatment of a large number of different materials (ref. 6.7). In certain applications, for example the micromachining of electronic components, good focusing is required and hence we would wish to use a short focal length lens. This, however, may be impracticable on a production line because of the limited depth of field. We must provide sufficient depth of field to allow for vibrations and inaccuracy in positioning in the vertical sense. The depth of field is given by

$$Z \simeq \frac{4\lambda f^2}{D^2} \simeq \frac{r_s^2}{\lambda}$$

and therefore increases as the square of the spot radius. We must then reach an acceptable compromise between sufficiently large depth of field and small focal area.

The selection of a laser for a given application involving laser 'heating' depends very much on the nature of the application. For many cutting tasks it may be advantageous to use a CW laser; for continuous output, the highest powers are produced by CO_2 lasers for which values up to 100 kW have been quoted. For welding operations, a pulsed laser may be preferred. In this case, because of the very short pulses which can be produced, we find that Q-switched Nd:glass lasers generating pulses with peak powers of about 10^{11} W are commercially available.

The focusing properties of laser radiation are also important in low power applications, two of which represent the first laser based devices to be used by the public at large. The first of these is the 'point of sale' device used to price items in supermarkets and to provide for automatic stock information upgrading. Products have a coded label, consisting of a series of parallel bars of varying widths, placed on them. This is scanned by a laser. The light reflected from the bars is detected, thereby identifying and pricing the product and adding the price to the bill.

The second application is in the preparation and readout of some home video-disk systems. The information is imprinted on the video-disk in digital form by forming small pits in the surface of the disk with a laser. These pits are subsequently read by a low power laser to provide a video signal for playback on a television set. The system has a number of advantages over tape or stylus pick up systems including the absence of wear, the high information density that can be accommodated by the closely focused laser beam and the fact that warped disks can be played equally well. Also high quality 'frozen' images can be selected at will and held indefinitely. Similar arrangements are used in compact disk (CD) audio systems, and are becoming increasingly popular for archival and computer storage (ref. 6.6).

6.5.1.6 TUNABILITY

We saw in Chapter 5 and the earlier part of this chapter that some lasers can be tuned to emit radiation over a range of wavelengths. With dye lasers, for example, the range of tunability can be large. Indeed dye lasers can be tuned over most of the visible spectrum, and by harmonic generation this range can be extended into the ultraviolet. On the other hand, optical parametric amplifiers used with a primary laser source can up-convert into the range 1–25 μm in the infrared (section 3.9.1).

Laser tunability leads to applications in photochemistry, high resolution and Raman spectroscopy and isotope separation (ref. 6.8).

6.6 Measurement of distance

The main methods of measuring distance using lasers are; (i) interferometry; (ii) beam modulation telemetry, and (ii) pulse time of flight (ref. 6.9a).

6.6.1 Interferometric methods

We saw in Chapter 1 that if the wavefront from a light source is divided into two parts which then traverse different distances before being recombined, then an interference fringe pattern is produced. The irradiance distribution in the pattern is characteristic of the point-for-point path differences between the two parts of the beam. Thus, if one of the path lengths is changed the fringe pattern will move across the field of view and the change in path length can be measured in terms of the fringe shift.

The classical method for measuring distance (or changes of distance) in this way is the Michelson interferometer and nearly all other methods are variations of this instrument. The Michelson interferometer, which is shown in Fig. 6.16, consists of a beam splitter, two plane mirrors and an observing telescope. The wavefront from the laser source is divided by the beam splitter B; the two parts then proceed to the plane mirrors M_1 and M_2 and are reflected back to B. Some of the light is reflected by the beam splitter and some is transmitted as shown so that the beam splitter serves to recombine the beams and interference fringes can be seen through the telescope. We may regard the fringes as being produced in the thin film formed between the mirror M_1 and M_2' which is the reflection of mirror M_2 in B. Thus if M_1 and M_2' are exactly parallel, that is M_1 and M_2 are exactly perpendicular to each other, a system of circular fringes will be seen as explained in Chapter 1. On the other hand, if one of the mirrors is tilted slightly, then a system of straight-line fringes is formed.

We saw in Chapter 1 (eq. (1.23)) that for thin film interference a bright fringe is formed when

$$p\lambda = 2D \cos \theta \simeq 2D \quad \text{(if } \theta \text{ is small)}$$

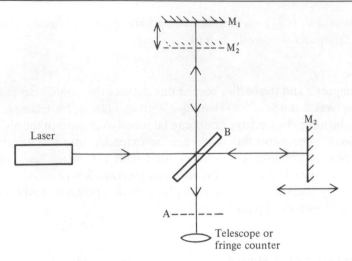

Fig. 6.16 Michelson inteferometer: M_1 is a fixed mirror, M_2 is movable and M_2' is its image in the beamsplitter B. Interference fringes can be seen through the telescope, or alternatively the number of fringes crossing an aperture placed in position A can be detected using a photodetector and counted electronically.

where D is the optical thickness of the film. Hence if one of the mirrors, M_2 say, is moved D will change and the fringe pattern will move. Specifically, if D changes by $\lambda/2$ a complete fringe will pass a reference point in the field of view. Therefore we can measure the distance moved by M_2 in terms of fringe shifts, that is in terms of the wavelength λ of the light used. To measure an unknown distance one simply aligns M_2 with one end and counts the fringe shift as it is moved until it coincides with the other end of the distance being measured.

Optical interferometry predates the laser by many years, of course, but the technique was always limited by the coherence limitations of the light sources available. Distances of a few centimeters could be measured at best. With an He–Ne laser, however, coherence lengths of many meters are available so that, in principle, we can measure up to such distances with an accuracy of a fraction of a wavelength. Fringe displacements of 0.01 of a fringe, equivalent to $\lambda/200$ can be detected. We must remember, however, that the distances measured are optical path lengths which include the refractive index of the air. Changes in refractive index, due to pressure and temperature variations and atmospheric turbulence, result in random fringe shifts and thereby limit the distance which can be measured and the accuracy attainable. Accuracies of about 1 part in 10^6 can be achieved quite readily. In practice the plane mirrors are replaced by cube corner retroreflectors (Fig. 6.17). These have the property of reflecting an incident beam back along a direction parallel to its incident path thereby simplifying the alignment of the instrument. The lateral displacement involved prevents the returning light from entering the laser cavity and thereby creating an undesirable modulation of the laser output. The large number of fringes which cross the field of view is counted electronically using, for example, a silicon photodiode as a detector.

Fig. 6.17 Schematic diagram of a cube corner reflector. The incident light reflects once off each side face and emerges along a path almost exactly parallel to that of the incident light.

The technique is widely used in machine tool control, length standard calibration and for seismic and geodetic purposes.

6.6.2 Beam modulation telemetry

As we mentioned above, owing to fluctuations in the density of the atmosphere, interferometric distance measuring methods are limited to distances not exceeding about 100 m. For greater distances, techniques involving amplitude modulation of the laser beam are useful. The beam from an He–Ne or GaAs laser is amplitude modulated and projected to the 'target' whose distance is to be measured. The light reflected from the target is received by a telescope and sent to a detector. The phase of the modulation of the reflected beam is different from that of the emitted beam because of the finite time taken for the light to travel to the target and return to the telescope. The phase shift ϕ is given by

$$\phi = \frac{2\pi}{\lambda}(2n_gL) \tag{6.10}$$

where L is the target distance and n_g is the group index of refraction of the atmosphere. The value of n_g for the 632.8 nm He–Ne laser wavelength is $n_g = 1.00028$ for dry air at 15°C, 760 torr and 0.03% CO_2. Corrections for varying atmospheric temperature and pressure are available. These corrections are difficult to apply for measurements in the field, however, and one must attempt to average n_g over the entire path traversed by the light.

Figure 6.18 shows the schematic diagram of a beam modulation system. The light is amplitude modulated at a given frequency f, collimated and transmitted to the target. Reflected light is collected by the telescope and focused onto the detector (the presence of a retroreflector on the target is a great help). A phase detector compares the relative phase of the reflected beam with that of the original beam.

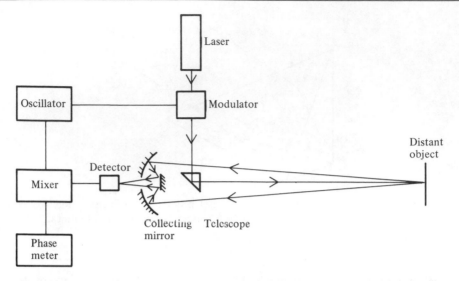

Fig. 6.18 Schematic diagram of a beam modulation distance measurement system.

The phase difference can be written as

$$\phi = (p + q)2\pi$$

where p is an unknown integer and q is a fraction less than unity. The phase comparison gives q but not p. To find p, the measurement must be repeated with different values of the modulation frequency. Having found ϕ then L can be determined from eq. (6.10). The narrow bandwidth of the laser light enables high discrimination against stray light so that the system can be used in daylight with a high signal-to-noise ratio, while the small beam divergence allows a high degree of selectivity in the target being examined.

Beam modulation telemetry units have been developed for which the accuracy is better than 1 mm at distances up to 1000 m and 1 part in 10^6 for greater distances. Such devices have been used for the measurement of large structures such as dams and bridges, as airborne instruments for land profiling and for geodetic surveying.

6.6.3 Pulse echo techniques

We can measure large distances by timing the round trip transit time for a very short pulse reflected from a distant target. The system consists of a pulsed laser, preferably Q-switched, a telescope to collect the reflected light, a photodetector and an accurate timer. The narrowly collimated beam of the laser makes it possible to measure the range to specific targets and the technique is finding military application as a range finder. Accuracy of the order of ±5 m in 10 km has been achieved.

A novel application has been in measuring the distance to the moon. Using retroreflectors left on the surface of the moon during the Apollo 11, 14 and 15 space missions, the lunar distance has been measured to an accuracy of ±15 cm.

This technique, which is often known as *optical radar* or '*lidar*' (*light detection and* ranging), has been extended to atmospheric studies. By measuring the amount of backscattered light, the presence of air turbulence can be detected and the amounts of various atmospheric pollutants such as CO_2 and SO_2 can be measured (ref. 6.9b).

6.7 Holography

Although holography was developed prior to the laser (the first hologram was recorded by Gabor in 1948, ref. 6.10), the requirement of holography for light with a high degree of spatial and temporal coherence has closely linked the development of holography to that of lasers. Holography is a method of recording information from a three-dimensional object in such a way that a three-dimensional image may subsequently be reconstructed; the phenomenon is often known as wavefront reconstruction. A great deal of work has been done on holography and its applications; a selection of texts on the subject is given in ref. 6.11.

Figures 6.19(a) and (b) illustrate the basic principles. A photographic plate is exposed simultaneously to waves of light scattered by the 'object' and to waves of light from a 'reference' source. The reference beam, shown here in Fig. 6.19(a) as a plane parallel beam, may be of any reproducible form and is derived from the same laser source as the light illuminating the object. Because of their high degree of mutual coherence the two sets of waves produce an interference pattern on the plate, which is recorded in the photographic emulsion and forms a *hologram*.

The photographic plate is now processed and illuminated with only the reference beam present as shown in Fig. 6.19(b). Most of the light from the reference beam passes straight through the hologram; some of it, however, is diffracted by the interference pattern in the emulsion. By the normal diffraction grating equation (eq (1.28)), light of wavelength λ will experience constructive interference at angles such that $\lambda = d \sin \theta$, where d is the local fringe spacing of the interference fringes whose exact shape and distribution depends on the shape of the object and the wavefronts reflected from it. Thus the constructive interference of these diffracted waves reconstructs the original wavefronts from the object and to an observer the wavefronts appear to be coming from the object itself. These wavefronts constitute what is termed the *virtual image*. However, just as a diffraction grating gives diffracted orders on either side of the 'straight-through' position, the hologram generates a second image; this image, which is usually inferior in quality to the virtual image, is called the *real image*.

The hologram serves as a 'window' on the object scene, which has been illuminated by the laser, through which the object can be viewed from different angles. The range of views of the reconstructed object is limited only by the size and position of the hologram and a truly three-dimensional effect is created.

The mathematical analysis of holography is quite complicated, but the following simplified treatment will serve to illustrate the principles involved. We assume that the photographic plate is in the (x, y) plane and that we may represent the electric field of the wavefront reflected from the object in the (x, y) plane at time t by

$$\mathscr{E}_{\text{ob}} = U_0(x, y) \exp(-i\omega t)$$

Fig. 6.19 Schematic diagram showing the basic geometry for (a) making a hologram by recording the interference pattern produced by the interference of light reflected from the object, and the reference wavefronts and (b) reconstruction of the object wavefront. To the observer the reconstructed wavefront appears to be coming from the object itself and a virtual image is seen.

where $U_0(x, y)$ is the amplitude, which in general is complex. Similarly, the complex amplitude of the reference beam in the (x, y) plane at the same instant is $U_r(x, y)$. As the object and reference beams are coherent, the irradiance recorded on the photographic plate is given by adding the amplitudes and multiplying by the complex conjugate. Thus the irradiance is

$$I(x, y) = |U_0 + U_r|^2 = (U_0 + U_r)(U_0^* + U_r^*)$$

Therefore

$$I(x, y) = (U_0 U_0^* + U_r U_r^*) + (U_0 U_r^* + U_0^* U_r) \tag{6.11}$$

The first term in eq. (6.11) is the sum of the individual irradiances; the second term represents the interference which occurs and thus contains information in the form of amplitude and phase modulations of the reference beam.

The plate is now processed to form a transmission hologram. With correct processing, the transmission of the hologram is a constant (say T) times the irradiance function $I(x, y)$ given in Eq. (6.11). If the plate is now illuminated by the original reference beam only, the transmitted light will have a complex amplitude U_T which is proportional to U_r times the transmittance of the hologram $TI(x, y)$. Hence we may write

$$\begin{aligned}
U_T(x, y) &= U_r T I(x, y) \\
&= T[U_r(U_0 U_0^* + U_r U_r^*) + U_r^2 U_0^* + U_r U_r^* U_0] \tag{6.12}
\end{aligned}$$

As we mentioned above, the hologram behaves like a diffraction grating and produces a direct beam and a first order diffracted beam on either side of the direct beam. The first term in eq. (6.12) represents the direct beam. The last term is the one of greatest interest; $U_r U_r^*$ is constant so that the last term is essentially U_0, the object wavefront amplitude. Hence this diffracted beam represents a reconstruction of the wavefront from the original object and it forms the virtual image. The middle term represents the other diffracted beam and forms the real (conjugate) image.

This description can be verified by considering the simple case of an object comprising a single white line on a dark background. The hologram, in this case, turns out to be a simple periodic (or sinc) grating. The zero order of the diffracted light is the direct beam, whereas the first orders on either side comprise the virtual and real images.

6.7.1 Applications of holography

Undoubtedly, the full potential of the holographic technique is still to be realized although a number of applications are now firmly established (ref. 6.10). We shall describe one established and one potential application, namely holographic interferometry and computer memories respectively.

6.7.1.1 HOLOGRAPHIC INTERFEROMETRY

The determination of surface contours by conventional interferometry has been restricted to the examination of reflecting surfaces with simple shapes. This restriction is removed by holographic interferometry, which can be used for complicated shapes with diffusely reflecting surfaces. There are a number of recognizably different types of holographic interferometry which we now describe briefly.

Double exposure holographic interferometry is an important industrial process in which very small displacements or distortions of an object can be measured by counting interference fringes. The subject of the investigation is recorded as a hologram and before the plate is processed the subject is moved, distorted by stress or whatever and a second hologram recorded. After processing, each image can be reconstructed in the usual way. The two sets of wavefronts for the reconstructed images interfere and produce interference fringes over the full range of views of the subject obtainable through the hologram. A typical example of this technique is given in Fig. 6.20 which shows a circular membrane which has been deformed by a uniform pressure. The time between the two records may be anything from a fraction of a microsecond upwards, but the plate and subject must maintain the same relative positions except for the movement under investigation.

A variation of the technique is *real time holographic interferometry*, in which the interference fringes are viewed in real time. A hologram of the subject is recorded as above but in this case the plate is processed and replaced in its original place. The subject is now distorted and interference fringes can be observed through the holographic plate, changing as the distortion of the subject actually occurs. Although real time holographic interferometry provides a sensitive tool for measuring the strains of objects as they actually deform, it suffers from a number of problems. These include the difficulty of replacing the plate *exactly* in its original place and distortion of the photographic emulsion during processing. Figure 6.21 shows the fringe pattern in real time holographic interferometry as the object, a metal bar which is clamped at one end, is stressed.

The third technique, *time-average holographic interferometry*, is particularly useful

Fig. 6.20 Double exposure holographic interferogram showing the deformation of a circular membrane which has been caused by a uniform pressure. [Photograph courtesy of W. Braga and C. M. Vest, University of Michigan.]

Fig. 6.21 Illustration of real time or single exposure holographic interferometry. Interference of the actual object wave with the reconstructed object wave shows the deformation of the bar. The fact that the fringes are not horizontal indicates that the bar has suffered a twist in addition to bending. [From O'Shea, Callen and Rhodes, *Introduction to Lasers and their Applications* © 1977, Addison-Wesley, Reading, MA. Fig. 7.14. Reprinted with permission.]

for examining the spatial characteristics of low amplitude vibrations of an object. In most holographic situations, a general rule is that the object should remain stationary during the period of exposure. In the present case this rule is violated dramatically, for during the exposure the object is moving continuously. The resulting hologram may be regarded as the limiting case of a large number of exposures for many different positions of the surface. The fringes produced represent contour lines of equal amplitude of vibration of the surface. The brightest fringes occur at the nodes where the surface remains stationary. Elsewhere, providing the period of exposure covers many vibrations of the surface, there is a variation in irradiance due to the surface motion, with almost zero irradiance at the antinodes. An example of the application of time-average holographic interferometry to the analysis of the vibrations of a turbine blade is shown in Fig. 6.22.

6.7.2 *Holographic computer memories*

Holographic computer memories are being actively developed because, potentially, they have a very high storage capacity – theoretically up to 10^{10} bits mm^{-3} – with rapid

Fig. 6.22 Holographic reconstructions from time-average holograms showing flexural resonances (A and B) and torsional resonances (C and D) of a turbine blade. [From an article by Robert K. Erf in Robert K. Erf (Ed), *Holographic Non-Destructive Testing*, 1974; Courtesy Academic Press Inc. Ltd.]

access. A holographic memory records and reads out a large number of bits simultaneously as we can appreciate by considering Fig. 6.23. The information to be stored is formed into a two-dimensional array of bits by a device called a page composer. The page composer may be thought of as an array of light valves which may be open or closed corresponding to 'ones' and 'zeros' respectively. This array of up to perhaps 10^4 bits is then stored at one time in a particular location of the holographic memory. During recording, the light modulators allow maximum irradiance in both the signal and reference beams. To store a different array as another hologram on the storage medium the beam deflector moves the beams to the appropriate position.

Readout of data occurs when the hologram is addressed only by the reference beam. The first modulator is partly closed to reduce the irradiance of the light reaching the holographic storage medium, while the second modulator is closed to cut off the signal beam. The deflector directs the beam to the hologram to be read via the tracking mirror and an image representing the arrays of 'ones' and 'zeros' is produced. This image is focused by a lens and projected onto the detector array. Each bit originally stored in the page composer is incident on one photodetector in the array. The stored information is thus converted into electrical signals.

The storage of data in this way offers a number of advantages compared to a bit-orientated memory. The information about a particular array is distributed in a holographic fringe pattern across the entire hologram. Thus the hologram is not

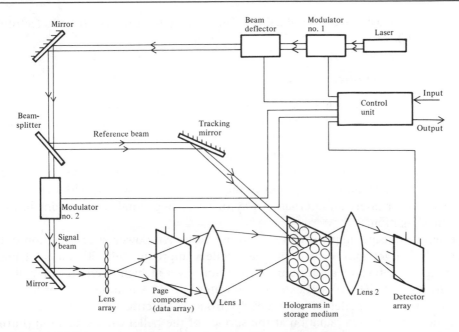

Fig. 6.23 Schematic diagram of a holographic memory system.

sensitive to small scratches or particles of dust which might otherwise cause the loss of bits of data in a bit-orientated memory.

Secondly the information is recovered essentially in parallel. A large number of bits are all read simultaneously allowing a very high readout rate. The requirements on light beam deflection are also reduced. Each position to which the beam is deflected represents 10^4 bits of data, thus a 10^8 bit memory requires only 10^4 separate locations. This lies within the capabilities of present light beam deflectors. Addressing can be done using electro-optical deflectors which have access times in the region of microseconds.

A third advantage is that the holographic recording and construction is insensitive to the exact position of the reference or reading beam on the hologram. The hologram can be moved but the focused spots remain stationary. This means that holographic memory systems are easier to align and less subject to problems of vibration than other optical memories.

Finally it is possible to record several holograms in different planes of one thick sensitive material provided that different reference beam directions are used to record each hologram. The appropriate hologram is then read with a readout beam aligned at precisely the same angle as the original reference beam in the recording process.

Despite these advantages, the availability of suitable storage media has to date limited the development of holographic memories. Although photographic emulsion is satisfactory, it provides a permanent store only, which cannot be easily updated. Ideally the storage medium should be stable and alterable, have low write energy and high readout efficiency. Potentially useful read/write/erase optically sensitive materials include thermoplastic-photo-conducting layers, magneto-optic MnBi and electrophotochromic KCl.

6.8 Laser induced nuclear fusion

For many years, research has been directed towards a system for producing controlled thermonuclear reactions to generate energy. Nuclear fusion of light elements occurs within a very high temperature plasma such as exists in the sun. Until recently, laboratory experiments aimed at reproducing these conditions were based entirely on magnetic confinement of the plasma and these have advanced considerably in recent years using the Tokamak concept. Since the early 1970s, however, with the advent of very high power lasers an alternative way of producing suitable conditions has been under investigation.

The basic concept simply involves focusing very high power laser radiation onto a target. The target may consist of glass pellets approximately 50 μm in diameter containing a mixture of deuterium and tritium gases at high pressures or pellets of frozen heavy water (D_2O) and extra heavy water (T_2O). A number of laser beams are directed onto the pellet simultaneously from symmetrically arrayed directions. Absorption of the laser radiation at the surface of the pellet causes ablation (burning off) of the outer material and an implosion of the contents. The implosion is caused by a compressional wave driving radially into the material from the periphery thereby squeezing the pellet into a very dense core. Very high temperatures, in excess of 10^8 K, are produced within this core and at these temperatures the velocities of the deuterium and tritium atoms are so great that the electrostatic repulsion of the positive nuclei is overcome and the atoms undergo fusion. A typical fusion reaction is:

$$\,^3_1H \; + \; \,^2_1H \rightarrow \,^4_2He \; + \; \,^1_0n \; (14 \text{ MeV}) \tag{6.13}$$

The reaction yields a helium atom and an energetic neutron. For many such reactions to take place within the compressed pellet, the high temperature must be maintained for about 1 ps and the compression must be in the region of 10^4 : 1. These conditions require enormous laser pulse energies.

Neutron numbers in excess of 10^9 per pulse have been observed but this is considerably lower than that required to reach 'scientific breakeven', which is defined as the level at which the thermonuclear energy generated equals the laser energy input. Calculations indicate that a laser input pulse of perhaps 10^{14} W with a subnanosecond duration may be necessary to achieve breakeven and also that laser induced fusion may be most efficient in the wavelength range 300–600 nm. A series of lasers with steadily increasing capability has been built over the last few years for this research. These include the Shiva Nd: glass laser at the Lawrence Livermore Laboratory, California, which has 20 amplifier chains each delivering about 1 TW in a 100 ps pulse to the fuel pellets. The individual amplifier chains may have quite contorted geometries to equalize their lengths to ensure that they all deliver their energy to the target at precisely the same instant. The energy supplied to the amplifier chains is derived from a single Nd: glass master laser. A system called Nova, which is twenty times more powerful than Shiva was completed in 1985. Nova can produce pulses of 10^5J in 300 ps (i.e. 3×10^{14}W). It is hoped that Nova will enable scientific breakeven to be attained and to show, at least in principle, that power generation by laser induced fusion is feasible (ref. 6.12).

Although neodymium based lasers are attractive for initial fusion studies, it is unlikely that they will be useful for practical power generation. The main reason for this is that they cannot be cooled rapidly enough to allow the 100 pulses-per-second repetition rate which may be required. Gas lasers can be cooled quickly by convection and the use of CO_2 lasers is being studied at the Los Alamos Scientific Laboratory. However, the CO_2 laser wavelength may be less favorable than shorter wavelengths. Hence the excimer lasers discussed in sec. 5.10.3.3, which produce many wavelengths in the 300–600 mm range, are viewed as possible candidates for the next generation of fusion lasers.

Problems

6.1 If the halfwidth of the He–Ne 632.8 nm transition is 1500 MHz, what must be the length of the laser cavity to ensure that only one longitudinal mode oscillates? Estimate the accuracy to which the temperature must be controlled if the frequency stability is to be better than 10^8 Hz. Take the coefficient of expansion of the laser tube to be $10^{-6} \text{C}^{\circ-1}$.

6.2 If a cavity mode burns holes in the gain versus velocity curve for the 632.8 nm He–Ne transition at the spectral halfwidth points (halfwidth = 1500 MHz), what is the velocity of the atoms involved in the hole burning?

6.3 If one of the mirrors in the Michelson interferometer moves with velocity v, show that the rate at which the fringes cross the field of view is $2v/\lambda$. Show that this result can be obtained by calculating the frequency of the beats generated between the light reflected from the stationary mirror and the moving mirror. (The Doppler shift of the frequency of light reflected from moving objects is the basis of many laser velocimeters – see ref. 6.4e, pp 306–15 and 334.)

6.4 Calculate the Doppler frequency shift for light ($\lambda = 500$ nm) reflected from an object moving at 20 m s^{-1}; what implications does your answer have for the frequency stabilization of lasers used in Doppler velocimeters?

6.5 Explain how a laser may be used to measure the width of a narrow slit from the Fraunhofer diffraction pattern produced by the slit. The second minimum in the diffraction pattern, which is formed in the focal plane of a lens of focal length 0.5 m, is 2 mm from the central maximum. What is the width of the slit (take $\lambda = 632.8$ nm)? Discuss how this technique may be extended to the measurement of the diameter of a thin wire and hence used in controlling the diameter of the wire during production.

6.6 A CW argon laser emits 1 W at $\lambda = 488$ nm; if the beam divergence is 0.5 mrad and the diameter of the beam at the output mirror is 2 mm, calculate the brightness of the laser. To what *photometric* brightness (or luminance) does this correspond?

6.7 Calculate the pulse width produced by a mode locked Nd: glass laser assuming that the laser cavity is 0.2 m long and that there are 3000 participating longitudinal modes. What is the time separation of the pulses?

6.8   If the halfwidth of the 10.6 μm transition of a low pressure CO_2 laser is 60 MHz, calculate the coherence length of the laser. If the cavity length is 2 m, show that not more than one mode will oscillate. If we take the width of the Fabry–Perot resonances as an (over)estimate of the spectral width of the mode, calculate the coherence length – take the mirror reflectance R to be 0.95. (See ref. 1.1c, pp 90–96, for example.)

6.9   What is the total energy release in the nuclear reaction given in eq. (6.13)? Why does this value differ from that given in the text for the energy of the neutron? (You will need a table of nuclear masses.)

References

6.1   (a) G. M. S. Joynes and R. B. Wiseman, 'Techniques for single mode selection and stabilisation in helium–neon lasers,' in H. G. Jerrard (Ed.), *Electro-optics/Laser International '80 UK* (Conference Proceedings) IPC Science & Technology Press, 1980, p. 163.

(b) G. E. Moss, 'High power single-mode HeNe laser', *Appl. Opt.*, **10**, 1971, 2565.

6.2   E. E. Basch (ed.), *Optical Fiber Transmission*, Howard W. Sams., Indianapolis, 1987, Chapter 9, section 4.

6.3   A. Yariv, *Introduction to Optical Electronics*, Holt, Rinehart & Winston, New York, 1971, pp 120–5.

6.4   (a) J. Wilson and J.F.D. Hawkes, *Lasers: Principles and Applications*, Prentice Hall, New York, 1987.

(b) M. Ross (ed.), *Laser Applications* Vol. 1, Academic Press, New York, 1972.

(c) S. S. Charschan (ed.), *Lasers in Industry*, Van Nostrand–Reinhold, New York, 1972.

(d) J. E. Harry, *Industrial Lasers and their Applications*, McGraw-Hill, Maidenhead, 1974.

(e) J. F. Ready, *Industrial Applications of Lasers*, Academic Press, New York, 1978.

(f) *Laser Focus*, Advanced Technology Publications, Newton, Mass., (a controlled circulation publication available to those working in laser-related fields).

(g) *Electro-Optical Systems Design*, Kiver Publications, Chicago (a controlled circulation publication).

6.5   M. V. Klein and T. E. Furtak, *Optics* (2nd edn), Wiley, New York, 1986, Chapter 8.

6.6   M. Anson, 'Laser speckle vibrometry: a technique for analysis of small vibrations', *Proc. Physiol. Soc. (Lond.)*, **300**, 1980, 8P.

6.7   (a) A. Yariv, *op. cit.*, Chapter 5.

(b) J. E. Harry, *Industrial Lasers and their Applications*, McGraw-Hill, Maidenhead, 1974, pp 111–40.

(c) J. F. Ready, *op. cit.*, Chapters 13–16.

6.8   R. N. Zare, 'Laser separation of isotopes', *Scientific American*, Feb. 1977.

6.9   (a) A. Yariv, *op. cit.*, Chapter 4.

(b) J. F. Ready, *op. cit.* Chapter 11.

6.10   D. Gabor, 'A new microscopic principle', *Nature,* **4098**, 1948, 777.

6.11   (a) G. W. Stroke, *An Introduction to Coherent Optics and Holography* (2nd edn), Academic Press, New York, 1969.

(b) R. H. Collier, C. B. Burkhardt and L. H. Lin, *Optical Holography*, Academic Press, New York, 1972.

(c) H. M. Smith, *Principles of Holography*, Wiley-Interscience, New York, 1975.

(d) Yu I. Ostrovsky, *Holography and its Applications*, Mir Publications, Moscow, 1977.

(e) M. Wenyon, *Understanding Holography*, Arco Publishing Company, New York, 1978.

6.12 (a) R. O. Goodwin, W. F. Hagen, J. F. Holzricher, W. W. Simmons and J. B. Trenholme, 'Livermore builds the Nova laser', *Laser Focus*, **17**(5), 1981, 58–64.

(b) J. L. Emmett, J. Nuckolls and L. Wood, "Fusion power by laser implosion", *Scientific American*, **230**, 1974, 24–37.

(c) R. O. Goodwin & W. W. Simmons "How Industry helped build Nova, the world's largest high precision optical project", *Laser Focus* **21** 1985 pp 78–88.

Chapter 7

Photodetectors

Introduction

The optical detectors in this chapter may be classified as either *thermal* or *photon* devices. In thermal detectors, the absorption of light raises the temperature of the device and this in turn results in changes in some temperature-dependent parameter (for example electrical conductivity). As a consequence, the output of thermal detectors is usually proportional to the amount of energy absorbed per unit time by the detector and, provided the absorption efficiency is the same at all wavelengths, is independent of the wavelength of the light. In photon detectors, on the other hand, the absorption process results directly in some specific quantum event (such as the photo-electric emission of electrons from a surface) which is then 'counted' by the detection system. Thus the output of photon detectors is governed by the rate of absorption of light quanta and not directly on their energy. Furthermore, all the photon processes considered here require a certain minimum photon energy to initiate them. Since the energy of a single photon is given by $E = h\nu = hc/\lambda$ (see eq. (1.2)), photon detectors have a long wavelength 'cut-off', that is a maximum wavelength beyond which they do not operate.

A problem encountered with photon detectors operated in the infrared is that the photon energies involved become comparable with the average thermal energies ($\approx kT$) of atoms in the detector itself. A relatively large number of quantum events may then be generated by thermal excitation rather than by light absorption and will thus constitute a source of noise. The obvious way to reduce this noise signal is to reduce the temperature of the detector; indeed most photon detectors operating above a wavelength of 3 μm or so must be cooled to liquid nitrogen temperatures (77 K) or below.

7.1 Thermal detectors

To gain an insight into the performance characteristics of thermal detectors, we consider the behavior of the simple model shown in Fig. 7.1. The incoming radiation is

Fig. 7.1 Model of a thermal detector used to derive the frequency response characteristics. The incoming radiation causes the instantaneous temperature of the sensing element to be $T_s + \Delta T$. The element is connected via a conducting link (of conductance G) to a heat sink which remains at the temperature T_s.

absorbed within the sensing element of heat capacity H. This is connected to a heat sink, at constant temperature T_s, via a heat conducting link which has a thermal conductance G. If the instantaneous rate of heat absorption is given by W, then during a small time interval δt the heat absorbed is $W\delta t$. If we let the temperature of the element be $T_s + \Delta T$, then during the same time interval the amount of heat lost through the thermal link is $G\Delta T\delta t$. The difference between these two represents the amount of heat available to raise the temperature of the element. Hence we may write:

$$W\delta t - G\Delta T\delta t = H\delta(\Delta T)$$

If we take the limit $\delta t \to 0$ we obtain

$$W = H\frac{d(\Delta T)}{dt} + G\Delta T \tag{7.1}$$

Now suppose that W has a time dependence given by $W = W_0 + W_f \cos(2\pi ft)$ where $W_0 \ll W_f$ and also that ΔT can similarly be written $\Delta T = \Delta T_0 + \Delta T_f \cos(2\pi ft + \phi_f)$. By substituting these relations into eq. (7.1), it may be verified by the reader (see Problem 7.2) that ΔT_f is given by

$$\Delta T_f = \frac{W_f}{(G^2 + 4\pi f^2 H^2)^{\frac{1}{2}}} \tag{7.2}$$

For good sensitivity, it is obviously desirable to have as large a value of ΔT_f as possible; inspection of eq. (7.2) shows that this implies small values for both H and G. This may be achieved by using thin absorbing elements of small area (to reduce H) which have minimal support (to reduce G). We may thus expect high sensitivity devices to be rather fragile.

Looking now at the frequency characteristics, we may rewrite eq. (7.2) as

$$\Delta T_f = \frac{W_f}{G(1 + 4\pi f^2 \tau_H^2)^{\frac{1}{2}}}$$

where τ_H, the *thermal time constant*, is given by:

$$\tau_H = \frac{H}{G} \tag{7.3}$$

For good response at a frequency f we require:

$$\tau_H \ll \frac{1}{2\pi f}$$

Thus once H has been fixed (from size considerations) then G cannot be made too small, otherwise the response time may become unduly long. Typical values for τ_H found in practice usually range from 10^{-3} s upwards, although smaller values can be achieved.

The limiting sensitivity of thermal detectors is governed by temperature fluctuations within the detector, which arise from random fluctuations in the energy flow rate out of the element. It may be shown (ref. 7.1) that the root mean square (r.m.s.) fluctuations in the power (ΔW_f), flowing through a thermal link, which have frequencies between f and $f + \Delta f$ can be written

$$\Delta W_f = (4kT^2G)^{\frac{1}{2}}\Delta f \tag{7.4}$$

The smallest value of G obtainable is when energy exchange takes place by means of radiative exchange only. In this case, the minimum detectable power at room temperature for a 100 mm^2 area detector is about 5×10^{-11} W (see Problem 7.3). The best detectors available approach to within an order of magnitude of this figure at room temperature, and the performance of some can be improved further by cooling.

The receiving element is often in the form of a thin metal strip with a suitably absorbant surface coating such as 'gold black' (an evaporated film of gold which is uniformly absorbant at wavelengths from the UV well into the IR). Mounting the element in a vacuum enclosure gives increased stability due to isolation from air movement, although window transmission can then be a problem if a wide wavelength range is desired. To overcome drift in the output caused by changes in the ambient temperature, we can take the difference in output of two identical detectors in close proximity, only one of which is exposed to the incident radiation.

When very large amounts of radiation are encountered (such as in the outputs from high power lasers), more massive detector elements are used; these are often in the form of stainless steel disks or cones.

Because of their relative unimportance in the field of optoelectronics we deal only briefly with a few of the better known types of thermal detector; for a more detailed discussion, the reader is referred to ref. 7.1.

7.1.1 Thermoelectric detectors

Thermoelectric detectors use the principle of the thermocouple (i.e. the Seebeck effect) whereby the heating of one junction between two dissimilar metals relative to the other

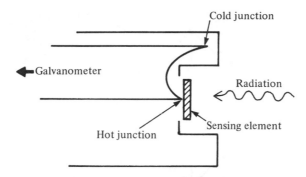

Fig. 7.2 Thermoelectric detector: the temperature change of the sensing element induced by the absorption of radiation is detected using a thermocouple, one junction of which is attached to the sensing element, the other is shielded from the radiation.

causes a current to flow round the circuit which is proportional to the temperature difference between the junctions. In thermoelectric detectors one junction is used to sense the temperature rise of the receiving element whilst the other is maintained at ambient temperature, as shown in Fig. 7.2. A rather more sensitive detector may be made by connecting several thermocouples together in series; the device is then known as a *thermopile*. Efficient operation calls for materials with large electrical conductivities (to minimize Joule heating effects) and also small thermal conductivities (to minimize heat conduction losses). These two requirements are usually incompatible, and a compromise has to be reached. Although metals are most often used for the junction materials, certain heavily doped semiconductors can offer improved sensitivity, but are generally less robust and give rise to constructional problems. The usefulness of thermoelectric detectors lies in their simplicity and their rugged construction.

7.1.2 The bolometer

In the bolometer the incident radiation heats a fine wire or metallic strip causing a change in its electrical resistance. This may be detected in several ways, for example the element may be inserted into one arm of a Wheatstone bridge (Fig. 7.3), or in place of the photoconductor in the circuit of Fig. 7.18. Care must be taken to ensure that any currents flowing through the element are sufficiently small not to raise its temperature by a significant amount. The main parameter of interest in assessing the performance is the temperature coefficient of resistance α, which is given by

$$\alpha = \frac{1}{\varrho} \frac{\mathrm{d}\varrho}{\mathrm{d}T}$$

where ϱ is the resistivity of the material and T the temperature. The resistivity of metals increases with increasing temperature, and hence for these α will be a positive quantity. Platinum and nickel are the most commonly used, and both have α values of about 0.005 K^{-1}. Greater sensitivity may be achieved by using semiconducting elements, which are sometimes called *thermistors*. These consist of oxides of manganese,

Fig. 7.3 Wheatstone bridge circuit incorporating a bolometer radiation sensing element. When the resistance values are such that $R_1/R_2 = R_3/R_4$ then the current i_G through the galvanometer is zero. If, however, the sensing element resistance changes slightly then a current will flow which is proportional to the resistance change.

cobalt or nickel and they have α values of about -0.06 K^{-1} (for these materials, α is dependent on temperature). The negative sign arises because of the characteristic decrease in resistivity with increasing temperature of semiconductors above a certain temperature (see eq. (2.24) and the discussion following eq. 2.36)).

Carbon resistance bolometers cooled to liquid helium temperature (4.2 K) have proved successful in far-infrared astronomy where very sensitive detectors are required.

7.1.3 Pneumatic detectors

The receiving element in a pneumatic detector is placed inside an airtight chamber. Radiation falling on the element causes the air temperature inside the chamber to rise and hence the air pressure to increase. This pressure increase may be detected in several ways. In one of the most sensitive detectors, the Golay cell, a wall of the chamber has a hole in it covered by a flexible membrane silvered on its outside surface. This acts as a mirror whose focal length depends on the pressure within the chamber (see Fig. 7.4). A beam of light originating from a source S passes through a grating, is then reflected from the flexible mirror to re-pass through the grating, and finally is directed onto a detector D. When no radiation is being absorbed within the chamber, the optics are arranged so that an image of the transmitting region of the grating is superimposed on a nontransmitting region and there is then no output from D. However, if the mirror changes its curvature slightly, light will be transmitted through the grating and recorded by D. The output from D is then proportional to the amount of radiation absorbed within the chamber. Golay cells are available which can detect radiation powers down to 10^{-11} W; they are, however, rather fragile and difficult to set up.

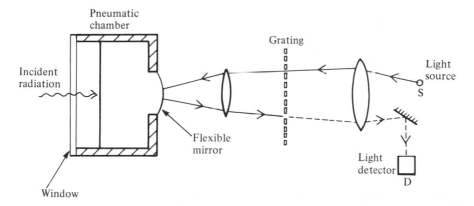

Fig. 7.4 Schematic diagram of a Golay cell detector. A beam of light originating from a source S passes through a grating. It is then reflected from a flexible mirror which forms part of the wall of a pneumatic chamber. The beam subsequently re-passes through the grating and is directed onto a light detector D. Radiation absorbed within the chamber causes pressure fluctuations which in turn cause the curvature of the flexible mirror to change.

7.1.4 Pyroelectric detectors

Pyroelectric detectors are a comparatively recent development, and while they do not have the same sensitivity as the Golay cell they can be made with very rapid response times and are very robust. The incident radiation is absorbed in a ferroelectric material which has molecules with a permanent electrical dipole moment. Below a critical temperature (the *Curie temperature* T_c) the dipoles are partially aligned along a particular crystallographic axis giving rise to a net electrical polarization of the crystal as a whole. When the material is heated, the increased thermal agitation of the dipoles decreases the net polarization, which eventually becomes zero above T_c as shown in Fig. 7.5.

The most sensitive material in use is triglycine sulfate (TGS), but this has an inconveniently low Curie temperature of only 49 °C and more commonly used materials are ceramic based, such as lead zirconate, which have Curie temperatures of several hundred degrees centigrade.

The detector consists of a thin slab of ferroelectric material cut such that the spontaneous polarization direction is normal to the large area faces. Transparent electrodes are evaporated onto these faces and connected together via a high value load resistor (of up to 10^{11} Ω) as shown in Fig. 7.6(a). A temperature change of the ferroelectric causes the spontaneous polarization to vary and hence also the amount of captive surface charge on the faces. Changes in the surface charge induce corresponding changes in the charge on the electrodes, thus causing a current to flow through the load resistor. This in turn results in a changing voltage signal appearing across the load resistor. Radiation of constant irradiance will not cause any change in the charge stored on the electrodes and will consequently not give rise to an output signal. The frequency

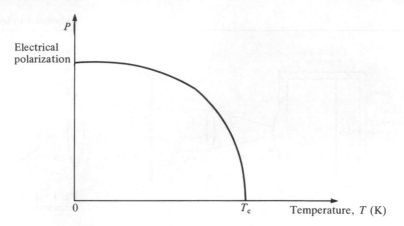

Fig. 7.5 Spontaneous electrical polarization versus temperature for a ferroelectric material (schematic diagram). The polarization falls to zero at the Curie temperature T_c.

Fig. 7.6 (a) Pyroelectric detector. A slab of ferroelectric material is sandwiched between two electrodes (one being transparent). The electrodes are connected by a load resistor R_L. Radiation absorbed within the ferroelectric material causes it to change its polarization. The induced charge on the electrodes changes and current flows through R_L causing a voltage signal to appear across R_L. (b) Equivalent circuit and typical impedance matching circuitry for a pyroelectric detector. The varying amounts of charge stored on the electrodes are equivalent to a current generator feeding into the electrode capacitance C. The load resistor R_L is in parallel with C. Since R_L is usually very high ($\approx 10^9\ \Omega$), an impedance matching circuit is often employed to reduce the signal source impedance. Shown here is a typical circuit using a JFET. The output impedance in this case is then R.

response of the pyroelectric detector is considered in detail in Problem 7.4. At low frequencies, the output rises from zero to reach a plateau when $f > 1/2\pi\tau_H$, where τ_H is the thermal time constant given by eq. (7.3).

At higher frequencies, the electrode capacitance C acts as a signal shunt across the load resistor R_L and the output voltage falls to $\sqrt{\frac{1}{2}}$ of its maximum value at a cut-off frequency f_c given by

$$f_c = \frac{1}{2\pi R_L C} \tag{7.5}$$

As the voltage output is proportional to R_L (again see Problem 7.4) there is a trade-off between sensitivity and frequency response. Typically, a detector with a frequency bandwidth of 1 Hz at an operating frequency of 100 Hz can detect radiation powers of about 10^{-8} W.

Because of the comparatively large values of the load resistor encountered in pyro-electric detectors, an impedance matching circuit is usually built into the detector. A source follower circuit using a JFET is commonly used as shown in Fig. 7.6(b).

Pyroelectric detectors can be made with response times in the nanosecond region and with a wavelength response extending out to 100 μm. They have proved very useful as low cost, robust infrared detectors in such uses as fire detection and intruder alarms.

7.2 Photon devices

7.2.1 Photoemissive devices

When radiation with a wavelength less than a critical value is incident upon a metal surface, electrons are found to be emitted; this is called the photoemissive or photo-electric effect (see section 2.6). When a photon of energy $h\nu$ enters the metal it may be absorbed and give up its energy to an electron. Provided the electron is able to reach the surface and has enough energy to overcome the surface potential barrier (given by $e\phi$, where ϕ is the surface work function) it may escape and photoelectric emission takes place as illustrated in Fig. 7.7.

If the electron is initially at the Fermi level, its kinetic energy E on emission is given by:

$$E = h\nu - e\phi \tag{7.6}$$

However, the electron may be initially below the Fermi level and may also suffer inelastic collisions before emission; hence eq. (7.6) represents the maximum energy available to the emitted electrons. No electrons at all will be emitted when $h\nu < e\phi$ (or, in terms of wavelength, $\lambda > hc/e\phi$). If the probability of inelastic collisions of the excited electrons is high, then only a fraction of them may be able to escape. The ratio of the number of emitted electrons to the number of absorbed photons is called the *quantum yield* or *quantum efficiency*.

Pure metals, however, are rarely used as practical photocathodes since they have low quantum efficiencies ($\sim 0.1\%$), and high work functions. (Caesium has the lowest value for ϕ, which is 2.1 eV.) We may divide practical photoemissive surfaces into two

Fig. 7.7 Electron energy level diagram at a metal–vacuum interface illustrating the photoelectric effect. To escape from inside a metal, an electron must gain at least an energy $e\phi$, where ϕ is the work function expressed in electron volts.

groups (a) the older *classical* types and (b) the newer *negative electron affinity* (NEA) types. The former consist of a thin evaporated layer containing compounds of alkali metals (usually including Cs) and one or more metallic elements from group V of the periodic table (e.g. Sb). They are often designated by an 'S' number. To some extent we may regard them as semiconductors, and hence most electrons must gain an energy of at least $E_g + \chi$ (where E_g is the energy gap and χ is the electron affinity) to escape from the surface (see Fig. 2.17(b)). For example, in the material NaKCsSb ('S20') $E_g \simeq 1$ eV and $\chi \simeq 0.4$ eV, and hence it should have a threshold at a photon energy of about 1.4 eV, which is indeed close to that observed.

It is possible, however, to reduce the effective value of χ, as far as the bulk electrons are concerned, provided that *band bending* takes place at the surface. This is found to occur when there are states within the energy gap at the semiconductor surface. If we suppose that a large number of these are close to the valence band, they will fill with holes and lead to a local depletion in the hole population. The resulting uncovering of negatively charged acceptor ions in the vicinity will lead to the formation of a depletion region at the surface very similar to that formed within the p material at a p–n junction. The potential drop across the depletion region leads to band bending as shown in Fig. 7.8. If we write the potential drop as V_s, then the effective electron affinity χ_{eff}, as far as the electrons in the bulk are concerned, is given by

$$\chi_{\text{eff}} = \chi - V_s$$

If $V_s > \chi$ then we have a *negative electron affinity* and the effective work function for bulk electrons is just E_g.

In practice, NEA photocathodes are formed by evaporating very thin layers of caesium or caesium oxide onto the semiconductor surface. The resulting band structure is more complex than shown in Fig. 7.8 but the essential features remain. Photocathodes using GaAs have been used successfully, these operate with quite high

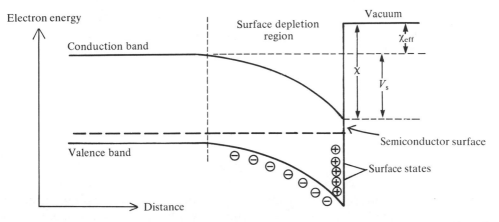

Fig. 7.8 The effective electron affinity of a semiconductor may be altered if band bending takes place at the surface. Here holes trapped in surface states cause a surface depletion region to be formed. The potential drop V_s across the depletion region reduces the effective electron affinity for bulk electrons from χ to $\chi - V_s$.

quantum efficiencies up to a wavelength corresponding to the energy gap of GaAs (≈ 0.9 μm). So far, it has not proved possible to make NEA photocathodes which operate much above 1.1 μm with any reasonable quantum efficiencies. There is no doubt, however, that the NEA photocathodes will prove to be superior to the 'classical'

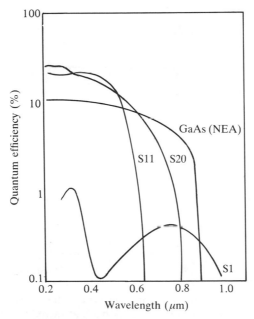

Fig. 7.9 Quantum efficiency versus wavelength for a number of the more common photocathode materials.

types in the near-infrared, although the lower cost of the classical types will ensure that they will continue to be used in the visible region. The quantum efficiencies of a number of the more common photocathode materials are shown as a function of wavelength in Fig. 7.9.

7.2.2 Vacuum photodiodes

In the vacuum photodiode, a photoemissive surface (the *photocathode*) is placed inside a vacuum tube with another electrode (the anode) placed nearby and biased positively with respect to it, as shown in Fig. 7.10. When the photocathode is illuminated, the emitted electrons will be collected by the anode and a current will flow in the external circuit. If the bias voltage is large enough (in practice a few hundred volts), all the emitted electrons will be collected and the current will be almost independent of bias voltage and also will be proportional to the light intensity.

Generally, the output of simple photodiodes is relatively small and requires amplification, although photodiodes with a high current capability can be useful when examining high intensity laser pulses. There are, however, two methods by which useful internal gain may be achieved within the device itself. In the *gas filled phototube* the tube contains a gas such as argon under low pressure (≈ 1 torr or less). Photoelectrons on their way from the cathode to the anode will collide with gas atoms, and provided they are sufficiently energetic may ionize them and so generate further electrons. The overall current gain may be typically of the order of 10 and will obviously depend fairly

Fig. 7.10 Schematic diagram of a photoelectric cell. Electrons released from the photocathode surface by the incident radiation are attracted to the anode, thus causing a current to flow round the external circuit. A voltage will then appear across the load resistor R_L.

critically on the anode voltage. The other, and much more popular, method of obtaining internal gain is achieved in the *photomultiplier*.

7.2.3 Photomultipliers

In the photomultiplier, the photoelectrons are accelerated towards a series of electrodes (called *dynodes*) which are maintained at successively higher potentials with respect to the cathode. On striking a dynode surface, each electron causes the emission of several secondary electrons which in turn are accelerated towards the next dynode and continue the multiplication process. Thus, if on average δ secondary electrons are emitted at each dynode surface for each incident electron, and if there are N dynodes overall, then the total current amplification factor between the cathode and anode is given by

$$G = \delta^N \tag{7.7}$$

Considerable amplification is possible; if we take, for example, $\delta = 5$ and $N = 9$, we obtain a gain of 2×10^6. The variation of δ with interdynode voltage for a typical dynode material is illustrated in Fig. 7.11.

Four of the most common photomultiplier dynode configurations are illustrated in Fig. 7.12. Three of them (venetian blind, box and grid, and linear focused) are used in 'end on' tubes. These have a semitransparent cathode evaporated onto the inside surface of one end of the tube envelope. The photoelectrons are emitted from the opposite side to that of the incident radiation. Obviously, in this arrangement the thickness of the photocathode is very critical. If it is too thick, few photons will penetrate to the electron emitting side, whilst if it is too thin few photons will be absorbed.

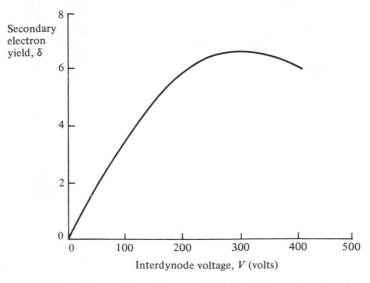

Fig. 7.11 Variation of secondary electron yield δ with interdynode voltage for Be−Cu dynodes.

Fig. 7.12 Dynode structures of four common types of photomultiplier: (a) venetian blind, (b) box and grid, (c) linear focused and (d) circular cage focused. Typical trajectories of an electron through the systems are also shown.

In the venetian blind type, electrons strike a set of obliquely placed dynode slats at each dynode stage; the electrons are attracted to the next set of slats by means of the interdynode potential applied between a thin wire grid placed in front of the slats. This arrangement is compact, relatively inexpensive to manufacture and is very suitable for large area cathodes. The box and grid type (Fig. 7.12(b)) is somewhat similar in performance. In both of these, very little attempt is made to focus the electrons, which is in contrast to the linear focused and circular cage focused types (Figs. 7.12(c) and (d)), where some degree of electron focusing is obtained by careful shaping and positioning of the dynodes.

The focused types have somewhat higher electron collection efficiencies and a much better response to high signal modulation frequencies (we discuss frequency response later in this section). The circular cage focused type is very compact and usually used in conjunction with a side window geometry. In this, the photocathode material is deposited on a metal substrate within the glass envelope and the photoelectrons are emitted from the same side of the cathode as that struck by the incident radiation.

The dynode potentials are usually provided by means of the circuit shown in Fig. 7.13. Care must be taken to ensure that the voltage between the cathode and the first dynode is large enough to maintain proportionality between cathode current and cathode illumination. Usually a voltage value is recommended for a particular tube, and in some circumstances (for example when examining fast pulses) it may be preferable to use a Zener diode in place of the fixed resistor R_k in Fig. 7.13 to keep the voltage at this value.

The intermediate stages usually operate satisfactorily over quite a wide voltage range provided the voltage is distributed uniformly. To maintain this uniformity, the current flowing down the dynode chain must be considerably larger (say 100 times) than the anode current. If high anode currents are likely, then the last few stages may also be biased using Zener diodes.

The photomultiplier responds to light input by delivering charge to the anode. This

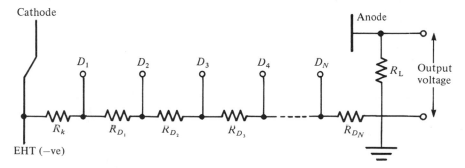

Fig. 7.13 Dynode biasing circuit using a linear resistor chain. The EHT voltage supply is applied across a resistor chain (R_k, R_{D1}, R_{D2}, etc.) which acts as a potential divider and maintains the dynodes (D_1, D_2, D_3, etc.) at increasingly higher positive potentials relative to the cathode. When an amplified signal current pulse arrives at the anode, it flows through the anode load resistor R_L causing a voltage to appear across it.

charge may be allowed to flow through a resistor R_L or to charge a capacitor; the corresponding voltage signal then provides a measure of the input signal. If individual pulses need to be examined (as for example when using photon counting techniques, which we discuss in section 7.2.5), then it is important to ensure that the response time of the external circuitry is less than that of the pulse rise time. This usually implies a low value for the load resistor.

7.2.3.1 SPEED OF RESPONSE

The electrons take a finite time (the *electron transit time*) to traverse the dynode chain from cathode to anode, although in fact this time will not be the same for all the electrons. There are two main reasons for the spread in transit times; first, the electrons have a spread in velocities when they are ejected from the cathode and subsequent dynodes; secondly, they traverse slightly different paths (in different electric fields) through the photomultiplier. It is this spread in transit times that limits the ability of the photomultiplier to respond faithfully to a fast optical pulse. Transit time spread may be reduced by using fewer dynodes and designing these so that electrostatic focusing gives the electrons very similar paths in nearly identical fields (as for example in the linear focused type shown in Fig. 7.12(c)). Overall tube gain may be maintained when using fewer dynodes if the interdynode voltages are relatively high and if the dynode surfaces have high secondary electron emission coefficients. Dynode surface materials are now available which have much higher values of δ than that of the hitherto commonly used Be–Cu. Photomultiplier tubes with rise times of about 2 ns and transit times of about 30 ns are now readily available. In terms of the structures illustrated in Fig. 7.12, the order of increasing speed of response is: box and grid, venetian blind, circular focused and linear focused.

7.2.4 Noise in photomultipliers

All electrical systems suffer from noise, i.e. a randomly varying output signal unrelated to any input signal. In a photomultiplier there are several sources of noise which we now discuss.

7.2.4.1 DARK CURRENT

Even when no radiation is falling onto the photocathode surface, thermionic emission gives rise to a 'dark current' which often constitutes the main source of noise in photoemissive devices. The thermionic emission current i_T for a cathode at temperature T of area A and work function ϕ is given by the Richardson–Dushman equation (ref. 7.2):

$$i_T = aAT^2 \exp\left(-\frac{e\phi}{kT}\right) \tag{7.8}$$

Here a is a constant which for pure metals has the value 1.2×10^6 A m^{-2} K^{-2}. Thermionic emission may obviously be reduced by reducing the temperature and indeed this is often essential, especially for low work function photocathodes, although other sources of noise may dominate at low temperatures (for example electrons may be emitted from the photocathode by radioactive bombardment). If the dark current were absolutely constant, it would be a relatively simple matter to subtract it from the total output; however, it is itself subject to random fluctuations due to the statistical nature of thermionic emission. The root mean square variation in the current is given by the same equation as for shot noise which is discussed next.

7.2.4.2 SHOT NOISE

Shot noise is encountered whenever there is current flow and arises directly from the discrete nature of the electronic charge. Thus when a current flows past any point in a circuit the arrival rate of electrons will fluctuate slightly and gives rise to fluctuations in the current flow at that point. It may be shown (ref. 7.3) that the magnitude of the r.m.s. current fluctuations Δi_f with frequencies between f and $f + \Delta f$ is given by:

$$\Delta i_f = (2ie\Delta f)^{\frac{1}{2}} \tag{7.9}$$

where i is the current flowing, which in the case of the photomultiplier is the sum of the dark current and the signal current leaving the photocathode.

The presence of dark current shot noise sets a limit on the minimum detectable signal. We assume that if an optical signal results in an electrical output signal that is smaller than the noise signal, then it cannot be detected without further processing. We define the *responsivity R_λ* of the photomultiplier as i/W, where W is the optical power falling on the photocathode (see section 7.3). Therefore, the minimum detectable signal power in the presence of a thermionic dark current i_T is given in watts by:

$$W_{\min} = \frac{(2i_Te\Delta f)^{\frac{1}{2}}}{R_\lambda} \tag{7.10}$$

Example 7.1 Minimum detectable signal for a photomultiplier

To calculate the minimum signal power from eq. (7.10) we need values for i_T, R_λ and Δf. The dark current may be estimated from eq. (7.8). If we assume a cathode area of 1000 mm^2 together with a work function ϕ of 1.25 eV, we then have:

$$i_T = 1.2 \times 10^6 \times 10^{-3} \times (300)^2 \exp(-1.25/0.025)$$
$$= 2 \times 10^{-14} \text{ A}$$

Next we calculate the responsivity R_λ. We assume a quantum efficiency η of 0.25 at a wavelength of 0.5 µm, and so we have that

$$R_\lambda = \frac{\eta e\lambda}{hc} = \frac{0.25 \times 1.6 \times 10^{-19} \times 0.5 \times 10^{-6}}{6.6 \times 10^{-34} \times 3 \times 10^8} = 0.1 \text{ A W}^{-1}$$

Finally, if we take a bandwidth of 1 Hz, then from eq. (7.10) we obtain

$$W_{\min} = \frac{(2 \times 2 \times 10^{-14} \times 1.6 \times 10^{-19} \times 1)^{\frac{1}{2}}}{0.1} = 8 \times 10^{-16} \text{ W}$$

7.2.4.3 MULTIPLICATION NOISE

It is found that the current noise at the anode is always greater than that expected from shot noise alone, (i.e. $G(2ie\Delta f))^{\frac{1}{2}}$. The reason for this is a statistical spread in the secondary electron emission coefficient about the mean value δ which causes the anode current noise to be increased by a factor $(\delta/(\delta - 1))^{\frac{1}{2}}$ (ref. 7.4). This factor is only appreciable when δ is near to unity; for a typical value of $\delta = 4$ we see that the noise current is increased by some 15%.

7.2.4.4 JOHNSON (OR NYQUIST) NOISE

Johnson noise arises because of the thermal agitation of charge carriers within a conductor; the random nature of this motion results in a fluctuating voltage appearing across the conductor. The r.m.s. value of this voltage, ΔV_f, having frequency components between f and $f + \Delta f$ across a resistance R at a temperature T is given by (see ref. 7.5):

$$\Delta V_f = (4kTR\Delta f)^{\frac{1}{2}} \tag{7.11}$$

In a photomultiplier, such a noise voltage will appear across the anode load resistor (R_L in Fig. 7.13). In practice, Johnson noise is often smaller than the dark current shot noise. For example, with a load resistor of $10^4 \ \Omega$ at 300 K and a bandwidth Δf of 1 Hz then $V_f = 1.3 \times 10^{-8}$ V. On the other hand, taking a dark current of 10^{-16} A gives a shot noise of about 6×10^{-18} A (eq. (7.9)). If the photomultiplier gain is 10^7 then the shot noise voltage signal appearing across the load resistor is about 6×10^{-7} V.

7.2.5 Photon counting techniques

A useful technique for dealing with very low level signals is photon counting. If the cathode quantum efficiency is unity, then each photon striking the photocathode gives rise to a single current pulse at the anode. It has been found for certain photomultiplier designs that the magnitude of these current pulses lies within fairly closely defined limits, whereas the noise pulses have a much wider amplitude distribution. A typical pulse height distribution containing both signal and noise is shown in Fig. 7.14.

A photon counting system basically consists of a fast pulse counter coupled to a pulse height discriminator. Pulses are ignored by the counter when they lie outside certain preset limits (e.g. the discriminator levels shown on Fig. 7.14). Thus, by only accepting those signals that lie within the fairly well-defined signal peak, a significant improvement in signal-to-noise ratio is possible. If the light signal being detected is 'chopped' on and off then a further reduction in background noise level may be obtained by subtracting the count rate when the signal is off from the count rate when both signal and noise counts are present.

7.2.6 Image intensifiers

As their name implies, image intensifiers are designed to boost very low intensity optical images to the point where they become useful. They can also act as wavelength

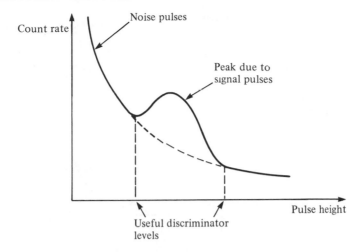

Fig. 7.14 Amplitude distribution of pulses from a photomultiplier designed for photon counting. Signal pulses have a relatively restricted range of pulse heights, whereas noise pulses show a much wider range of pulse heights. Discriminator levels can therefore be set to accept most of the signal pulses while a large proportion of the noise pulses will be rejected.

'downconverters', i.e. they can convert near-infrared radiation into visible radiation. The primary image is formed on a photocathode surface and the resulting photoelectron current from each point on the image is then intensified either by increasing the energy of the individual electrons (in the so-called *first generation* types) or by increasing the actual numbers of electrons (in *second generation* types). The electrons subsequently fall onto a cathodoluminescent phosphor screen to produce the intensified image. A schematic diagram of a typical first generation type is shown in Fig. 7.15.

The electrons may be focused onto the screen by either electrostatic or magnetic means; usually the former is used for reasons of simplicity. An accelerating potential is applied between the cathode and the phosphor screen, thus increasing the electron energy from a few electron volts to 10 keV or so. Luminance gains of up to 2000 may be achieved quite readily, while higher gains are possible by cascading two or more units with fiber optic coupling between them.

Second generation devices make use of the so-called *microchannel plate* to achieve gain through electron multiplication. The microchannel plate consists of a slab of insulating material ($\sim 500\ \mu$m thick) with a high density of small diameter ($\sim 15\ \mu$m) holes or *channels* in it. The inner surfaces of the channels are made slightly conducting and a potential (~ 1 kV) is applied between opposite faces of the slab. Electrons entering one of the channels are accelerated down it and strike the walls soon after entering, since the axis of the channel is slightly inclined to the electron trajectory. As in the photomultiplier, secondary electrons are generated by the impact and the process is repeated along the channel as illustrated in Fig. 7.16.

Focusing may be achieved most simply by placing the microchannel plate in close proximity to both the photocathode and the phosphor screen; this arrangement gives a

Fig. 7.15 Schematic diagram of a first generation image intensifier. Electrons are released from the photocathode and accelerated towards the anode which is coated with a phosphor layer. Electrostatic focusing elements ensure that electrons released from a certain spot on the photocathode are all focused onto a corresponding spot on the phosphor screen. Both the photocathode and the phosphor screen are curved, and fiber optical coupling can be used to convert flat images into curved images and vice versa. On striking the phosphor screen, light is generated by cathodoluminescence.

Fig. 7.16 Electron multiplication in a microchannel plate image intensifier. When the primary electrons enter the channel and strike the walls, secondary electrons are emitted that in turn generate further secondaries. The channel thus acts like a miniature photomultiplier tube. On emerging from the channel, the electrons generate light by striking a phosphor screen.

very compact device. Alternatively, to achieve yet higher gain, the microchannel plate may be placed just in front of the screen in a first generation image intensifier. In some instances an electronic output may be required rather than a visual one, in which case the luminescent screen may be replaced by an array of electron detectors such as silicon diodes.

7.2.7 Photoconductive detectors

As we showed in Chapter 2, an electron may be raised from the valence band to the conduction band in a semiconductor where the energy gap is E_g by the absorption of a photon of frequency v provided that

$$hv \geqslant E_g$$

or in terms of wavelength

$$\lambda \leqslant \frac{hc}{E_g} \tag{7.12}$$

We define the *bandgap wavelength* λ_g as the largest value of wavelength that can cause this transition. That is

$$\lambda_g = \frac{hc}{E_g} \tag{7.13}$$

As long as the electron remains in the conduction band, it will cause an increase in the conductivity of the semiconductor. This is the phenomenon of *photoconductivity*, which is the basic mechanism operative in photoconductive detectors. For convenience, we suppose the semiconductor material to be in the form of a slab of width W, length L and thickness D with electrodes on opposite ends, as shown in Fig. 7.17. An external potential across the electrodes is usually provided by the simple circuit of Fig. 7.18.

Fig. 7.17 Geometry of slab of photoconductive material. The slab of length L, width W and thickness D has electrodes on opposite faces; radiation falls onto the upper surface.

Fig. 7.18 Photoconductor bias circuit. The photoconductor is placed in a series circuit comprising a voltage source, a load resistor R_L, and the photoconductor itself. Changes in the resistance of the photoconductor cause changes in the voltage appearing across R_L. If only the a.c. component of this voltage is required, then a blocking capacitor C may be placed as shown.

Any change in the conductivity of the detector results in an increased flow of current round the circuit which will increase the potential across the load resistor R_L. This may then be detected using a high impedance voltmeter. If we are only interested in the time-varying part of the incident radiation, then a blocking capacitor C may be inserted in the output line to remove any d.c. component. The optimum size for R_L in a particular situation is determined by the fractional change in the resistance of the photodetector when under maximum illumination. If this is small (say $< 5\%$) then it may be shown (Problem 7.6) that the largest sensitivity is obtained when $R_L = R_D$, where R_D is the photodetector resistance. On the other hand, if it is relatively large, then linearity of output can only be maintained if the potential drop across the load resistor always remains small compared with the potential drop across the photo-conductor. This requires that $R_L \ll R_D$. It is therefore obviously advantageous, as far as the output voltage is concerned, for R_D to be high.

We suppose that the radiation falling normally onto the slab is monochromatic and of irradiance I_0. The transmitted irradiance I is determined by α, the *absorption co-efficient*, via the equation

$$I = I_0 \exp{(-\alpha D)} \qquad (7.14)$$

For wavelengths longer than the bandgap wavelength (λ_g), the absorption coefficient is comparatively small. For wavelengths below λ_g, however, α increases rapidly and can attain values in the region of 10^6 m^{-1} or greater. Figure 7.19 shows the variation of α with wavelength for some important semiconductor detector materials.

Values of α of the order of 10^6 m^{-1} imply that most electron–hole pairs will be generated within a few microns of the semiconductor surface, although for wavelengths nearer the band edge this figure may be considerably larger. For simplicity, we assume that the thickness of the slab is sufficiently large to ensure that practically all of the radiation falling on the slab is absorbed within it (this implies that $D \gg \alpha^{-1}$). Thus, the total number of electron–hole pairs generated within the slab per second is $\eta I_0 WL/h\nu$

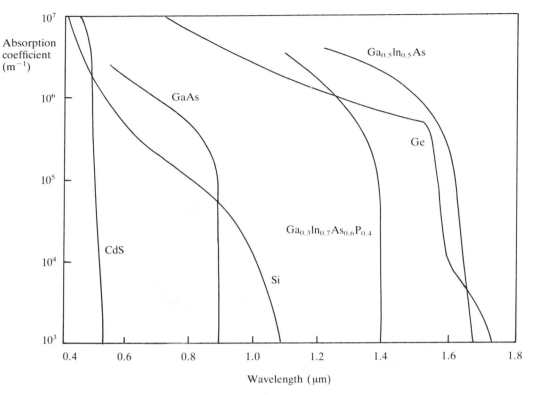

Fig. 7.19 Variation of the optical absorption coefficient α with wavelength for several semiconductor materials.

where η is the quantum efficiency of the absorption process. The average generation rate r_g of carriers per unit volume is then given by:

$$r_g = \frac{\eta I_0 WL}{h\nu WLD}$$

or

$$r_g = \frac{\eta I_0}{h\nu D} \tag{7.15}$$

As we saw in section 2.7, the recombination rate r_r of excess carriers depends on the densities of the excess carrier populations Δn, Δp (where for charge neutrality $\Delta n = \Delta p$) and on the minority carrier lifetime τ_c via the equation:

$$r_r = \frac{\Delta n}{\tau_c} = \frac{\Delta p}{\tau_c} \tag{7.16}$$

In equilibrium the recombination rate must equal the generation rate, and therefore

$$\Delta n = \Delta p = r_g \tau_c \tag{7.17}$$

We may write the conductivity σ of a semiconductor material as (see eq. (2.24))

$$\sigma = ne\mu_e + pe\mu_h$$

Hence under illumination the dark conductivity will increase by an amount $\Delta\sigma$ given by

$$\Delta\sigma = \Delta ne\mu_e + \Delta pe\mu_h$$
$$= r_g\tau_c e(\mu_e + \mu_h) \tag{7.18}$$

The application of a voltage V across the electrodes (Fig. 7.17) will result in a photoinduced current Δi where:

$$\Delta i = \frac{WD}{L}\,\Delta\sigma V$$

Using eq. (7.18) we obtain

$$\Delta i = \frac{WD}{L}\,r_g\tau_c e(\mu_e + \mu_h)V \tag{7.19}$$

We may define an effective quantum efficiency parameter, known as the *photoconductive gain G*, as the ratio of the rate of flow of electrons per second from the device to the rate of generation of electron–hole pairs within the device. That is

$$G = \frac{\Delta i}{e}\,\frac{1}{r_g WDL}$$

or, using eq. (7.19),

$$G = \frac{\tau_c(\mu_e + \mu_h)V}{L^2} \tag{7.20}$$

Unlike the quantum efficiency for, say, the photoelectric effect, G may be larger than unity. It may be increased by increasing V and decreasing L, although at high values of the electric field the current tends to saturate owing to space charge effects (see Problem 7.8). High values of the gain will also be favored by large values of τ_c, although this implies that the response time will be correspondingly poor. There is a good deal of evidence that in some materials (such as CdS) impurity energy levels exist within the energy gap into which carriers may fall but which do not cause recombination; the carrier is merely released by thermal excitation at some later time. Such levels are termed *traps* or *sensitization centers*. Whilst a carrier is held in a trap, a carrier of the opposite type must be present in the semiconductor to maintain charge neutrality – thus the presence of traps further increases the gain. This increase in gain, however, will once again be at the expense of the response time. Under fairly intense background illumination levels and at relatively elevated temperatures, the traps in most materials tend to be full; consequently they have little effect on the photosignal. Significant effects, however, may be observed at low temperatures and low background illumination levels.

Figure 7.20 shows an idealized wavelength response curve for a photoconductive detector expressed in terms of output per unit of incident energy and assuming a

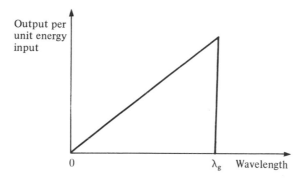

Fig. 7.20 Idealized wavelength response curve for a photoconductive detector. Beyond λ_g the output falls to zero since the photons have insufficient energy to excite carriers across the bandgap.

constant quantum efficiency η for carrier generation when $\lambda < \lambda_g$. In practice, η is found to fall off at short wavelengths. We may explain this by the increase in the absorption coefficient with decreasing wavelength (Fig. 7.19) which results in carriers being generated increasingly closer to the semiconductor surface where there is often a much higher probability of radiationless transitions taking place than in the bulk. Careful surface passivation techniques are needed to obtain good short wavelength response characteristics.

7.2.8 Noise in photoconductive detectors

The main source of noise in photoconductive detectors arises from fluctuations in the rates of generation and recombination of electron–hole pairs, and is termed *generation–recombination* noise. Both optical and thermal excitation processes contribute to generation noise. The relative importance of the thermal processes is strongly dependent on the size of the bandgap, since the probability for thermal excitation of a carrier across the gap is approximately proportional to the factor $\exp\left(-E_g/2kT\right)$ (see the discussion following eq. (2.36)). Thus for detectors capable of operating out to relatively long wavelengths, and which consequently have small energy gaps, thermal generation noise is likely to be large unless the temperature is reduced. As a rough rule of thumb, the operating temperature should be such that $T < E_g/25k$. Provided that the thermal generation noise has been reduced to negligible proportions the r.m.s. noise current fluctuations Δi_f within the frequency range f to $f + \Delta f$ resulting from generation–recombination is given by (ref. 7.6)

$$\Delta i_f(\text{generation–recombination}) = \left(\frac{4ieG\Delta f}{1 + 4\pi^2 f^2 \tau_c^2}\right)^{\frac{1}{2}} \tag{7.21}$$

Here G is the photoconductive gain, τ_c the minority carrier lifetime and i the total current flowing.

The noise current given by eq. (7.21) is almost independent of frequency when $f \ll 1/(2\pi\tau_c)$, whilst for $f > 1/(2\pi\tau_c)$ the noise current declines with increasing

Fig. 7.21 Schematic diagram of the noise spectrum for a photoconductive detector. The frequency f_1 below which '$1/f$' noise becomes predominant is about 1 kHz. Generation–recombination noise has a flat spectrum at frequencies below $1/(2\pi\tau_c)(=f_2)$ where τ_c is the minority carrier lifetime. Above f_2, generation–recombination noise falls with increasing frequency as $1/f$. At the highest frequencies ~ 1 MHz) Johnson noise associated with the circuit load resistor will eventually become predominant.

frequency. At sufficiently high frequencies, the dominant noise mechanism will be Johnson noise.

It is also found that at frequencies less than about 1 kHz a relatively little understood source of noise known as *flicker*, or '$1/f$', noise becomes predominant. Flicker noise is present in most semiconductor devices and, although the cause has not been established with certainty, there appear to be some definite links with trap distributions that are metastable at the device operating temperature. Empirically it is found that the r.m.s. noise current due to flicker noise may be written as

$$\Delta i_f (\text{flicker}) = i \left(B \frac{\Delta f}{f} \right)^{\frac{1}{2}}$$

where B is a constant ($\approx 10^{-11}$). A complete noise spectrum for the photoconductive detector is shown schematically in Fig. 7.21.

7.2.9 Characteristics of particular photoconductive materials

Cadmium sulfide (CdS) and cadmium selenide (CdSe) are both used for low cost, visible radiation sensors, for example in light meters for cameras. These devices usually have high photoconductive gains (some 10^3 to 10^4) but poor response times (about 50 ms). The response time is in fact strongly dependent on the illumination level, being much reduced at high levels, a behavior indicative of the presence of traps. A typical construction is shown in Fig. 7.22. A film of the material in polycrystalline form is

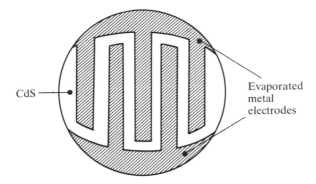

CdS

Evaporated
metal
electrodes

Fig. 7.22 Top view of typical electrode geometry of a CdS photoconductive cell.

deposited on an insulating substrate and the electrodes are formed by evaporating a suitable metal, such as gold, through a mask to give the comb-like pattern shown. This geometry, which results in a relatively large area of sensitive surface and a small interelectrode spacing, helps to give a high sensitivity to the device (see eq. (7.19)).

Lead sulfide (PbS) is a well known near-infrared detector material with a useful wavelength response from 1 to 3.4 μm. By varying the growth conditions, the detector characteristics can be varied widely with the expected trade-off between gain and frequency response. A typical response time would be about 200 μs. In the region of 2 μm, it is one of the most sensitive photodetectors. The wavelength response may be extended to 4 μm by cooling to −30 °C (unlike most other photoconductive materials, the energy gap of PbS decreases with decreasing temperature). However, cooling also has the effect of reducing both the overall gain and the frequency response. The internal dark resistance of these detectors is usually quite high (≈ 1 MΩ). This is an advantage when using the bias circuit of Fig. 7.18, since then comparatively large values for the bias resistor R_L can be used, resulting in relatively high output voltage signals.

Indium antimonide (InSb) detectors are usually formed from single crystals and tend to have low impedances (≈ 50 Ω). Consequently, their output voltages tend to be small. They have a wavelength response extending out to 7 μm and exhibit response times of around 50 ns. Operation at room temperature is possible, but a much improved noise performance results on cooling to liquid nitrogen temperatures (77 K), although the peak wavelength response then shifts to 5 μm.

Mercury cadmium telluride ($Hg_xCd_{1-x}Te$) may be thought of as an alloy composed of the semi-metal HgTe and the semiconductor CdTe. Semi-metals have overlapping valence and conduction bands and may be regarded as having a *negative* bandgap (for HgTe the bandgap is 0.3 eV). Consequently, depending on the composition of the alloy, a semiconductor can be formed with a bandgap varying between zero and 1.6 eV (the bandgap of pure CdTe). Detectors are available whose peak sensitivities lie in the range 5 to 14 μm. This region is of particular importance, since it covers the peak emission wavelengths of bodies at or somewhat above ambient temperatures and also corresponds to a region of good atmospheric transmission (see Fig. 9.6). Usually cooling to 77 K or below is necessary for a satisfactory noise performance, but detectors operating at the lower end of this wavelength range are sometimes used in conjunction with a thermoelectric cooler. Photoconductive gains of up to 500 are possible, while at low temperatures and under low illumination the effects of traps can increase the gain yet further. As with indium antimonide, device resistances tend to be low.

Doped semiconductors, rather than using band-to-band transitions it is also possible to use transitions from impurity levels within the bandgap to the appropriate band edge. Typical among these are zinc and boron doped germanium detectors whose wavelength response can extend from 20 to 100 µm, though cooling to 4 K (liquid helium temperatures) is essential to reduce background noise.

7.2.10 Junction detectors

7.2.10.1 PHOTODIODES

When a *p–n* junction is formed in a semiconductor material, a region depleted of mobile charge carriers is formed with a high internal electric field across it known as the depletion region (see the discussion in section 2.8.1). If an electron–hole pair is generated by photon absorption within this region, the internal field will cause the electron and hole to separate as shown in Fig. 7.23.

We may detect this charge separation in two ways. If the device is left on open circuit, an externally measurable potential will appear between the *p* and *n* regions. This is known as the *photovoltaic* mode of operation. On the other hand, we may short circuit the device externally (or more usually operate it under reverse bias) in which case an external current flows between the *p* and *n* regions. This is known as the *photo-conductive* mode of operation.

The junction will also respond to electron–hole pairs generated away from the depletion region provided they are able to diffuse to the edge of the depletion region

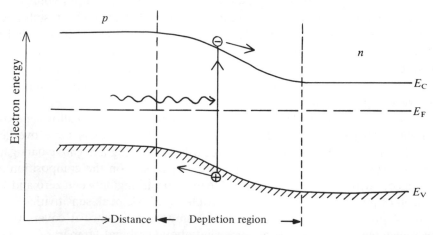

Fig. 7.23 Electron energy level diagram illustrating the generation and subsequent separation of an electron–hole pair by photon absorption within the depletion region of a *p–n* junction.

before recombination takes place. From the discussion in Chapter 2, it is evident that only carriers generated within a minority carrier diffusion length or so of the edge of the depletion region are likely to be able to do this; nevertheless, this does increase the sensitive volume of the device.

In operation, we may represent the photodiode by a constant current generator (the current flow i_λ being generated by light absorption) with an ideal diode across it (to simulate the effect of the $p\text{--}n$ junction), as shown in Fig. 7.24. The internal characteristics of the cell may be better modeled by the introduction of a shunt resistor (R_{sh}), a shunt capacitor (C_d) and a series resistor (R_s). If we assume a quantum efficiency η for the photon absorption process and also that all the incident radiation is absorbed within the cell, then we may write

$$i_\lambda = \frac{\eta I_0 A e \lambda}{hc} \tag{7.22}$$

where I_0 is the light irradiance falling on a cell of area A.

We now develop expressions for the sensitivities of the photodiode when operated in the two different modes. For simplicity, we assume that operation is at fairly low optical modulation frequencies, so that the effects of the shunt capacitance may be neglected (i.e. $i_c = 0$).

Considering the current to flow as shown in Fig. 7.24, we then have

$$i_\lambda = i_d + i_{sh} + i_{ext} \tag{7.23}$$

also

$$V_{ext} = V_d - i_{ext} R_s \tag{7.24}$$

Photovoltaic mode   Photoconductive mode

Fig. 7.24 Photodiode equivalent circuit. In operation, the photodiode can be represented by a photogenerated current source i_λ feeding into an ideal diode. The internal cell characteristics are better modeled by the introduction of a shunt resistor (R_{sh}), a shunt capacitor (C_d), and a series resistor (R_s). In the photovoltaic mode, an external high value resistor (R_L) is connected across the output and the voltage across this measured. In the photoconductive mode, an external bias (V_{app}) is applied in conjunction with a series load resistor (R_L). The current flowing through R_L is monitored by measuring the voltage across it.

and

$$V_d = i_{sh}R_{sh} \tag{7.25}$$

We now take the current–voltage behavior of the diode to be given by eq. (2.51), i.e.

$$i_d = i_0 \left[\exp\left(\frac{eV_d}{kT}\right) - 1 \right]$$

where i_0 is the diode reverse bias leakage current.

In the *photovoltaic mode*, the external current flow is very small; hence, taking $i_{ext} = 0$, eqs. (7.23) and (7.24) become $i_\lambda = i_d + i_{sh}$ and $V_{ext} = V_d$ respectively. Substituting for i_d and i_{sh} in the former we obtain

$$i_\lambda = i_0 \left[\exp\left(\frac{eV_d}{kT}\right) - 1 \right] + \frac{V_d}{R_{sh}}$$

Rearranging gives

$$\exp\left(\frac{eV_d}{kT}\right) = 1 + \frac{i_\lambda}{i_0} - \frac{V_d}{i_0 R_{sh}}$$

Under normal operating conditions $i_\lambda \gg i_0$, whereas V_d is usually the same order of magnitude as $i_0 R_{sh}$. For example, typical values might be $i_0 \approx 10^{-8}$ A, $R_{sh} \approx 10^8$ Ω and $V_d \approx 0.6$ V. Thus, to a first approximation we may write

$$\exp\left(\frac{eV_d}{kT}\right) = \frac{i_\lambda}{i_0}$$

It then follows that

$$V_d \, (= V_{ext}) = \frac{kT}{e} \ln\left(\frac{i_\lambda}{i_0}\right)$$

Substituting for i_λ from eq. (7.22) we have

$$V_{ext} = \frac{kT}{e} \ln\left(\frac{\eta I_0 e\lambda A}{hci_0}\right) \tag{7.26}$$

Hence the external voltage should be a *logarithmic* function of the incident light irradiance.

The *solar cell* is basically a p–n junction detector operated under conditions such that it can deliver power into an external load. The full i–V characteristics of a p–n junction under illumination are sketched in Fig. 7.25. The solar cell operates in the quadrant where i is negative and V is positive. Most power will be delivered when the product iV is a maximum and this determines the optimum load impedance across the cell. At present, most commercially available cells are made from silicon which can convert up to 15% of the solar radiation falling on them into electrical energy. In bright sunlight, each cell can deliver current densities of the order of 80 Am^{-2} at a voltage of about 0.6

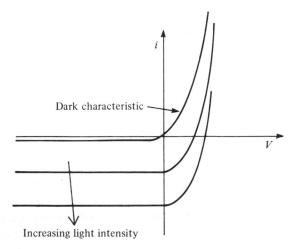

Fig. 7.25 Current–voltage characteristics of a p–n junction solar cell under various levels of illumination. The dark characteristic is that of an ordinary p–n junction diode. Under increasing levels of illumination the curve is progressively shifted downwards.

V. As a means of power generation, they are not yet economic except for certain specialist uses (e.g. low power requirements in remote areas and as a power source for satellites). Cheaper processing technology for silicon or the development of thin film evaporated cells may well change this in the near future.

In the *photoconductive mode*, a relatively large reverse bias (≈ 10 V or more) is usually applied across the diode (Fig. 7.24). Since the diode current saturates at i_0 for relatively small values of reverse bias (i.e. a few tenths of a volt), eq. (7.23) can be written as

$$i_\lambda = i_0 + i_{sh} + i_{ext}$$

Now $i_0 \approx 10$ nA, whilst from eq. (7.25)

$$i_{sh} = V_d/R_{sh} \approx 10 \text{ V}/100 \text{ M}\Omega \approx 100 \text{ nA}$$

so that when i_λ is of the order of microamps and above we may write $i_{ext} = i_\lambda$. Substituting from eq. (7.22) we then have:

$$i_{ext} = \frac{\eta I_0 A e \lambda}{hc} \tag{7.27}$$

Hence, in the photoconductive mode the external current flowing is directly proportional to the incident light irradiance.

In addition to its inherently linear response, the photoconductive mode usually offers the advantages of faster response, better stability and greater dynamic range. The main drawback is the presence of a dark current ($i_0 + i_{sh}$) which, as in the photomultiplier, gives rise to shot noise (section 7.2.4.2) and limits the ultimate sensitivity of the device. Both modes of operation are subject to generation noise (section 7.2.8), but recombination noise is absent since the charge carriers are separated in the depletion region before they can recombine.

The most common semiconductor material used for photodiodes is silicon. This has an energy gap of 1.14 eV and provides excellent photodiodes with quantum efficiencies up to 80% at wavelengths between 0.8–0.9 μm. A typical construction is shown in Fig. 7.26. It should be noted that electrical contact to the semiconductor material is always made via a metal–n^+ (or $-p^+$) junction; this is found to be the most convenient way of providing an ohmic contact (see section 2.8.6). It is also possible to make detectors that are illuminated from the side (i.e. parallel to the junction); this type of construction results in good sensitivity for wavelengths close to the bandgap limit where the optical absorption coefficient is relatively small. Detection efficiency may be increased by providing antireflection coatings on the front surfaces of the detectors consisting of a λ/2 coating of SiO_2. A typical spectral response is shown in Fig. 7.27, while Fig. 7.28 illustrates the electric field variation within the structure of Fig. 7.26 when under reverse bias.

If we assume we have an abrupt junction, that the external bias V is much larger than the internal junction potential and that $N_a \gg N_d$, then the depletion layer widths can be written as (see eqs. (2.60), (2.59) and (2.55))

$$x_n = \left(\frac{2\varepsilon_0\varepsilon_r V}{eN_d}\right)^{\frac{1}{2}}$$

$$x_p = \left(\frac{2\varepsilon_0\varepsilon_r VN_d}{eN_a^2}\right)^{\frac{1}{2}}$$

$$(7.28)$$

Since we have a p^+–n structure it follows that $x_n \gg x_p$. For efficient detection the electron–hole pairs should be generated either inside or within a diffusion length or

Fig. 7.26 Typical silicon photodiode structure for photoconductive operation. A junction is formed between heavily doped p-type material (p^+) and fairly lightly doped n-type material so that the depletion region extends well into the n material. The p^+ layer is made fairly thin. Metallic contacts can be made directly to the p^+ material but to obtain an ohmic contact to the n material an intermediate n^+ layer must be formed.

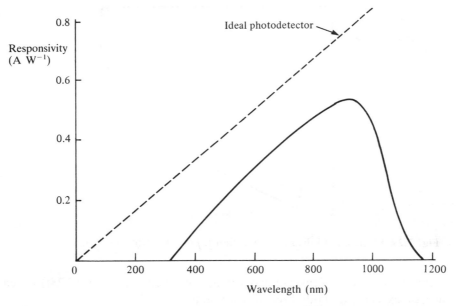

Fig. 7.27 Typical current responsivity of a silicon photodiode. Also shown is the responsivity of an ideal photodiode with unit quantum efficiency.

so of the depletion region. At short wavelengths, where the absorption coefficient is relatively high, they will be generated close to the surface. Consequently, to achieve a good short wavelength response the p^+ region should be made as thin as possible. Conversely, at the upper wavelength range of the detector the absorption coefficient is relatively small and a wide depletion region is necessary for a good response. This

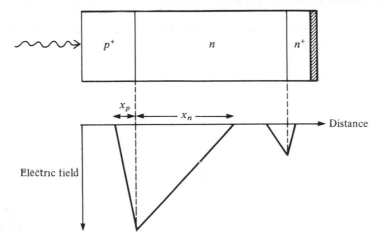

Fig. 7.28 Electric field distribution within the p^+-n junction diode shown in Fig. 7.26, assuming an abrupt diode structure.

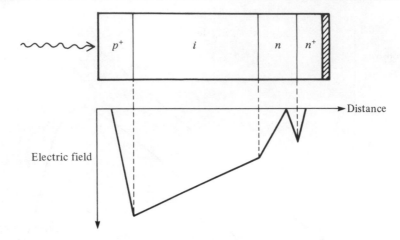

Fig. 7.29 Electric field distribution within a *pin* structure.

implies that large values of reverse bias voltage are needed (see eq. (7.28)) which may approach or exceed the diode breakdown voltage.

A structure that results in a good long-wavelength response with only relatively modest bias levels is the so-called *pin* (or PIN) structure, illustrated in Fig. 7.29.

The intrinsic (i) region has a high resistivity (i.e. low values of N_d and N_a) so that only a few volts of reverse bias are needed to cause the depletion region to extend all the way through to the *n* region and thus provide a large sensitive volume. In practice, the bias is maintained considerably higher than the minimum value and the intrinsic region then remains fully depleted of carriers even at high light levels.

7.2.10.2 RESPONSE TIME OF PHOTODIODES
Three main factors limit the speed of response of photodiodes, as follows.

Diffusion time of carriers to the depletion region. Carrier diffusion is inherently a relatively slow process, the time taken for carriers to diffuse a distance *d* may be written (ref. 7.7):

$$\tau_{diff} = \frac{d^2}{2D_c} \tag{7.29}$$

where D_c is the minority carrier diffusion coefficient.

Example 7.2 Diffusion time of carriers in Si

We consider the time taken for electrons to diffuse through a 5 μm thick layer of *p*-type silicon. Taking $D = 3.4 \times 10^{-3}$ m^{2}s^{-1}, we obtain from eq. (7.29) that

$$\tau_{diff} = (5 \times 10^{-6})^2/(2 \times 3.4 \times 10^{-3}) = 3.7 \times 10^{-9} \text{ s}$$

To ensure that as few carriers as possible are generated outside the depletion region, we require that at the wavelengths being used

$$W = x_n + x_p \gtrsim \frac{1}{\alpha(\lambda)}$$

where $\alpha(\lambda)$ is the absorption coefficient at wavelength λ. At wavelengths near the bandgap limit and at fairly low reverse bias levels this inequality may not hold and the speed of detection of some of the optically generated carriers will then be limited by diffusion. In these circumstances, the response to a narrow optical pulse will be as illustrated in Fig. 7.30. Those carriers generated within the depletion region respond rapidly, whilst those generated outside it give rise to a 'slow' tail. The relative importance of the tail may be decreased by increasing the reverse bias or by using the *pin* structure where the depletion region is relatively wide in any case.

Drift time of carriers through the depletion region. In very high electric fields, the drift velocities of carriers in semiconductors tend to saturate. Provided the field within the depletion region exceeds the saturation value for most of its length (which is usually the case in practice), then we may assume that the carriers move with a constant velocity v_{sat}. The longest transit time will result when carriers are generated near to one edge of the depletion region. In this case, one of the carriers will have to traverse the full depletion layer width W. We may write the transit time as

$$\tau_{drift} = \frac{W}{v_{sat}} \tag{7.30}$$

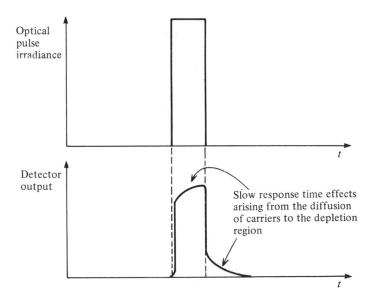

Fig. 7.30 Response of a p–n junction detector to a narrow optical pulse. The effects arising from the relatively long time it takes for carriers generated away from the depletion region to diffuse to the junction are shown.

Example 7.3 Depletion region transit time in Si photodiodes

In fields above 10^7 V m^{-1}, holes in silicon have a saturation velocity of about 10^5 m s^{-1}. Thus the transit time with a depletion layer 5 μm thick is given by eq. (7.30) to be $5 \times 10^{-6}/10^5$ or 5×10^{-11} s.

Junction capacitance effects. A diode under reverse bias exhibits a voltage-dependent capacitance caused by the variation in stored charge at the junction. For example, an abrupt junction diode has a capacitance given by (see eq. (2.63))

$$C_{\mathrm{j}} = \frac{A}{2} \left[\frac{2e\varepsilon_{\mathrm{r}}\varepsilon_0}{(V_0 - V)} \left(\frac{N_{\mathrm{d}}N_{\mathrm{a}}}{N_{\mathrm{d}} + N_{\mathrm{a}}} \right) \right]^{\frac{1}{2}}$$

If we assume that the external bias V is large compared to the zero bias junction potential V_0, and that we have a p^+–n junction, then this expression simplifies to

$$C_{\mathrm{j}} = \frac{A}{2} \, (2e\varepsilon_{\mathrm{r}}\varepsilon_0 N_{\mathrm{d}})^{\frac{1}{2}} V^{-\frac{1}{2}} \qquad (7.31)$$

In practice, junctions are rarely abrupt; however, it still remains true that the capacitance decreases with increasing reverse bias. For example, in a linearly graded junction we have $C_{\mathrm{j}} \propto V^{-1/3}$.

We see from Fig. 7.24 that at high frequencies the diode capacitance acts as a shunt across the output resistance network and reduces the output.

Usually $R_{\mathrm{sh}} \gg R_{\mathrm{L}}$ and $R_{\mathrm{s}} \ll R_{\mathrm{L}}$, hence as far as the diode current, i_λ, is concerned, the diode capacitance C_{j} is in parallel with the load resistor R_{L}. It is easy to show (Problem 7.10) that the output voltage of the device as a function of the modulation frequency f is given by

$$V_0 = \frac{i_\lambda R_{\mathrm{L}}}{(1 + 4\pi^2 f^2 C_{\mathrm{j}}^2 R_{\mathrm{L}}^2)^{\frac{1}{2}}}$$

The electrical bandwidth, Δf_{el} is defined as the frequency range over which the output is above $\sqrt{\frac{1}{2}}$ of its maximum value (see Appendix 5). Thus in the present instance we have $4\pi^2 \Delta f_{\mathrm{el}}^2 C^2 R_{\mathrm{L}}^2 = 1$, i.e.

$$\Delta f_{\mathrm{el}} = \frac{1}{2\pi R_{\mathrm{L}} C_{\mathrm{j}}} \qquad (7.32)$$

The bandwidth may obviously be improved by reducing C_{j}. Inspection of eq. (7.31) shows that this may be achieved by reducing the diode area A, reducing the doping level N_{d} or increasing the reverse bias voltage V. Unfortunately, there are difficulties associated with each of these courses of action. For example, there is a limit as to how small A can be made without encountering problems associated with focusing an incident beam onto a small area. In addition, a reduction in N_{d} and an increase in V

Example 7.4 Frequency bandwidth limitations arising from diode capacitance

First, we use eq. (7.31) to estimate the diode capacitance. We assume the typical values: $A = (1 \text{ mm}^2)$, $\varepsilon_r = 11.7$, $N_d = 10^{21} \text{ m}^{-3}$ and $V = 10 \text{ V}$. We then obtain:

$$C_j = \frac{10^{-6}}{2} [2(1.6 \times 10^{-19})(11.7)(8.85 \times 10^{-12})(10^{21})]^{\frac{1}{2}}(10)^{-\frac{1}{2}} \simeq 30 \text{ pF}$$

To match the input impedance of fast oscilloscopes, etc. the load impedance for high speed applications is usually 50 Ω; thus, for a diode with a capacitance of 30 pF and a 50 Ω load resistance, we calculate from eq. (7.32) that $\Delta f_{el} = 100$ MHz.

---

both cause the depletion layer width to increase (eq. (7.28)), and this has the effect of increasing the drift transit time (eq. (7.30)). The bulk resistance of the device (and hence R_s) will also tend to increase if N_d is reduced. To achieve the largest bandwidths, therefore, a compromise is reached whereby the response time associated with the diode capacitance is made equal to that associated with the drift of carriers across the depletion region. Silicon photodiodes with bandwidths greater than 1 GHz (i.e. response times less than 1 ns) are readily available.

The silicon photodiode is one of the most popular of all-purpose radiation detectors in the wavelength range 0.4 to 1 μm. It has the virtues of high quantum efficiency, small size, good linearity of response, large bandwidth, simple biasing requirements and relatively low cost.

Photodiodes made from materials other than silicon are available, but they have not yet achieved the same standards of performance. Germanium photodiodes can be used up to a wavelength of 1.8 μm but they have rather low quantum efficiencies and somewhat high leakage currents (and hence high noise currents), although cooling to 77 K greatly improves their performance. This is unfortunate, since the wavelength region from 1 to 1.6 μm is of great interest from the point of view of optical communications (see Chapter 9), and suitable low noise high efficiency detectors are currently being sought that will cover this region. Recently, interest has been concentrated on diodes made from both ternary (e.g. GaInAs and HgCdTe) and quaternary (e.g. GaInAsP) compounds.

7.2.10.3 AVALANCHE PHOTODIODE

As explained above, most fast photodiodes are designed for use with a 50 Ω load impedance and the voltage output ($= i_{ext}R_L$) often requires considerable amplification. Useful internal amplification of the photocurrent is achieved in the avalanche photodiode. In this device, a basic p–n structure is operated under very high reverse bias. Carriers traversing the depletion region therefore gain sufficient energy to enable further carriers to be excited across the energy gap by impact excitation. The process is illustrated in Fig. 7.31.

Fig. 7.31 Illustration of the principle of operation of an avalanche photodiode. An electron having reached the point A has sufficient energy above the conduction band bottom to enable it to excite an electron from the valence band into the conduction band (C → D). In so doing it falls from A to B.

An electron having reached point A on the diagram has enough energy above the conduction band bottom such that it can collide with an electron from the valence band and raise it to the conduction band (C → D) (the minimum energy required to initiate this process was discussed in section 4.3 when dealing with cathodoluminescence). This generates a new electron–hole pair. In so doing, the electron will of course lose an equivalent amount of energy and move from A to B. These newly generated carriers may both subsequently generate further electron–hole pairs by the same process, although in general the cross section for hole impact excitation is much smaller than that for electrons. The excitation cross sections vary rapidly with electric field $\mathscr{E}$ according to a relation of the form $\exp(-A/\mathscr{E})$ where A is a constant.

Current gains in excess of 100 are readily obtainable, although as might be expected the current gain is very sensitive to the value of the bias voltage, as illustrated in Fig. 7.32. If the bias voltage is made too large then a self-sustaining avalanche current flows in the absence of any photoexcitation and this sets an upper limit to the voltage that may be used. To ensure a uniformly high avalanche breakdown voltage and uniform gain throughout the diode it is necessary to provide a highly uniform field profile across the device. A number of designs have been proposed to achieve this; one type, for example, uses the guard-ring structure shown in Fig. 7.33. The guard ring restricts the avalanche region to the central illuminated part of the cell.

A number of precautions are necessary when using the avalanche photodiode. For example, the rapid variation of gain with bias voltage shown in Fig. 7.32 requires the use of a very stable power supply if constant gain is to be maintained. Also, care must be taken if the simple bias circuit of Fig. 7.18 is used, since the relatively large output current may cause a significant voltage drop across the bias resistor and the diode series resistance, giving rise to nonlinearities in the output. A further difficulty is that the gain is highly dependent on temperature. This may be overcome by having two identical diodes, one of which is masked from the incident radiation and is operated near

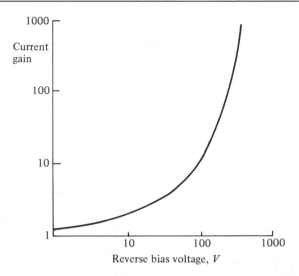

Fig. 7.32 Typical variation of current gain with reverse bias voltage for an avalanche photodiode.

breakdown at constant current. A change in temperature will alter the bias voltage at the control diode and this change may be used to regulate the operating bias applied to the signal diode.

If each photogenerated carrier always gave rise to the same number of secondary carriers (say M) then the r.m.s. noise current due to shot noise would be given by

$$\Delta i_f = M(2ie\Delta f)^{\frac{1}{2}}$$

where i is the total current flowing (see eq. (7.9)). The multiplication process is a statistical one, however, and any primary carrier may generate more or less than the average value. When this factor is taken into account then the true r.m.s. noise current may be written:

Fig. 7.33 Silicon avalanche photodetector with guard-ring structure. The guard ring is a region of comparatively low doping, and hence the depletion region extends an appreciable distance into it. Thus, in the vicinity of the guard ring the total depletion layer thickness is greater, and hence the maximum electric field strengths are less, than in the central region.

$$\Delta i_f \text{ (avalanche)} = M(2ie\Delta f \, F(M))^{\frac{1}{2}} \tag{7.33a}$$

$F(M)$, the *excess noise factor*, is given by (ref. 7.8)

$$F(M) = M \left[1 - \left(1 - \frac{1}{r} \right) \left(\frac{M-1}{M} \right)^2 \right] \tag{7.33b}$$

where r is the ratio of the electron to hole ionization probabilities. In silicon, r tends to have relatively large values (i.e. about 50), whereas for germanium r is about unity. This implies that avalanche photodiodes made from germanium are inherently more noisy than those made from silicon.

Avalanche photodiodes are very popular candidates for use in fiber optical communication systems (see Chapter 9). It remains to be seen, however, whether suitable materials giving good low noise performance in the region beyond 1.1 μm (the limit of silicon detectors) will be developed.

7.2.10.4 SCHOTTKY PHOTODIODES

In a *Schottky* photodiode a thin metal coating (usually gold) is applied to an n type silicon substrate (Fig. 7.34(a)). The energy band structure in the region of the junction is shown in Fig. 7.34(b). It can be seen from this that when an electron–hole pair is generated within the depletion region, the electron and hole will be separated by the action of the internal field as in a p–n junction photodiode. The main advantage of the Schottky photodiode is that the surface metal layer can be made sufficiently thin to transmit blue and near-ultraviolet radition, thus giving an enhanced sensitivity in this region.

7.2.10.5 PHOTOTRANSISTORS

The *phototransistor* is another device, like the avalanche photodiode, where the current flow from a p–n junction detector is internally amplified. The construction is basically

Fig. 7.34 Basic structure of a Schottky photodiode (a) and the energy level behavior in the region of the junction (b). A potential barrier of height φ–χ is formed between the metal and semiconductor.

that of a junction transistor, with the base region exposed to the incident radiation. Normally no external connection is made to the base (see Fig. 7.35(a)). To understand the operation of the device, we consider the external currents to be as shown in Fig. 7.35(b). The base current i_b will be supplied by the photogenerated current.

We must have

$$i_c = i_e - i_b$$

where i_c and i_e are the collector and emitter currents respectively. The collector current has two components: (i) the normal diode reverse saturation current i_{co} and (ii) that part of the emitter current that manages to cross to the collector. (The current is carried by minority carrier diffusion across the base, and not all the minority carriers leaving the emitter will reach the collector.) We write this latter current as αi_c where α is slightly less than unity (α is known as the common base current gain).

Thus

$$i_{co} + \alpha i_e = i_e - i_b$$

whence

$$i_e = \frac{i_b + i_{co}}{(1 - \alpha)}$$
$$= (i_b + i_{co})\left(\frac{\alpha}{1 - \alpha} + 1\right)$$
$$= (i_b + i_{co})(h_{fe} + 1)$$

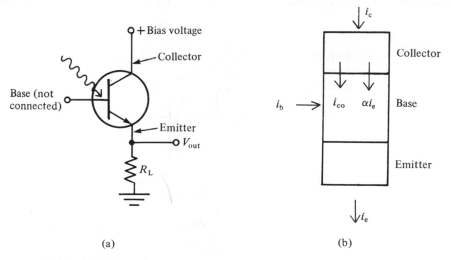

(a)   (b)

Fig. 7.35 External connections made to *npn* phototransistor (a). Light absorbed within the base region causes an emitter current to flow through the load resistor R_L and thus a signal voltage will appear across it. (b) shows the currents assumed to be flowing in the phototransistor.

where $h_{fe}(= \alpha/(1 - \alpha))$ is known as the *common emitter current gain* of the transistor. Typical dark values for h_{fe} in phototransistors are about 100. With no incident radiation $i_b = 0$ and the current flowing, $i_{co}(h_{fe} + 1)$, is the dark current of the device. This is obviously larger than for comparable p–n junction devices when the dark current in this notation is just i_{co}.

When illuminated there will be a base current of magnitude i_λ where, from eq. (7.22), $i_\lambda = \eta(I_0 A e\lambda/hc)$. The external current flowing is now $(i_\lambda + i_{co}) \times (1 + h_{fe})$ which, if $i_\lambda \gg i_{co}$, is equal to $i_\lambda(1 + h_{fe})$. Thus the device gives us internal *gain*, and has a responsivity lying between that of a *pin* photodiode and an avalanche photodiode. There are two main disadvantages, however. First, at low light levels, and hence low base currents, h_{fe} can drop to quite low values. Secondly, the frequency response tends to be relatively poor (typically less than 200 kHz), mainly because of the time constant associated with the capacitance and resistance of the base junction.

7.2.10.6 VIDICONS AND PLUMBICONS

The vidicon is a generic name for a family of devices that rely on the phenomenon of photoconductivity to convert an optical image into an electrical signal. Figure 7.36 shows a typical structure. The optical image is formed on a thin target of semiconducting material (antimony trisulfide is commonly used) that has a transparent conducting layer (usually SnO_2) on the side facing the incident radiation. This conducting layer is connected to a potential of some +50 V above ground via a bias resistor. The other side of the semiconductor target is scanned with an electron beam in the same way as in a CRT. In operation, the target acts rather like a 'leaky' capacitor. When not illuminated its resistance will be high, and charge will accumulate on opposite faces. The electron beam side will charge up to around cathode potential (0 V), whilst the other will charge

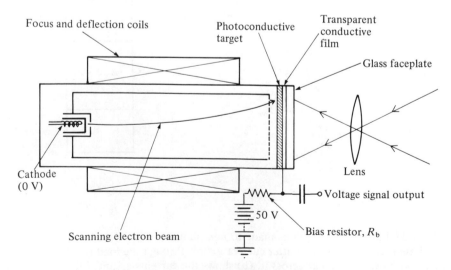

Fig. 7.36 Basic construction of a vidicon tube.

up to around +50 V. Under illumination, however, the resistivity of the target material will be much reduced and the charge on the 'capacitor' will leak away (i.e. the capacitor will discharge itself) whenever the scanning beam is not incident on the area in question. When the beam does return to the 'discharged' area it will recharge the beam side and a corresponding amount of opposite charge must be supplied via the bias resistor and external bias supply circuit to the opposite side. The amount of charge flowing will be dependent on how discharged the 'capacitor' has become, which, in turn, is directly related to the amount of light falling on the target. The output voltage signal is obtained by taking the voltage across the bias resistor.

One of the problems with the vidicon is its relatively high dark current, which gives rise to poor signal-to-noise ratios at low light irradiances. A device which exhibits very low dark currents is the plumbicon. This is essentially identical with the vidicon except for the nature of the photosensitive layer. In the plumbicon, this consists of a thin film *pin* structure formed from lead oxide PbO (Fig. 7.37(a)). The transparent SnO_2 layer acts as the *n*-type contact while the other surface has an excess of oxygen, which causes it to be *p* type. The region in between (typically 15 μm thick) is effectively an intrinsic semiconductor. With no illumination, the potentials across the device cause it to be in reverse bias (and hence very little dark current will flow). Any illumination of the film generates electron-hole pairs, these will then flow to opposite sides of the structure and reduce the amount of stored charge. As in the vidicon, when the electron beam recharges the beam side a corresponding amount of charge is drawn through the bias resistor from the external supply. The plumbicon is widely used in colour TV studio cameras.

The energy gap of PbO is about 2 eV so that the red sensitivity of the device is poor;

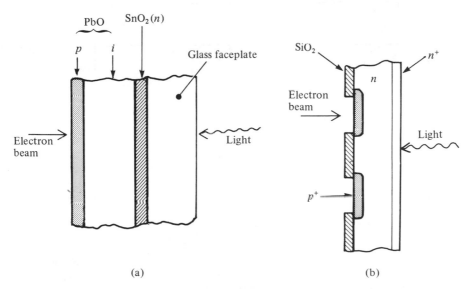

Fig. 7.37 Target structures for vidicon type imaging tubes using (a) a lead oxide based *pin* structure (the plumbicon) and (b) a discrete array of silicon *p–n* junction diodes.

this can be improved by adding a thin layer of PbS ($E_g \simeq 0.4\,\text{eV}$) to the electron beam side of the target. Any red light not absorbed by the PbO is then absorbed by the PbS.

A further development has been the replacement of the lead oxide target layer by an array of silicon diodes (Fig. 7.37(b)). Dark currents are very small, and the devices exhibit very uniform sensitivities over the range 0.45 to 0.85 μm with a good tolerance to high light levels.

7.2.11 Detector arrays

In the previous section, we have seen how an optical image may be converted into an electrical signal by using a vidicon type device. An alternative approach is to use an array of discrete detectors. The image may be focused directly onto a two-dimensional array, or alternatively the image may be scanned past a one-dimensional array. The problem with either of these more direct approaches is that to obtain a scan of the image requires that the output of each detector must be read out sequentially. Given that an array may contain many thousands of detectors, the problem of correctly interconnecting them is not trivial. One way of partially avoiding these difficulties is to use a *charge-coupled device* (CCD) structure. The basic building block of this is the

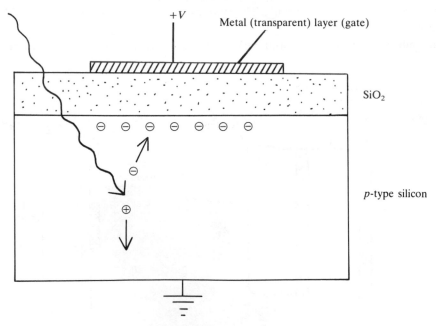

Fig. 7.38 MOS capacitor structure. When the gate is biased positively, photogenerated electron–hole pairs become separated and the electrons then become trapped at the SiO$_2$–Si boundary beneath the gate electrode.

metal-oxide–semiconductor (MOS) capacitor. These are formed by growing a layer of silicon dioxide (SiO_2) on a *p*-type silicon substrate; a metal electrode is then evaporated on top of the oxide layer (Fig. 7.38). The metal electrode is known as the *gate* and is biased positively with respect to the silicon. Photogenerated electron–hole pairs within the silicon will be separated, with the electrons being attracted to the surface of the silicon under the gate, and will remain there whilst the gate voltage is positive. The electrons are effectively trapped within a potential well formed under the gate contact. The amount of charge trapped in the well will be proportional to the total integrated light flux falling onto the device during the measurement period.

The problem now is to 'read out' the amount of trapped charge sequentially along a line of such detectors. This can be done by passing the charge at readout time from

Fig. 7.39 Basic CCD array composed of a line of MOS capacitors (a). The initial charge distribution within the potential wells when G_1 has a positive voltage applied (= V_g) and $G_2 = G_3 = 0$ is shown in (b). In (c) $G_1 = G_2 = V_g$, $G_3 = 0$ and the charge has now spread out under both G_1 and G_2. In (d), $G_1 = 0$, $G_2 = V_g$ and $G_3 = 0$. The charge that initially was under G_1 has now moved to the right to be under G_2. Note that no charge should be generated under the gates G_2 and G_3; these elements are therefore screened from incident light.

detector to detector. There are several ways of achieving this, the basic idea being illustrated in Fig. 7.39. The gate potentials are supplied from three voltage lines (L_1, L_2, L_3) each one being connected to every third electrode (G_1, G_2, G_3). We suppose that initially the potential of L_1 is at some positive value V_g, whilst L_2 and L_3 are at zero potential. Photogenerated charge will be trapped under the G_1 electrodes in proportion to the light falling on these elements (Fig. 7.39(b)). After a suitable integration time, the charge may be moved along the chain of MOS capacitors by applying a repeated sequence of potentials to the gate supply lines. Thus, suppose we apply a voltage V_g to L_2 whilst maintaining L_1 at V_g; the charge initially under L_1 will now be shared between G_1 and G_2 (Fig. 7.39(c)). Next we reduce the potential of L_1 to zero. All the charge that was initially under G_1 is then under G_2 (Fig. 7.39(d)). Continuing this cycle will progressively move the charge along the line of MOS capacitors from left to right. At the end of the line, the amount of charge arriving as a function of time then provides a sequential scan of the 'G_1' detector outputs.

For obvious reasons, this is known as a three-phase scheme. In practice, several other schemes are possible and ref. 7.9 may be consulted for more details. One problem with the device as described here is that a new light scan cannot be carried out until all of the charge has been transferred along the array. A faster scanning rate may be achieved with the layout indicated in Fig. 7.40. Here, a second CCD array (the *transport register*), which is shielded from incident radiation, lies alongside the first. Once a charge image has been built up in the sensing array it is then transferred 'sideways' to the transport register and can be read out sequentially along the transport register. This readout takes place at the same time as a new image is being built up. One-dimensional arrays of up to 2000 elements are available, the sensing elements being some 10 μm apart.

Two-dimensional arrays based on the above one-dimensional designs are also possible, and are known as *frame transfer* devices. Here the transfer registers feed into a *readout register* running down the edge of the device (Fig. 7.41(a)). The contents of each line are read out in sequence into the readout register so that the signal appearing at the end of the readout register represents a line-by-line scan of the image. One

Fig. 7.40 Basic arrangement of the detector array and transport register in a linear CCD optical sensing array. The directions of charge transport are indicated by arrows and shading indicates those areas that are shielded from incident radiation.

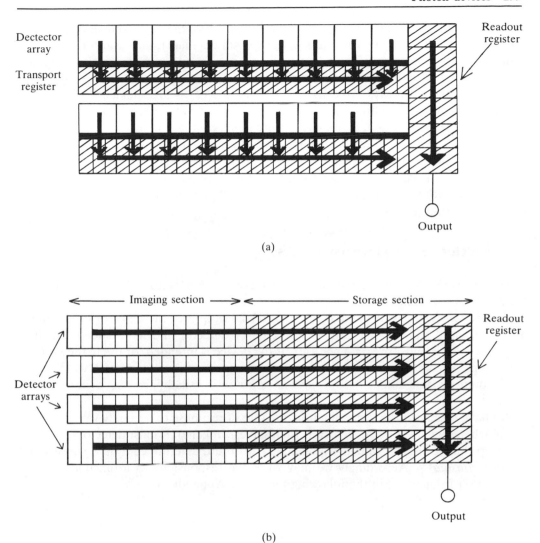

Fig. 7.41 Two schemes for reading out photogenerated charge from a two-dimensional CCD image sensor: frame transfer (a) and interline transfer (b). As in Fig. 7.40, charge flow is indicated by arrows and areas shielded from radiation by shading.

problem with this scheme is that the presence of the transfer register reduces the resolution possible by increasing the distance between the sensing arrays. It is possible to dispense with the transfer register altogether by adopting the design shown in Fig. 7.41(b)). Here, each array is made twice as long as before, with the second half of the array being shielded from the incident radiation. Then, after a charge image has been built up it is moved along each array into the shielded section where it can be stored until it can be transferred into the readout register as before. Again, readout takes place whilst a new image is being built up. This scheme is known as *interline transfer*. However, although it offers better resolution than frame transfer, it is also somewhat slower because of the relatively long time taken to transfer charge along the sensing array and into the storage section.

Two-dimensional arrays typically contain up to 512 × 512 elements with inter-element spacings similar to those of one-dimensional arrays (i.e. ∼ 10 μm). As yet, they are not quite fast enough for the requirements of domestic television. However, they can be used in other less demanding applications such as machine vision. In some of these, the analog image signal may have to be processed in some way using a digital computer. In this case, the output from the array is passed to a *frame grabber*. This digitizes and stores the signals representing one complete scan or frame of the image ready for further processing.

7.3 Detector performance parameters

We now briefly review the parameters that are commonly used to assess the performance of a detector.

The *responsivity R* is defined as the ratio of the output of the detector to its input. The units used will depend both on the type of detector and its intended use, but typically will be amps (or volts) per watt. If the device is intended for use in the visible, then amps per lumen are sometimes used. The responsivity will vary with wavelength and should be designated R_λ; the *spectral response* is usually given as a curve of R_λ versus λ.

We have discussed the factors limiting the rate of response to a rapidly varying input signal when dealing with the detectors themselves. In general, most detectors respond to a step change in input with an exponential rise or fall, determined by a *response time*, or *time constant* τ. Accordingly we may write the responsivity as a function of light modulation frequency f in the following form (see Appendix 5):

$$R(f) = \frac{R(0)}{(1 + 4\pi^2 f^2 \tau^2)^{\frac{1}{2}}} \tag{7.34}$$

The *cut-off frequency* f_c is defined as the frequency at which the responsivity falls to a value $R(0)/\sqrt{2}$ and is thus given by $f_c = 1/(2\pi\tau)$. Obviously, a detector can only give a faithful representation of an input signal pulse if its response time is short compared with the pulse width.

We have already discussed the principal sources of noise in most of the detectors described in this chapter. The presence of noise obviously limits the ability of the device to detect small signals. Conventionally, the minimum detectable signal is taken to be that which would generate an r.m.s. output equal to that generated by the noise. (The signal-to-noise ratio is then equal to unity.) An indication of the size of the minimum detectable signal is given by the *noise equivalent power* (NEP). This is defined as the power of sinusoidally modulated monochromatic radiation which would result in the same r.m.s. output signal in an 'ideal' noise-free detector as the noise signal encountered in the real detector. Statements of the NEP should be given along with a statement of the modulation frequency, detector bandwidth, detector temperature and detector area. The detector bandwidth is usually taken to be 1 Hz. It is useful to remember that most noise sources produce what is known as *white noise*, where the noise power within a bandwidth Δf is proportional to Δf. Because current and voltage are proportional to the square root of power, noise current and noise voltage are then proportional to $\Delta f^{\frac{1}{2}}$. (See for example eqs. (7.9), (7.11) and (7.21).)

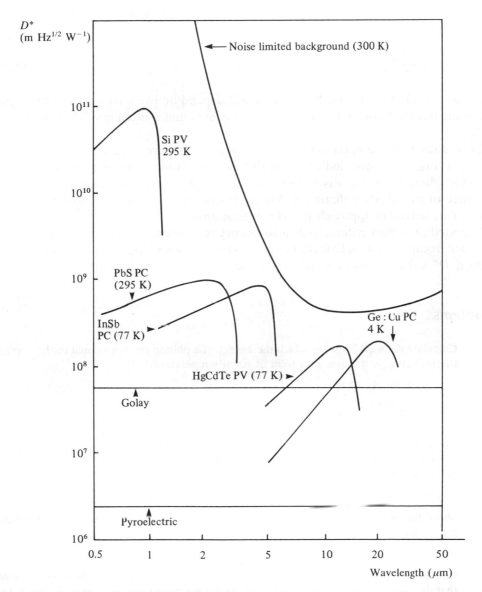

Fig. 7.42 Specific detectivity D^* as a function of wavelength for a number of representative photodetectors (PC = photoconductive, PV = photovoltaic).

If we assume that the noise power generated in a detector is proportional to its sensitive area A, then the noise current (or voltage) will vary as $A^{\frac{1}{2}}$. Thus we may define a unit NEP* which takes into account the effects of variable bandwidth and detector area where:

$$\text{NEP*} = \frac{\text{NEP}}{(A\Delta f)^{\frac{1}{2}}}$$

In fact it is the reciprocal of this quantity, known as the *specific detectivity* (D^*) that is commonly used, thus:

$$D^* = \frac{(A\Delta f)^{\frac{1}{2}}}{\text{NEP}} \tag{7.35}$$

The value of D^* for a particular detector will depend on the wavelength of the signal radiation and the frequency at which it is modulated and is often quoted in the format $D^*(\lambda, f)$.

Curves showing the variation of D^* with λ for several representative detectors are shown in Fig. 7.42. Also indicated on this diagram is the theoretical D^* curve for a detector where the noise arises from fluctuations in carrier generation caused by the presence of blackbody radiation at 300 K. It will be seen that, at their best, quite a number of detectors approach this theoretical limit.

Provided that other criteria such as frequency response are met, and that the noise is not background-radiation limited, then at a particular wavelength the detector with the highest D^* value is generally the best choice.

Problems

7.1 Calculate the wavelength at which the energy of a photon becomes equal to the average thermal energy of atoms in a solid at room temperature.

7.2 Starting from eq. (7.1), i.e.

$$W = H \frac{\mathrm{d}(\Delta T)}{\mathrm{d}t} + G\Delta T$$

which represents the energy balance condition in a thermal detector element, and assuming that W and ΔT may be written in the form $W = W_0 + W_f \cos(2\pi ft)$ and $\Delta T = \Delta T_0 + \Delta T_f \cos(2\pi ft + \phi_f)$, show that:

  (a) $\Delta T_0 = W_0/G$ and (b) $\Delta T_f = W_f/(G^2 + 4\pi^2 f^2 H^2)^{\frac{1}{2}}$

7.3 Show that the effective thermal conductance G_R, for a radiative link at absolute temperature T may be written:

$$G_R = 4\sigma A \varepsilon T^3$$

where σ is Stefan's constant, ε the receiver surface emissivity and A its area. Hence show that the noise power fluctuations $(P_f)_R$, in a detector of bandwidth Δf limited by radiative exchange is given by

$$(\Delta W_f)_R = 4(AKT^5\sigma\varepsilon\Delta f)^{\frac{1}{2}}$$

Assuming $\varepsilon = 1$, calculate $(P_f)_R$ for a detector of area 100 mm^2 and bandwidth 1 Hz at room temperature (300 K).

7.4   The equivalent circuit of a pyroelectric detector may be taken to be a current source i feeding into a parallel combination of a capacitor C and a load resistor R_L (see Fig. 7.6(b)).

If P is the dipole moment per unit volume of the pyroelectric material, show that the surface charge is PA where A is the surface area. Hence show that i may be written

$$i = A\,\frac{dP}{dT} \cdot \frac{dT}{dt}$$

where T is the temperature of the detector and dT/dt is its rate of increase of temperature with time due to the absorption of radiation.

By assuming the results of Problem 7.2, show that if the detector completely absorbs radiation of irradiance $W = W_0 + W_f \cos 2\pi ft$ then the voltage responsivity of the device R_v, can be written:

$$R_v = A\,\frac{dP}{dT}\frac{R_L}{G}\,\frac{2\pi f}{\left(1 + 4\pi^2 f^2\,\dfrac{H^2}{G^2}\right)^{\frac{1}{2}}} \cdot \frac{1}{(1 + 4\pi^2 f^2 C^2 R_L^2)^{\frac{1}{2}}}$$

Sketch the behavior of R_v with f.

7.5   Estimate the minimum photon flux required for a television scanning system using a photomultiplier tube with a quantum efficiency of 0.1. The signal bandwidth is 5 MHz and a signal-to-noise ratio of 100 is required.

7.6   Show that when the simple bias circuit shown in Fig. 7.18 is used with a photoconductive detector to detect small signal levels, then maximum voltage output signals across R_L are obtained when R_L is equal to the detector resistance R_D.

7.7   Show that the gain G of a photoconductive detector can be written as $G = \tau_c/\tau_d$ where τ_c is the minority carrier lifetime, and τ_d is given by

$$\frac{1}{\tau_d} = \frac{1}{\tau_n} + \frac{1}{\tau_p}$$

τ_n and τ_p being the times taken for electrons and holes to drift across the photodetector.

7.8   According to eq. (7.20), the gain G of a photoconductive detector may be increased by increasing the applied voltage. However, space charge limitations prevent the gain being increased indefinitely by this means. Derive an expression for the maximum useful gain by assuming that the applied voltage is limited by the condition that the photogenerated excess charge within the photoconductor volume is equal to that which can be stored on the area contacts to the photoconductor when these are viewed as forming a capacitor.

7.9   An ideal photodiode (of unit quantum efficiency) is illuminated with 10 mW of radiation at 0.8 μm wavelength; calculate the current and voltage output when the detector is used in the photoconductive and photovoltaic modes respectively. The reverse bias leakage current is 10 nA.

7.10   Show that the responsivity of a photodiode detector can be written as:

$$R(f) = \frac{R(0)}{(1 + 4\pi^2 f^2 C^2 R_L^2)^{\frac{1}{2}}}$$

where C is the diode capacitance and R_L the load resistance.

7.11 It is desired to make a silicon *pin* photodiode of area 1 mm^2 with as fast a response time as possible when used in conjunction with a 50 Ω load resistor. Estimate the thickness of the intrinsic region required. Take $\varepsilon_r = 11.8$ and v_s (the electron saturation velocity) = 10^5 m s^{-1}. Over what wavelength range would you expect the device to be most effective?

7.12 If the base width of a *pnp* phototransistor is 50 μm, estimate the frequency at which limitations due to finite carrier diffusion time across the base become apparent. (Take $D_h = 1.2 \times 10^{-3}$ m s^{-1}.)

7.13 Using Fig. 7.42, estimate the smallest detectable signal power at 2 μm using a lead sulfide cell with a sensitive area of 10^4 mm^2 and a frequency bandwidth of 100 Hz.

References

7.1 R. A. Smith, F. E. Jones and R. P. Chasmar, *The Detection and Measurement of Infra-Red Radiation* (2nd Ed), Oxford University Press, Oxford, 1968, section 5.9.

7.2 J. S. Blakemore, *Solid State Physics* (2nd Ed), Saunders, Philadelphia, 1974, section 3.3.

7.3 J. Pierce, 'Physical sources of noise', *Proc. IRE*, **44**, 1956, 601.

7.4 R. H. Kingston, *Detection of Optical and Infrared Radiation*, Springer-Verlag, Berlin, 1978, section 5.2.

7.5 R. King, *Electrical Noise*, Chapman and Hall, London, 1966, section 3.

7.6 R. H. Kingston, *op. cit.*, section 6.1.

7.7 B G. Streetman, *Solid State Electronic Devices* (2nd edn), Prentice Hall, Englewood Cliffs, N.J., 1980, section 7.8.2.

7.8 G. E. Stillman and C. M. Wolfe, 'Avalanche photodiodes', in R. K. Willardson and A. C. Beer, Eds., *Semiconductors and Semimetals*, Vol. 12, Academic Press, New York, 1977, Chapter 5.

7.9 B. G. Streetman, *op. cit.*. section 9.4.3.

Chapter 8

Fiber optical waveguides

Introduction

The phenomenon of total internal reflection at the interface between two dielectric media has been known for some considerable time. In 1870, Tyndall (ref. 8.1) demonstrated that as a result light could be guided within a water jet. However, although some theoretical studies were carried out in the early years of the present century, it was not until the mid-1960s that the idea of a communication system based on the propagation of light within circular dielectric waveguides was considered seriously (ref. 8.2). The main reasons for this delay were that initially only single dielectric rods were considered and these had to be made impracticably small to accommodate a single electromagnetic mode; furthermore, the modes would have penetrated into the air surrounding the rod thereby causing the losses to be high and making it difficult to support the guide.

These major problems were overcome with the development of cladded dielectric waveguides, which were first proposed in 1954 (ref. 8.3). A number of serious difficulties still remained, the most important of which was the high attenuation encountered in the fibers. These were typically of the order of 1000 dB km^{-1}, whereas today values below 1 dB km^{-1} can readily be achieved. It will be noticed that we have not used the optical absorption coefficient as a measure of optical attenuation, but rather the units dB km^{-1}. These are defined as follows. If a beam of power P_i is launched into one end of an optical fiber, and if P_f is the power remaining after a length L km has been traversed, then the attenuation is given by

$$\text{Attenuation} = \frac{10 \log_{10}(P_i/P_f)}{L} \text{ dB km}^{-1} \tag{8.1}$$

The propagation of light down dielectric waveguides bears some similarity to the propagation of microwaves down metal waveguides. A full mathematical treatment of either situation requires the solution of Maxwell's equations using the appropriate boundary conditions, which is beyond the scope of the present text. However it is possible to gain some insight by considering the propagation of individual light rays down a waveguide structure. The mechanism by which such rays are confined within the waveguide is that of total internal reflection, and we now discuss this phenomenon.

8.1 Total internal reflection

When an electromagnetic wave is incident upon the boundary between two dielectric media whose refractive indices are n_1 and n_2, then in general a portion of that wave is reflected and the remainder transmitted. Maxwell's equations require that both the tangential components of $\vec{\mathscr{E}}$ and $\mathscr{H}$ and the normal components of $\boldsymbol{D}$ $(= \varepsilon_r \varepsilon_0 \vec{\mathscr{E}})$ and $\boldsymbol{B}$ $(= \mu_r \mu_0 \mathscr{H})$ are continuous across the boundary. A detailed consideration of the consequences of applying these conditions is too lengthy to be reproduced here; however, this is quite straightforward and is given in many textbooks (see, for example, ref. 8.4). The resulting equations are known as *Fresnel's equations* and are summarized below. We suppose the wave to be incident on the interface at an angle θ_i to the normal and that the reflected and transmitted waves are at angles θ_r and θ_t respectively, as shown in Fig. 8.1. These angles are related by the equations:

$$\theta_i = \theta_r$$

and

$$\frac{\sin \theta_i}{\sin \theta_t} = \frac{n_2}{n_1}$$

(8.2a)

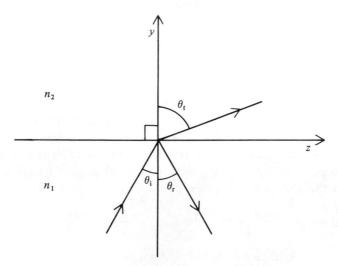

Fig. 8.1 Illustration of the behavior of a light ray incident on the boundary between two media with refractive indices n_1 and n_2 where $n_2 < n_1$. In general, a transmitted and a reflected beam are produced. The angles of incidence, transmission and reflection are θ_i, θ_t and θ_r respectively.

Fresnel's equations deal with the magnitudes of the transmitted and reflected electric fields ($\mathscr{E}_t$, $\mathscr{E}_r$) relative to the incident field $\mathscr{E}_i$. We must distinguish between the cases where the electric field vector is parallel to and perpendicular to the plane of incidence: we denote these by $\mathscr{E}^{\parallel}$ and $\mathscr{E}^{\perp}$ respectively:

$$\left.\begin{aligned}\frac{\mathscr{E}_r^{\perp}}{\mathscr{E}_i^{\perp}} &= \frac{n_1 \cos\theta_i - n_2 \cos\theta_t}{n_1 \cos\theta_i + n_2 \cos\theta_t} \\[2mm] \frac{\mathscr{E}_r^{\parallel}}{\mathscr{E}_i^{\parallel}} &= \frac{n_1 \cos\theta_t - n_2 \cos\theta_i}{n_1 \cos\theta_t + n_2 \cos\theta_i}\end{aligned}\right\} \tag{8.2b}$$

and

$$\left.\begin{aligned}\frac{\mathscr{E}_t^{\perp}}{\mathscr{E}_i^{\perp}} &= \frac{2n_1 \cos\theta_i}{n_1 \cos\theta_i + n_2 \cos\theta_t} \\[4mm] \frac{\mathscr{E}_t^{\parallel}}{\mathscr{E}_i^{\parallel}} &= \frac{2n_1 \cos\theta_i}{n_1 \cos\theta_t + n_2 \cos\theta_i}\end{aligned}\right\} \tag{8.2c}$$

We will be particularly interested in the situation when $n_1 > n_2$ since it is then possible to have total internal reflection (see section 4.6.4), i.e. when $\theta_i > \theta_c$ where:

$$\theta_c = \sin^{-1}(n_2/n_1) \tag{8.3}$$

there is then no transmitted wave in the second medium.

Further confirmation that total internal reflection occurs may be obtained as follows: we have that

$$\cos\theta_t = (1 - \sin^2\theta_t)^{\frac{1}{2}}$$

and substitution for $\sin^2\theta_t$ from eq. (8.2a) gives

$$\cos\theta_t = \left[1 - \left(\frac{n_1}{n_2}\right)^2 \sin^2\theta_i\right]^{\frac{1}{2}}$$

when $\theta_i > \theta_c$, $\sin\theta_i > n_2/n_1$ and $\cos\theta_t$ then becomes wholly imaginary.

We may therefore write

$$\cos\theta_t = \pm iB \quad (\theta_i > \theta_c) \tag{8.4}$$

where

$$B = \left[\left(\frac{n_1}{n_2}\right)^2 \sin^2\theta_i - 1\right]^{\frac{1}{2}}$$

We must take the negative sign in eq. (8.4) for reasons which will be explained later (see eq. (8.8) and the subsequent discussion). Inspection of the right-hand sides of eqs. (8.2b) shows that they both have modulus unity, thus verifying that when $\theta_i > \theta_c$ the irradiances of the reflected and the incident beams are the same and thus there is no transmitted beam. Since the electric field ratios are now complex quantities, however, a *phase shift* is introduced between the incident and reflected beams. We now calculate this phase shift. The $\mathscr{E}^{\perp}$ component of eqs. (8.2b) may be written in the form:

$$\frac{\mathscr{E}_r^{\perp}}{\mathscr{E}_i^{\perp}} = \frac{A + iB}{A - iB} \tag{8.5}$$

where

$$A = \frac{n_1}{n_2} \cos \theta_i$$

Hence we may write

$$\frac{\mathscr{E}_r^{\perp}}{\mathscr{E}_i^{\perp}} = \frac{\cos \psi + i \sin \psi}{\cos \psi - i \sin \psi} = \frac{\exp(i\psi)}{\exp(-i\psi)} = \exp(2i\psi) \tag{8.6a}$$

where

$$\tan \psi = B/A$$

$$= \frac{n_2 \left[\left(\frac{n_1}{n_2}\right)^2 \sin^2 \theta_i - 1 \right]^{\frac{1}{2}}}{n_1 \cos \theta_i}$$

or

$$\tan \psi = \frac{\left[\sin^2 \theta_i - \left(\frac{n_2}{n_1}\right)^2 \right]^{\frac{1}{2}}}{\cos \theta_i} \tag{8.6b}$$

Similarly it may be shown that

$$\frac{\mathscr{E}_r^{\parallel}}{\mathscr{E}_i^{\parallel}} = \exp(2i\delta) \tag{8.7a}$$

where

$$\tan \delta = \left(\frac{n_1}{n_2}\right)^2 \tan \psi \tag{8.7b}$$

The phase changes on reflection for $\mathscr{E}^{\perp}$ and $\mathscr{E}^{\parallel}$ are thus given by 2ψ and 2δ respectively; in both bases $\mathscr{E}_r$ *leads* $\mathscr{E}_i$ in phase.

Example 8.1 Phase shift on reflection

We take a glass/air interface with $n_1 = 1.5$, $n_2 = 1$ and calculate the phase shifts introduced when $\theta_i = 60°$. From eq. (8.6b) we have

$$\tan \psi = \frac{\left[\sin^2(60°) - \left(\frac{1}{1.5}\right)^2 \right]^{\frac{1}{2}}}{\cos(60°)} = 1.106$$

Hence $\psi = 47.9°$. From eq. (8.7b) we also have $\delta = 68.1°$.

Although all the energy in the beam is reflected when $\theta_i > \theta_c$, there is still a disturbance in the second medium whose electric field amplitude decays exponentially with distance away from the boundary. No energy is conveyed away from the surface provided the second medium extends an infinite distance from the boundary. We may derive an expression for this decay by considering the phase factor $\mathcal{P}$ of the transmitted wave, which at a point r we may write as

$$\mathcal{P} = \exp i(\omega t - \mathbf{k}_t \cdot \mathbf{r})$$

where k_t is the wavevector associated with the transmitted wave. Reference to Fig. 8.2 shows that we may write r as $z \sin \theta_t + y \cos \theta_t$ and hence we have

$$\mathcal{P} = \exp i[\omega t - k_t(z \sin \theta_t + y \cos \theta_t)]$$

$$= \exp i\left[\omega t - \frac{2\pi n_2}{\lambda_0}(z \sin \theta_t + y \cos \theta_t)\right]$$

where λ_0 is the wavelength of the radiation in vacuum. If we substitute expressions for $\sin \theta_t$ and $\cos \theta_t$ from eqs. (8.2a) and (8.4), we obtain

$$\mathcal{P} = \exp i\left[\omega t - \frac{2\pi n_2}{\lambda_0}\left(z \frac{n_1}{n_2}\sin \theta_i + (\pm iB)y\right)\right]$$

$$= \exp\left(\pm B \frac{2\pi n_2}{\lambda_0}y\right)\exp i\left(\omega t - \frac{2\pi n_1 \sin \theta_i}{\lambda_0}z\right) \qquad (8.8)$$

Thus, in the y direction the wave either grows or decays exponentially with distance. The former situation is obviously a nonphysical solution and we must choose $\cos \theta_t = -iB$ in eq. (8.4).

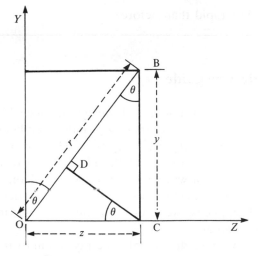

Fig. 8.2 Illustration of the relationship between the rectangular coordinates y and z and the distance r measured from the origin O. We have OB = OD + DB = OC $\sin \theta$ + BC $\cos \theta$. Hence, in terms of the coordinates, $r = z \sin \theta + y \cos \theta$.

The decay with distance in the second medium is given by the factor $F(y)$ where:

$$F(y) = \exp\left(-\frac{2\pi n_2}{\lambda_0} By\right)$$

$$= \exp\left\{-\frac{2\pi n_2}{\lambda_0}\left[\left(\frac{n_1}{n_2}\right)^2 \sin^2\theta_i - 1\right]^{\frac{1}{2}} y\right\} \tag{8.9}$$

Usually, $F(y)$ decays rapidly with y (see Example 8.2). However, when θ_i is very close to θ_c then $[(n_1/n_2)^2 \sin^2\theta_i - 1]^{1/2}$ will be close to zero and the disturbance may extend an appreciable distance into the second medium. That part of the wave which is in the second medium is referred to as the *evanescent wave*.

Example 8.2 Field penetration into the less dense medium during total internal reflection

If we again take a glass/air interface, $n_1 = 1.5$, $n_2 = 1$, with $\theta_i = 60°$, from eq. (8.9) we obtain:

$$F(y) = \exp\left(-5.21\frac{y}{\lambda_0}\right)$$

Thus, in a distance equal to the wavelength, the electric field vector will fall by a factor $\exp(-5.21)$ or 5.5×10^{-3}. If, however, we take a value for θ_i which is much closer to θ_c ($= 41.8°$) say $42°$, then we have:

$$F(y) = \exp\left(-0.54\frac{y}{\lambda_0}\right)$$

The decay is now less rapid than before.

8.2 Planar dielectric waveguide

We now turn to the simplest form of optical waveguide, one which consists of a slab of dielectric of thickness d and refractive index n_1 sandwiched between two semi-infinite regions both of refractive index n_2, where $n_1 > n_2$. The central slab is called the *core* whilst the surrounding regions are called the *cladding* regions. A ray of light may readily propagate down such a waveguide in a zigzag fashion, as shown in Fig. 8.3, provided that total internal reflection occurs at the boundaries (i.e. $\theta_i > \theta_c$). In the interests of notational simplicity, we refer to the interior angle in a waveguide as θ for the remainder of this chapter.

Although in Fig. 8.3 we have drawn only one ray, we must realize that in reality an 'infinite' number of such rays, all slightly displaced from each other, will also be propagating down the guide. In fact, the rays merely represent lines drawn normally to the plane wavefronts as shown in Fig. 8.4, and all points along the same wavefront must

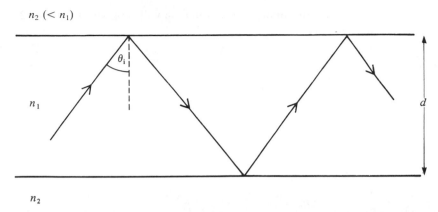

Fig. 8.3 Zigzag path of a light ray down a planar dielectric waveguide that results when the angle of incidence at the boundary θ_i is greater than the critical angle θ_c.

have identical phases. If we consider the particular wavefront, FC, drawn in Fig. 8.4, we see that it intersects two of the upwardly traveling portions of the same ray at the points A and C. Unless the phase at the point C differs from that at A by a multiple of 2π, we obtain destructive interference making it impossible for light to propagate properly down the guide. Moving along the ray path between the points A and C involves a total phase change of $(AB + BC)2\pi n_1/\lambda_0 - 2\phi$. The first term here arises from the pathlength $AB + BC$, whilst the second allows for the two phase changes on reflection at the points B and C. The fact that these two terms have opposite signs may appear puzzling until we remember that the basic equation of a wave $A = A_0 \cos(\omega t - kr)$ implies that the phase at the point r_2 will be less than that at a point r_1 if $r_2 > r_1$. On the other hand, when a wave suffers total internal reflection its phase increases (see eqs. (8.6a) and (8.7a)). Depending on whether the electric field is

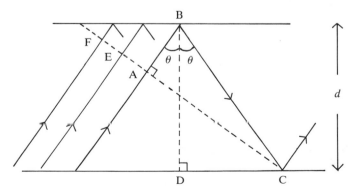

Fig. 8.4 Ray paths having the same internal angle θ within a planar waveguide. A wavefront is shown connecting the points F, E, A, C, which must therefore have the same phase.

perpendicular or parallel to the plane of incidence, ϕ will be given by either 2ψ or 2δ (see eqs. (8.6b) and (8.7b)).

From the triangle ABC we have that $AB = BC \cos 2\theta$, and hence $AB + BC = BC(\cos 2\theta + 1)$ or, since $\cos 2\theta = 2\cos^2\theta - 1$, $AB + BC = 2BC\cos^2\theta$. Also from the triangle BDC we have $BC \cos\theta = d$, so that, finally, $AB + BC = 2d\cos\theta$.

Thus the condition for the mode to propagate is:

$$\frac{4\pi n_1 d \cos\theta}{\lambda_o} - 2\phi = 2\pi m$$

or (8.10)

$$\frac{2\pi d n_1 \cos\theta}{\lambda_0} - \phi = \pi m$$

where m is an integer.

For each value of m there will be a corresponding value of θ (θ_m) that satisfies eq. (8.10). The difficulty is that ϕ depends on θ (via eqs. (8.6b) or (8.7b)), and we cannot obtain an explicit expression for θ in terms of m. The equation must be solved either graphically (see Problem 8.6) or numerically. There are, however, a number of interesting observations which can be made without having to solve the equation. Rearranging eq. (8.10) gives

$$\cos\theta_m = \frac{(m\pi + \phi)\lambda_0}{2\pi d n_1}$$ (8.11)

Bearing in mind that θ can only take values in the range θ_c to $\pi/2$, eq. (8.11) implies that small values of m must be associated with relatively large values of θ_m, and vice versa. However, the smallest value that θ_m can take is θ_c, which implies that there is a maximum value that m can take. We may estimate this by again rearranging eq. (8.10) to give

$$m = \frac{2dn_1 \cos\theta_m}{\lambda_0} - \frac{\phi}{\pi}$$ (8.12)

Since we must have $\sin\theta_m > n_2/n_1$ to maintain total internal reflection, $\cos\theta_m < [1 - (n_2/n_1)^2]^{\frac{1}{2}}$ and consequently

$$m \leqslant \frac{2dn_1}{\lambda_0}\left[1 - \left(\frac{n_2}{n_1}\right)^2\right]^{\frac{1}{2}} - \frac{\phi}{\pi}$$

or

$$m \leqslant \frac{2V}{\pi} - \frac{\phi}{\pi}$$ (8.13)

where

$$V = \frac{\pi d n_1}{\lambda_0}\left[1 - \left(\frac{n_2}{n_1}\right)^2\right]^{\frac{1}{2}} = \frac{\pi d}{\lambda_0}(n_1^2 - n_2^2)^{\frac{1}{2}}$$ (8.14)

V is sometimes known as the *normalized film thickness* or *normalized frequency*. It is perhaps unfortunate that the commonly used symbol for this quantity is identical with that for voltage; there should, however, be no confusion in the context in which it is used here.

Example 8.3 Number of modes in a guide

We may estimate the maximum *m* value that applies to a guide 100 μm wide where $n_1 = 1.53$, $n_2 = 1.50$ and for a vacuum wavelength of 1 μm. From eq. (8.14) we have

$$V = \pi \times 100 \times (1.53^2 - 1.50^2)^{\frac{1}{2}} = 94.7$$

Since ϕ cannot be greater than π, then eq. (8.13) gives the largest *m* value as approximately 60.

Each value of *m* is associated with a distinct wave pattern or *mode* within the waveguide. (We have met with the idea of modes previously when dealing with lasers in Chapter 5 and the present discussion is directly relevant to the possible modes in a semiconductor laser.) If $m > 2(V/\pi) - \phi/\pi$, the condition for total internal reflection will not be met. It is still possible in these circumstances for a mode to propagate down the guide, since some of the radiation will be partially reflected at each 'bounce' from the core–cladding interface although it will be subject to very high attenuation (see Problem 8.7). When the total internal reflection condition is not met, the mode is said to be beyond 'cut-off'. All guided modes are below cut-off.

We recall that when we dealt with a wave incident on an interface (section 8.1) two different situations were considered: (a) when $\mathscr{E}$ was in the plane of incidence; and (b) when $\mathscr{E}$ was perpendicular to the plane of incidence. These two situations involve different phase shifts when total internal reflection takes place (eqs. (8.6a) and (8.7a)) and hence they give rise to two independent sets of modes. Because of the directions of $\mathscr{E}$ and $\mathscr{H}$ with respect to the direction of propagation down the waveguide, the two types of mode are called *transverse magnetic* (TM) and *transverse electric* (TE), which correspond to the $\mathscr{E}^{\parallel}$ and $\mathscr{E}^{\perp}$ situations respectively. Thus, a TM mode involves the field components $\mathscr{H}_x$, $\mathscr{E}_y$ and $\mathscr{E}_z$ whilst the TE mode involves the components $\mathscr{E}_x$, $\mathscr{H}_y$ and $\mathscr{H}_z$ (Fig. 8.5(a) and (b)). The mode numbers *m* are incorporated into this nomenclature by referring to the TM_m and TE_m modes, etc.

Returning to eq. (8.13) we see that if $2V < \phi$ no mode whatsoever can be propagated. However, for any value of *V* it is always possible to find an angle θ such that the corresponding value of ϕ is less than $2V$, and consequently at least one mode can always be propagated. (This is not so if the guide has media of differing refractive indices above and below the central layer.)

As we have seen, a given mode can be regarded as being made up from an 'infinite'

Fig. 8.5 Ray with $\mathscr{E}$ in the plane of incidence (a), gives rise to the three field components $\mathscr{H}_x$, $\mathscr{E}_y$ and $\mathscr{E}_z$ corresponding to a TM mode. A ray with $\mathscr{H}$ in the plane of incidence (b), gives rise to $\mathscr{E}_r$, $\mathscr{H}_y$ and $\mathscr{H}_z$ and corresponds to a TE mode.

collection of rays traveling down the guide with the same value of internal angle θ. If we consider any particular point within the guide, only two of these rays can pass through it. One will be directed 'upwards' the other directed 'downwards'. Since, in general, there will be a phase difference between these two, they will interfere and thereby give rise to a variation in the field amplitude across the guide. Each mode will be characterized by a different field variation.

This phase difference, as a function of position within the guide, may be determined from Fig. 8.6. This shows two rays meeting at a point C, a distance y above the centre of the guide. The line AC represents a wavefront, and hence the phases at the points A and C must be the same. Thus the phase difference between the two rays meeting at C, $\Delta\phi(y)$, arises from a path difference of AB + BC together with a phase change of ϕ on reflection at B.

We have that AB = BC $\cos 2\theta$, so that AB + BC = 2BC $\cos^2\theta$. Since we also

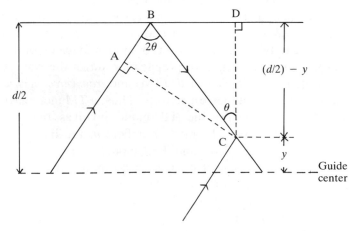

Fig. 8.6 Two 'oppositely' directed rays meeting at a point C a distance y above the guide center.

have BC = {(d/2) − y}/cos θ, it follows that

$$AB + BC = 2\{(d/2) - y\} \cos \theta$$

Thus we may write:

$$\Delta\Phi(y) = 2\left(\frac{d}{2} - y\right)\frac{2\pi n_1 \cos \theta_m}{\lambda_0} - \phi$$

By substituting for $\cos \theta_m$ from eq. (8.11) we obtain

$$\Delta\Phi(y) = 2\left(\frac{d}{2} - y\right)\frac{2\pi n_1}{\lambda_0}\frac{(m\pi + \phi)\lambda_0}{2\pi d n_1} - \phi$$

or

$$\Delta\Phi(y) = \pi m - \frac{2y}{d}(\pi m + \phi) \tag{8.15}$$

Now the resultant of two waves with a phase difference $\Delta\Phi(y)$ may be written as

$$\mathscr{E}_0\{\cos(\omega t) + \cos(\omega t + \Delta\Phi(y))\}$$

or

$$2\mathscr{E}_0 \cos\left[\omega t + \frac{\Delta\Phi(y)}{2}\right]\cos\left[\frac{\Delta\Phi(y)}{2}\right]$$

The effective amplitude of the electric field is thus given by $2\mathscr{E}_0 \cos(\Delta\Phi(y)/2)$ which, from eq. (8.15), can be written as

$$2\mathscr{E}_0 \cos\left[\frac{\pi m}{2} - \frac{y}{d}(\pi m + \phi)\right]$$

At the center of the guide ($y = 0$) this will have the value $\pm 2\mathscr{E}_0$ or zero, depending on whether m is even or odd. At the guide edges ($y = d/2$) the amplitude is $2\mathscr{E}_0 \cos(\phi/2)$.

From eqs. (8.6b) and (8.7b), we see that as θ_m moves closer to θ_c, both ψ and δ (and hence ϕ) tend to zero. Hence the closer θ_m is to θ_c (or the higher the mode number m), the higher will be the relative value of the mode field at the edge of the cladding. Since the rapidity of decay of the field within the cladding also lessens as we move to higher modes (i.e. as θ_m approaches θ_c – see the discussion on evanescent waves in section 8.1), we may conclude that higher order modes carry along with them a higher proportion of mode energy within the cladding. This trend may be observed in Fig. 8.7, which shows the variation in the electric field amplitude across a planar waveguide for the first four TE modes. In fact, for the particular waveguide parameters chosen, these are the *only* TE modes allowed to propagate. Fig. 8.8 shows the ray paths corresponding to the four modes of Fig. 8.7.

Because of the different ray angles it is evident that the modes will traverse the same length of guide in different times. This phenomenon is known as *inter-modal dispersion*. If several modes are excited by a narrow pulse of radiation, then inter-modal dispersion will cause the pulse to broaden as it travels down the guide. From the

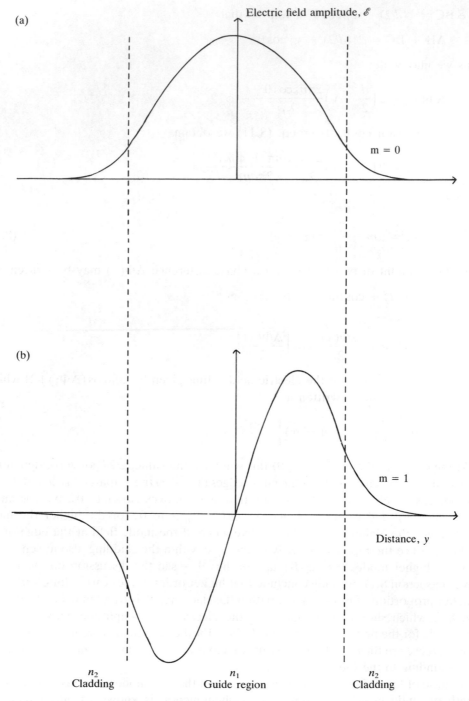

Fig. 8.7 Transverse electric field amplitudes for the four lowest modes in a symmetric planar waveguide, where $n_1 = 1.5$, $n_2 = 1$, $d = 2$ µm and $\lambda_o = 1.2$ µm.

Electric field amplitude, $\mathscr{E}$

(c)

m = 2

(d)

m = 3

Distance, y

n_2
Cladding

n_1
Guide region

n_2
Cladding

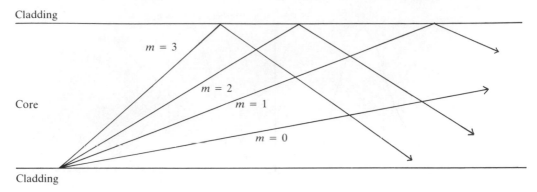

Core

Cladding

Fig. 8.8 Internal guide angles corresponding to the four TE_m modes shown in Fig. 8.7.

point of view of optical communications, this result is of great importance since it provides a limitation to the bandwith of the signal that can be carried down the guide. If, for example, we launch two optical pulses into a waveguide close together in time, inter-modal dispersion may cause the pulses to become broadened to such an extent that they overlap and become difficult to resolve into two separate pulses (see Appendix 4 for a further discussion of this point).

To estimate the magnitude of inter-modal dispersion, we consider the effective velocity of a ray with an internal angle θ down the guide. The velocity along its path is c/n_1, whilst the component of velocity along the guide axis is $(c/n_1)\cos(90° - \theta)$, i.e. $c\sin\theta/n_1$. Hence it will traverse a length L of fiber in a time τ given by $\tau = Ln_1/(c\sin\theta)$. Now the values that θ can take are restricted to lie between θ_c and $90°$, and so the maximum possible difference $\Delta\tau$ between the times taken for any two rays to traverse a distance L is given by

$$\Delta\tau = \frac{Ln_1}{c}\left(\frac{1}{\sin\theta_c} - \frac{1}{\sin(90°)}\right) = L\frac{(n_1 - n_2)n_1}{n_2c} \tag{8.16}$$

For example, over a distance of 1 km and with $n_1 = 1.53$ and $n_2 = 1.50$ we would obtain a maximum transit time difference of about 100 ns.

It should be emphasized that the ray model is itself only an approximation; a full understanding of the propagation of light down a waveguide requires finding the appropriate solutions to Maxwell's equations. A detailed treatment is contained in ref. 8.5.

One or two points arise from the more detailed treatment which are worthy of further consideration here. As with a plane wave, the mode fields depend on z and t as $\exp i(\omega t - \beta z)$, where β is known as the mode propagation constant (for plane waves the corresponding parameter is usually written as k, see eq. (1.6), and is known as the wavevector). The mode phase velocity is then ω/β, but the velocity with which energy can travel down the guide is given by the group velocity $\partial\omega/\partial\beta$ (see eq. (1.8)). The variation of ω with β for a typical planar waveguide is shown in Fig. 8.9(a). At large values of ω (or β) far from cut-off, the group velocity tends towards the value

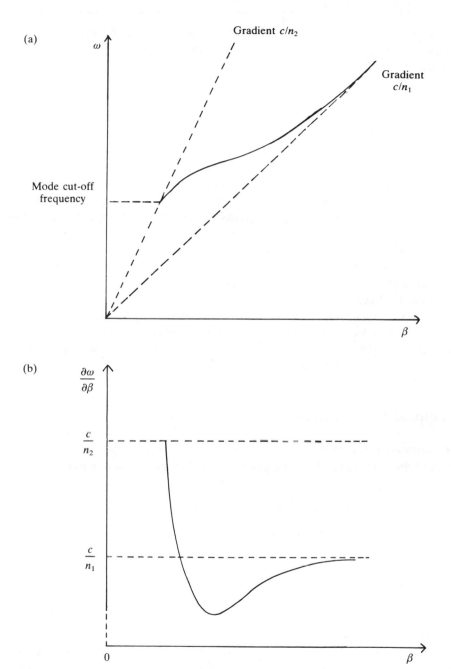

Fig. 8.9 Schematic relationship between ω and β (a), and between ∂ω/∂β and β (b) for a particular mode in a planar dielectric waveguide.

c/n_1. As we move back down the curve towards cut-off, the group velocity slowly decreases at first, to be followed by a final rapid increase to the value c/n_2 as cut-off is approached (Fig. 8.9(b)).

We can understand most of this behavior using the simple ray picture. Far from cut-off, a mode has an internal guide angle that approaches 90°, and hence its effective velocity down the guide should indeed approach c/n_1. As we move towards cut-off, the ray angle decreases and, because of the increased pathlengths involved, the effective velocity down the guide decreases. However, the final rapid increase in velocity cannot be explained using this simple picture. The discrepancy arises essentially from our neglect of the phase change on reflection. This may be incorporated into the ray model by allowing the ray to penetrate some way into the cladding at each 'reflection' as shown in Fig. 8.10. Close to cut-off, the ray spends an increasing amount of time in the cladding where it travels faster than it would in the core, thus giving rise to an increase in the overall velocity down the guide.

The values for the mode velocities at and far from cut-off also agree with what might be expected in view of the behavior of the mode fields as shown in Fig. 8.7. Far from cut-off, modes have most of their energy contained within the core and hence it seems reasonable that they should have a velocity corresponding to that of a plane wave in a medium of refractive index n_1. Modes close to cut-off, on the other hand, have an appreciable fraction of their mode energy in the cladding and hence might be expected to have velocities approaching c/n_2.

8.3 Optical fiber waveguides

There are two main types of optical fiber waveguide in use at the present time, *step index* and *graded index*. These are characterized by the refractive index profiles shown

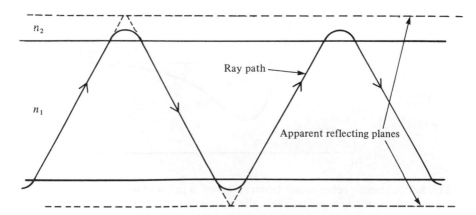

Fig. 8.10 Ray path in a waveguide illustrating the origin of the phase change on reflection in terms of penetration into the cladding. From within the guide, the ray behaves as if it were reflected from a plane a short distance inside the cladding.

Fig. 8.11 Refractive index profiles for (a) step index and (b) graded index fibers. In both cases, the cladding refractive index is n_2 whilst the maximum core refractive index is n_1.

in Fig. 8.11. First of all we consider the step index type, which has certain points of similarity to the planar waveguide.

8.3.1 Step index fibers

The path of a meridional ray, i.e. a ray that passes through the center of the guide, entering a step index fiber and undergoing total internal reflection is shown in Fig. 8.12. The angle α that the ray in the external medium (usually air) makes with the normal to the end of the guide is related to the internal angle θ by Snell's law, so that

$$\frac{\sin \alpha}{\sin (90° - \theta)} = \frac{n_1}{n_0}$$

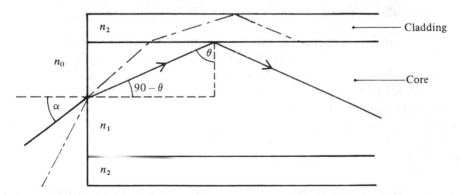

Fig. 8.12 Illustration of the path of a meridional ray (——) as it enters a circular step index waveguide from a medium of refractive index n_0. The ray is incident on the end of the fiber at an angle α to the normal. Inside the waveguide, the ray makes an angle θ with the normal to the guide axis. Also shown (—·—·) is the path of a ray that does not undergo total internal reflection at the core-cladding interface and that does not, therefore, correspond to a bound mode.

Hence

$$\sin \alpha = \frac{n_1}{n_0} \cos \theta$$

The maximum value that α can take, α_{max}, is thus determined by the minimum value that θ can take, which is of course the critical angle θ_c. Thus we have

$$n_0 \sin \alpha_{max} = n_1 (1 - \sin^2 \theta_c)^{\frac{1}{2}} = (n_1^2 - n_2^2)^{\frac{1}{2}}$$

The quantity $(n_1^2 - n_2^2)^{\frac{1}{2}}$ is known as the *numerical aperture* (NA) of the fiber, and hence

$$\alpha_{max} = \sin^{-1} \left(\frac{NA}{n_0} \right) \tag{8.17}$$

α_{max} is known as the *fiber acceptance angle* ($2\alpha_{max}$ is sometimes used and is called the *total acceptance angle*). It is often convenient to introduce the quantities

$$\Delta = (n_1^2 - n_2^2)/2n_1^2$$

and

$$\bar{n} = \frac{(n_1 + n_2)}{2}$$

Since for most optical fibers $n_1 \approx n_2$, Δ can also be written as

$$\Delta = (n_1 - n_2)/n_1$$

In terms of these

$$NA = (2\bar{n}n_1\Delta)^{\frac{1}{2}} \simeq n_1(2\Delta)^{\frac{1}{2}} \tag{8.18}$$

Also shown in Fig. 8.12 is a ray incident on the fiber end at an angle greater than α_{max}. This ray is refracted into the cladding and so does not correspond to a bound mode. It may, however, undergo total internal reflection at the cladding-air interface and thus be able to propagate down the fiber to some extent, although the attenuation is expected to be high. This is an example of a *cladding mode* (they may also be excited by direct illumination of the cladding at the end of the fiber). Although lossy, such modes can give rise to misleading results when short lengths of fiber are used to determine such parameters as fiber attenuation and numerical aperture (see section 8.6).

The meridional rays discussed so far give rise to guided modes which, as for the planar waveguide, may be designated transverse electric (TE) and transverse magnetic (TM). In this case, however, two integers. l and m, are required to completely specify the modes rather than just one (m) as before. This is essentially because the guide is now bounded in two dimensions rather than one. Thus we refer to the TE_{lm} and TM_{lm} modes.

Example 8.4 Calculation of typical fiber parameters

We consider a fiber in air with $n_1 = 1.53$, $n_2 = 1.50$ and $n_0 = 1$. The smallest value of the internal angle is θ_c, which is given by $\sin\theta_c = 1.50/1.53$, whence

$$\theta_c = 78.6°$$

We also have

$$\Delta = (1.53 - 1.50)/1.53 = 0.0196$$

and hence

$$NA = 1.53 \times (2 \times 0.0196)^{\frac{1}{2}} = 0.303$$

The value of α_{max} is then given by eq. (8.17) to be

$$\alpha_{max} = \sin^{-1}(0.303) = 17.6°$$

Meridional rays are not the only type that correspond to guided modes down fibers. There is another category called *skew rays* which describe angular 'helices' as they progress along the fiber. Figure 8.13 shows the projected path of such a ray. Because of the angles involved, components of *both $\mathscr{E}$ and $\mathscr{H}$* can be transverse to the fiber axis. Consequently, the modes originating from skew rays are designated as either HE_{lm} or EH_{lm}, depending on whether their magnetic or electric character dominates.

It is possible to launch some skew rays which, strictly speaking, do not correspond to bound modes, but which nevertheless can propagate for appreciable distances before being lost from the fiber. Such rays are called *leaky rays* (and the corresponding modes *leaky modes*).

As may be imagined, because of their rather involved geometry, skew rays are somewhat difficult to deal with. They have larger launch angles than do meridional rays, and thus increase the light gathering ability of the fiber. Strictly speaking, therefore, the maximum acceptance angle of the fiber will be greater than that given by eq. (8.18); however, for most purposes this extra complication is usually ignored.

In most practical waveguides, the refractive indices of core and cladding differ from each other by only a few percent and in this case it may be shown (ref. 8.6) that the full set of modes (i e. EH, HE and TE) can be approximated by a single set called *linearly polarized* (LP_{lm}) modes. The electric field intensity profiles of three such modes are illustrated in Fig. 8.14. An LP_{lm} mode in general has m field maxima along a radius vector and $2l$ field maxima round a circumference. On a ray picture, l is a measure of the degree of helical propagation: the larger its value the tighter the helix. The integer m, on the other hand, is related to the angle θ: the larger m the smaller the values of θ involved.

The number of modes that can be propagated may be related to a parameter V which is very similar to that used for planar waveguides and defined by eq. (8.14); here we put

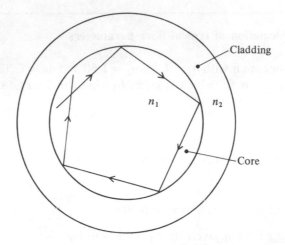

Fig. 8.13 Path of a skew ray in a circular step index fiber seen in a projection normal to the fiber axis.

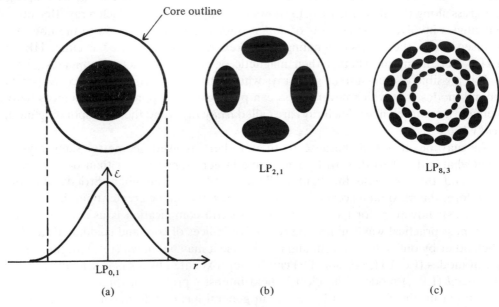

Fig. 8.14 Electric field distributions in three circular waveguide modes.

$$V = \frac{2\pi a}{\lambda_0}(n_1^2 - n_2^2)^{\frac{1}{2}} = \frac{2\pi a}{\lambda_0}(NA) \tag{8.19}$$

where a is the core radius.

When $V \gg 1$, the number of modes, N, that can propagate is given approximately by (ref. 8.7),

$$N \simeq \frac{V^2}{2} \tag{8.20a}$$

By substituting for V from eq. (8.19) and also making the approximation $\bar{n} \simeq n_1$ we obtain

$$N \simeq a^2 \left(\frac{2\pi}{\lambda_0}\right)^2 n_1^2 \Delta \tag{8.20b}$$

Example 8.5 Number of modes in a fiber

We consider a fiber where, as before, $n_1 = 1.53$, $n_2 = 1.50$ and $\lambda_0 = 1$ μm. If the core radius is 50 μm, then the value of V is given by eq. (8.19) as

$$V = \frac{2\pi \times 50 \times (1.53^2 - 1.50^2)^{\frac{1}{2}}}{1} = 94.72$$

Thus the number of modes N able to propagate is given approximately by

$$N \simeq (94.7)^2/2 = 4486$$

8.3.2 Inter-modal dispersion

As with the planar waveguide, different modes travel down the circular waveguide with differing velocities. Again, the mode fields depend on z and t according to the factor $\exp i(\omega t - \beta z)$, with the mode group velocity being given by $\partial\omega/\partial\beta$. Each mode has its own $\omega - \beta$ relationship, which is similar to that shown in Fig. 8.9(a). The phase velocities again vary from c/n_2 at cut-off to c/n_1 far from cut-off. However, not all modes have group velocities of c/n_2 at cut-off. This is a consequence of the fact that some modes (notably those corresponding to highly 'skewed' rays) reach cut-off before an appreciable fraction of the modal energy entends into the cladding.

In the interests of simplicity, we may make the assumption that when a large number of modes are present their group velocities cover the range from c/n_1 to c/n_2. An estimate, therefore, for the difference between the times taken for the slowest and fastest modes to travel a distance L along a fiber is given by

$$\Delta\tau^{\text{step}} = \frac{Ln_1}{c} - \frac{Ln_2}{c} = \frac{Ln_1\Delta}{c} \tag{8.21a}$$

or, using eq. (8.18),

$$\Delta\tau^{\text{step}} = \frac{L(NA)^2}{2cn_1} \tag{8.21b}$$

Example 8.6 Inter-modal dispersion in step index fibers

If we take a step index fiber with a numerical aperture of 0.2 and a core refractive index of 1.45, then over a distance of 1 km we should expect a maximum pulse broadening given by eq. (8.21b) of

$$\Delta\tau^{\text{step}} = \frac{10^3(0.2)^2}{2 \times 3 \times 10^8 \times 1.45} \text{ s km}^{-1}$$

i.e.

$$\Delta\tau^{\text{step}} = 4.6 \times 10^{-8}\,\text{s km}^{-1} \text{ or } 46\,\text{ns km}^{-1}$$

In practice, inter-modal dispersion is less of a problem than might be anticipated. The reason for this is that a real fiber is not perfectly uniform and also has small kinks or microbends: these tend to 'couple' the modes – i.e. the energy that is in one particular mode may be transferred to another (see section 8.4.1). Thus, although at one particular time a certain portion of the energy may be in a 'fast' mode and so be outdistancing the other modes, later on it may be in a slower one enabling the rest to catch up. The outcome is that after an initial distance (usually about 1 km), over which dispersion is proportional to the length of fiber traversed (L), the modes attain an 'equilibrium' relative energy content and the dispersion becomes proportional to L^q where $q \approx \frac{1}{2}$. A typical experimental result illustrating this is shown in Fig. 8.15.

8.3.3 Single mode fibers

It may be shown (ref. 8.7) that when V is less than 2.405 only the LP_{01} mode can propagate. (Strictly speaking, *two* orthogonal plane polarized LP_{01} modes can propagate; however, we assume we are dealing with only one of these.) In theory, the LP_{01} mode will propagate no matter how small the V value (i.e. no matter how small the core radius). As V decreases, however, the mode field will extend increasingly into the cladding. If the field becomes significantly large at the edge of the cladding, appreciable amounts of energy may be radiated away from the fiber causing the mode to be highly attenuated.

Thus the condition for only the LP_{01} mode to propagate is given by $V < 2.405$ or, in

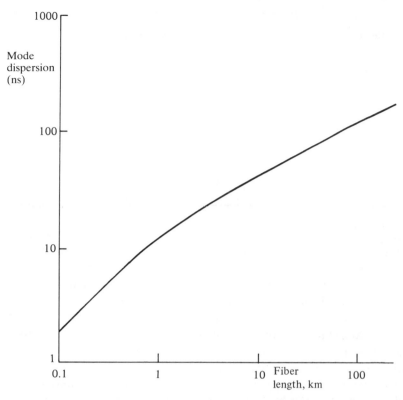

Fig. 8.15 Pulse spreading due to differences in mode velocity (inter-modal dispersion) as a function of fiber length. For short lengths (up to 1 km) the dispersion is proportional to the length traversed. With longer lengths the dispersion becomes approximately proportional to the square root of the length.

terms of the core radius

$$a < \frac{2.405\lambda_0}{2\pi(n_1^2 - n_2^2)^{\frac{1}{2}}}$$

or

$$a < \frac{2.405\lambda_0}{2\pi(\text{NA})}$$

Single mode fiber cores tend to be only a few microns in diameter (see Example 8.7) but it is an advantage from several points of view if they can be made with as large a core diameter as possible. From eq. (8.22) we see that this may be done by reducing the NA value, i.e. by making the core and cladding refractive indices very close together. This solution may present difficulties, however, since small fluctuations in the refractive indices may cause n_2 to exceed n_1 at some points along the fiber.

Example 8.7 Maximum radius of a single mode fiber

If we again take the typical values $n_1 = 1.53$, $n_2 = 1.50$ and $\lambda_0 = 1$ µm, we obtain from eq. (8.22) that the maximum radius a for single mode operation is such that

$$a < \frac{2.405 \times 10^{-6}}{2\pi(1.53^2 - 1.50^2)^{\frac{1}{2}}} \text{ m}$$

i.e.,

$$a < 1.27 \text{ µm}$$

We may also note that for a given core diameter the fiber will only be single mode when the wavelength of the radiation used is greater than a critical value. If a wavelength only just above this value is used, there is a danger that some parts of the fiber may not remain single mode because of small variations in the refractive indices. On the other hand, if the wavelength is well above the critical value then the mode may extend a long way into the cladding and thus be subject to high attenuation. To help counter this latter problem, the cladding material should be as high a quality as possible and should be reasonably thick (usually about 50 µm).

In an attempt to design single mode fibers with relatively large cores, a number of alternatives to the simple step index profile have been investigated. One of these is the 'W' index fiber. Here two cladding regions of refractive indices n_2 and n_3 surround the

Fig. 8.16 Refractive index profile in a W fiber.

core where $n_2 < n_3$ ($< n_1$), as shown in Fig. 8.16. In general, this structure allows a larger core radius to be used and also has the advantage of exhibiting reduced bending losses (see section 8.4.1). It is, however, more difficult to manufacture than a step index fiber.

One of the major advantages of using single mode fibers is that inter-modal dispersion no longer exists, and so they are capable of carrying very large bandwidth signals (other forms of pulse dispersion are still present, however, as will be seen in section 8.3.5). They are of importance in both optical fiber communications (Chapter 9) and optical fiber sensors (Chapter 10).

8.3.4 Graded index fibers

We turn now to the other main type of multimode fiber: the graded index fiber, whose refractive index profile was shown in Fig. 8.11. The main advantage of the graded index fiber is that mode dispersive effects are much less than in the step index type. However,

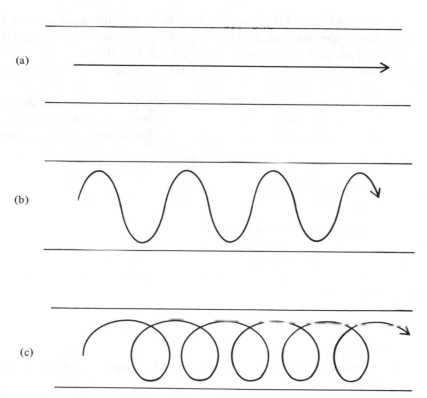

Fig. 8.17 Ray paths in a graded index fiber. We may distinguish between: the central ray (a); a meridional ray (b); and a helical ray avoiding the center (c).

the ray paths through the guide are more complicated and we may distinguish between three main categories: the *central* ray, the *meridional* rays and the *helical* rays. These are illustrated in Figs. 8.17(a)–(c). In the latter two cases, the rays follow smooth curves rather than the zigzags of step index fibers. These diagrams also provide a clue as to why inter-modal dispersion is relatively small. A helical ray, although traversing a much longer path than the central ray, does so in a region where the refractive index is less and hence the velocity is greater. Thus, to a certain extent the effects of these two factors can be made to cancel each other out resulting in very similar propagation velocities down the fiber for the two types of ray. Similar arguments apply to the meridional rays.

The variation in core refractive index is often expressed in the form:[†]

$$n(r) = n_1 \left[1 - 2\Delta \left(\frac{r}{a} \right)^\gamma \right]^{\frac{1}{2}} \qquad r < a$$

and (8.23)

$$n(r) = n_1 (1 - 2\Delta)^{\frac{1}{2}} \qquad r \geq a$$

where a is the core radius and $\Delta = (n_1 - n_2)/n_1$; here n_1 is the axial refractive index while n_2 is related to the index of the cladding. The parameter γ determines the core index variation. This particular choice of profile is usually made to simplify the mathematical treatment of wave propagation down the guide (ref. 8.8). We note that at $r = a$, since Δ is usually much less than unity, $n(a) \simeq n_1(1 - \Delta) = n_2$.

The guided modes are similar to those in the step index guide and may also be referred to as LP_{lm} modes. In fact the step index profile can readily be seen to be a particular case of the profile of eq. (8.23) with $\gamma = \infty$. The number of guided modes, N, is given by:

$$N = \left(\frac{\gamma}{\gamma + 2} \right) a^2 \left(\frac{2\pi}{\lambda_0} \right)^2 n_1^2 \Delta \tag{8.24}$$

which again reduces to the step index result (eq. (8.20b)) when $\gamma = \infty$.

It may be shown (ref. 8.9) that inter-modal dispersion is at a minimum when γ is close to the value $2(1 - 1.2\Delta)$ and that with a monochromatic source the time difference, $\Delta\tau^{\gamma opt}$ between the fastest and slowest modes over a fiber length L is given by the approximate expression

$$\Delta\tau^{\gamma opt} \approx \frac{Ln_1}{8c} \Delta^2 \tag{8.25}$$

A comparison with eq. (8.21a) shows that mode dispersion, as expected, is dramatically reduced; we have

---

[†] The profile parameter γ used here is usually referred to as α in most texts; we have made the change to avoid confusion with the exterior waveguide angle.

$$\frac{\Delta\tau^{\gamma opt}}{\Delta\tau^{\gamma=\infty}} \approx \frac{\Delta}{8} \quad \text{(i.e. much less than 1)} \tag{8.26}$$

Thus for a 1 km fiber which $n_1 = 1.53$ and $\Delta = 0.03$, the step index profile gives $\Delta\tau^{\gamma=\infty} = 150$ ns, whereas a graded index fiber of optimum profile yields $\Delta\tau^{\gamma opt} \simeq 0.56$ ns.

However, small deviations from the optimum profile, such as are likely to occur during manufacture, can easily worsen the mode dispersion obtained. Figure 8.18 shows the variation in mode dispersion as a function of γ. The extreme sensitivity of mode dispersion to slight variations in γ near the minimum is evident.

In practice, the situation is further complicated by the fact that all sources have a finite spectral width and in general the variation of refractive index with wavelength is different in the core from that in the cladding. This has the effect that now the optimum value of γ becomes (ref. 8.10):

$$\gamma_{opt} = 2\left[1 - Q - \Delta\left[\frac{(2-Q)(3-2Q)}{5-4Q}\right]\right] \tag{8.27}$$

where

$$Q = \frac{\lambda}{\Delta}\frac{d\Delta}{d\lambda}$$

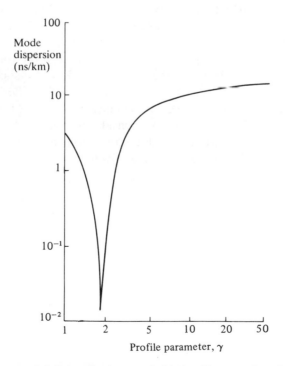

Fig. 8.18 Inter-modal dispersion in a graded index fiber as a function of the profile index γ. The curve has a very sharp minimum at a value of γ just less than two.

γ_{opt} depends on wavelength and on the fiber material; thus a fiber optimized for operation at one wavelength will not necessarily be optimized when operated at another. The variation of Δ with λ is referred to as *profile dispersion* and obviously must be taken into account when designing graded index fibers for small intermodal dispersion.

If we do have an optimized profile so that $\gamma \approx 2$, it is interesting to note from eq. (8.24) that the number of guided modes that a graded index fiber can support is about half that for a step index fiber ($\gamma = \infty$) with the same radius. This implies that the amount of energy that can be coupled into the graded index fiber is only about half of that which can be coupled into a comparable step index fiber. This conclusion has obvious repercussions in fiber optical communication systems and is taken up again in section 9.3.2.

8.3.5 Low dispersion fibers

The obvious way of avoiding inter-modal dispersion is to use single mode fibers. Even when this is done, however, there still remain three other sources of pulse dispersion that are present when the propagating radiation is not perfectly monochromatic. One of these, profile dispersion, was briefly mentioned at the end of the last section. The other two are *waveguide dispersion*, and *material dispersion*.

Material dispersion is due to the wavelength dependence of the refractive index and would also apply to a plane wave traveling in a medium of infinite extent. Waveguide dispersion arises because the mode velocity depends on wavelength via the parameter β (see Fig. 8.9(a)), and this would be true even if the refractive index were independent of wavelength. Provided the guide dimensions are such that the mode is not close to cut-off, waveguide dispersion is usually much smaller than material dispersion. Material dispersion is of special significance in silica-based fibers since it becomes zero at about 1.3 μm and we now consider it in more detail.

We saw in section 1.2 (eq. 1.8) that a pulse of radiation consisting of a finite spread of wavelengths travels with the *group velocity* v_g given by

$$v_g = \frac{d\omega}{dk}$$

Since $\omega = 2\pi v$ and $k = 2\pi/\lambda$

$$v_g = \frac{dv}{d\left(\dfrac{1}{\lambda}\right)} = -\lambda^2 \frac{dv}{d\lambda}$$

Using $v = c/\lambda n$ we have

$$v_g = -c\lambda^2 \left(-\frac{1}{\lambda^2 n} - \frac{1}{\lambda n^2} \frac{dn}{d\lambda} \right)$$

$$v_g = \frac{c}{n}\left(1 + \frac{\lambda}{n}\frac{dn}{d\lambda}\right) \tag{8.28}$$

If the wavelength spread is of width $\Delta\lambda$, then the spread in group velocity Δv_g is given by

$$\Delta v_g = \frac{dv_g}{d\lambda}\Delta\lambda$$

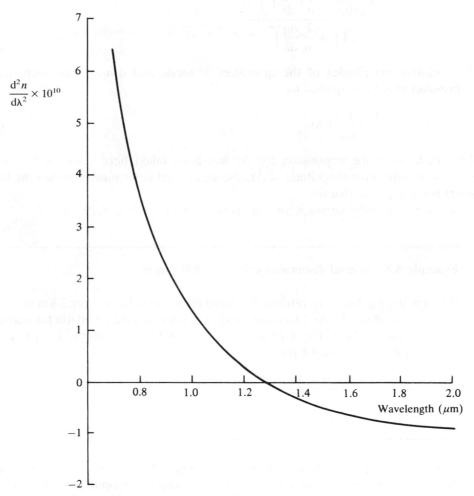

Fig. 8.19 Variation of the quantity $d^2n/d\lambda^2$ with wavelength for pure silica (SiO_2). The function becomes zero at about 1.3 µm, which results in very small values for the material dispersion near this wavelength. The addition of dopants displaces the curve slightly.

Using eq. (8.28) to evaluate $dv_g/d\lambda$ we have

$$\Delta v_g = \frac{c\lambda}{n^2}\left[\frac{d^2n}{d\lambda^2} - \frac{2}{n}\left(\frac{dn}{d\lambda}\right)^2\right]\Delta\lambda$$

Hence the spread in time $\Delta\tau$ of an initially very narrow pulse after traveling a distance L is given by

$$\Delta\tau = -\frac{L\Delta v_g}{v_g^2}$$

$$= -\frac{L\lambda\left[\frac{d^2n}{d\lambda^2} - \frac{2}{n}\left(\frac{dn}{d\lambda}\right)^2\right]\Delta\lambda}{c\left(1 + \frac{\lambda}{n}\frac{dn}{d\lambda}\right)^2}$$

The relative magnitudes of the quantities n, $dn/d\lambda$ and $d^2n/d\lambda^2$ are such that the expression may be simplified to

$$\Delta\tau = \left|\frac{-L\lambda}{c}\left(\frac{d^2n}{d\lambda^2}\right)\Delta\lambda\right| \tag{8.29}$$

The modulus of the expression for $\Delta\tau$ has been taken here since usually we are interested only in the magnitude of $\Delta\tau$, the sign in fact determines whether the long or short wavelengths arrive first.

A graph of $d^2n/d\lambda^2$ versus λ for pure silica is shown in Fig. 8.19.

Example 8.8 Material dispersion using an LED source

We now use eq. (8.29) to estimate material dispersion effects over 1 km of silica-based fiber when a GaAs LED operating at 850 nm and with a 50 nm linewidth is used as a source. From Fig. 8.19 we have that, at 850 nm, $d^2n/d\lambda^2 = 3 \times 10^{10}$ m^{-2}; thus, using eq. (8.29) we have

$$\Delta\tau = \frac{10^3 \times 850 \times 10^{-9} \times 3 \times 10^{10} \times 50 \times 10^{-9}}{3 \times 10^8}$$

$$= 4.3 \times 10^{-9} \text{ s}$$

Inspection of Fig. 8.19 shows that at about 1.3 μm, $d^2n/d\lambda^2$ becomes zero and hence at this wavelength the material dispersion term becomes very small indeed. However, a more interesting fact is that the *sign* of the dispersion term changes in passing through this point and that above 1.3 μm it has the opposite sign to the dispersion introduced by the waveguide dispersion and profile dispersion terms. It would seem, therefore, that it should be possible to obtain zero total dispersion in this region. The exact wavelength at which this occurs depends in a complicated way on the core diameter, refractive index profile and fiber composition (ref. 8.11) but it would appear perfectly feasible for the zero total dispersion point to be made to lie between 1.3 and 1.8 µm. This region is also of interest in that minimum absorption losses for silica fibers occur around 1.3 µm and 1.55 µm (see section 8.4.2). Zero overall dispersion is difficult to achieve in practice, however, because of the strict control that it is necessary to exercise over the core diameter and refractive index profile. To date, the lowest dispersions obtained in single mode fibers have been of the order of a few ps km^{-1} nm^{-1} (see for example ref. 8.12).

8.4 Losses in fibers

We may divide fiber losses into two categories (a) those which result from the distortion of the fiber from the ideal straight-line configuration and (b) those which are inherent in the fiber itself.

8.4.1 Bending losses

We may fairly readily appreciate why bends in fibers should give rise to losses by referring to Fig. 8.20. It will be seen that the part of the mode which is on the outside of the bend will need to travel faster than that on the inside to maintain a wavefront that is perpendicular to the propagation direction. Now each mode extends, in theory, an

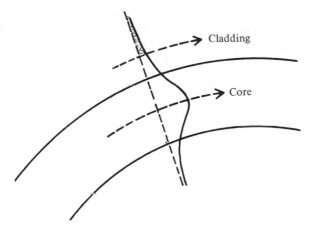

Fig. 8.20 Illustration of the mechanism of radiation loss in fibers at bends. To maintain a plane wavefront, a part of the mode (shown shaded on the diagram) may have to travel at velocities greater than that of light in the cladding. Since this is not possible, this portion of the mode must be radiated away.

infinite distance into the cladding despite the exponential decline of the electric field intensity within it (see eq. (8.9)). Consequently some part of the mode in the cladding will find itself attempting to travel at greater than the velocity of light in that medium. Since, according to Einstein's theory of relativity, this is not possible, the energy associated with this particular part of the mode must be radiated away. It is reasonable to deduce that the loss will be greater (a) for bends with smaller radii of curvature and (b) for those modes which extend most into the cladding. The loss can generally be represented by an absorption coefficient α_B, where (ref. 8.13)

$$\alpha_B = C \exp\left(-\frac{R}{R_c}\right)$$

Here C is a constant, R is the radius of curvature of the fiber bend and R_c is given by $R_c = a/(\mathrm{NA})^2$ where a is the fiber radius. It is evident that bends with radii of curvature of the order of magnitude of the fiber radius will give rise to heavy losses. However, the mechanical properties of the fiber usually cause the fiber to break long before such small-curvature bends can be achieved, so that losses from such 'macrobends' tends to be small.

What can give rise to appreciable losses is a continuous succession of small deformations or 'microbends' which can result when the fiber is pressed up against a surface which is not perfectly smooth (Fig. 8.21). To see why this might be, we consider the effect of a small angular change α in the direction of the fiber on a ray traveling down the fiber (Fig. 8.22). If the ray initially has an internal angle of θ, then after the discontinuity it will have become θ' ($= \theta - \alpha$). Effectively, therefore, the energy has been changed from one mode to another, i.e. we have *mode coupling*. In addition, if θ' turns out to be less than θ_c then the ray will be refracted out of the fiber. Since α is assumed to be small, only those modes with internal angles close to θ_c are in danger of being lost. A finite bend may be regarded as being made up of a whole series of such discontinuities. Thus we may conclude that, as far as bending loss is concerned, higher order modes are much more likely to be lost from the fiber than lower order modes. A more detailed analysis (ref. 8.14) shows that in the presence of a periodic deformation of period Λ, modes with propagation constants β_1, β_2 will be strongly coupled together provided that:

$$\Lambda^{-1} = \frac{|\beta_1 - \beta_2|}{2\pi} \tag{8.30}$$

It thus turns out that higher order modes are coupled with relatively small values of Λ, whilst adjacent lower order modes are coupled with larger Λ values. Here 'small'

Fig. 8.21 Small kinks or 'microbends' can be formed when a fiber is pressed against a surface that is not perfectly smooth.

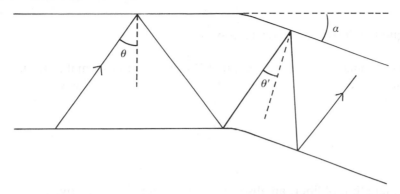

Fig. 8.22 In the presence of a small 'kink' of angle α in a fiber, the internal angles of the rays within the fiber will be altered. For the particular ray shown, θ changes to θ' where $\theta' = \theta - \alpha$.

means typically of the order of 1 mm, whilst 'large' means about 20 mm or more. Since coupling between higher order modes can cause loss because of coupling into non-guided modes, periodic deformations of the order of a few millimeters (i.e. micro-bends) are best avoided. Microbends can be caused by the bare fiber being pressed against a protective jacket or wound tightly on a drum. Unless some care is taken, microbending loss can add several decibels per kilometer to the attenuation of a fiber (ref. 8.15).

8.4.2 Intrinsic fiber losses

Losses intrinsic to fibers have two main sources: (a) scattering losses and (b) absorption losses.

Scattering losses. We have assumed, when discussing light propagation, that the fibers are homogeneous. Most fibers are made from glasses, however, which have a 'dis-ordered' structure. This disorder may be either structural or compositional in origin. In structural disorder, the same basic molecular units are present throughout the material but these are connected together in an essentially random way. In compositional disorder, on the other hand, the exact chemical composition varies from place to place. Whichever of these types of disorder is present, the net effect is a fluctuation in refractive index through the material. If the scale of these fluctuations is of the order of $\lambda/10$ or less, each irregularity acts as a point source scattering center. This type of scattering is known as *Rayleigh scattering* and is characterized by giving rise to an effective absorption coefficient that varies as λ^{-4} (ref. 8.16).

In detail, the absorption coefficient due to Rayleigh scattering (α_R) can be written

$$\alpha_R = \frac{8\pi^3}{3\lambda_0^4} n^8 p^2 \beta k T_F \tag{8.31}$$

where n is the refractive index, p the photoelastic coefficient, β the isothermal compressibility at the temperature T_F which itself is the temperature at which the disorder becomes effectively 'frozen in' as the glass cools.

Example 8.9 Rayleigh scattering loss

We may estimate the contribution that Rayleigh scattering makes to the attenuation of a silica fiber at a wavelength of 1 μm. Taking the values $n = 1.46$, $p = 0.286$, $\beta = 7 \times 10^{-11}$ m^2 N^{-1} and $T_F = 1400$ K, we obtain from eq. (8.31)

$$\alpha_R = \frac{8\pi^3}{3(10^{-6})^4} \times (1.46)^8 \times (0.286)^2 \times 7 \times 10^{-11} \times 1.38 \times 10^{-23} \times 1400$$

$$= 1.89 \times 10^{-4} \text{ m}^{-1}$$

Over a length L of fiber, an absorption coefficient of α_R will give rise to a loss of $-10 \log_{10}\{\exp(-\alpha_R L)\}$ dB. Over a kilometer of fiber, we should then expect a loss of

$$-10 \log_{10}\{\exp(-1.89 \times 10^{-4} \times 10^3)\} \text{ dB km}^{-1} \text{ or } 0.82 \text{ dB km}^{-1}.$$

Lower attenuations will be obtained at longer wavelengths. At 1.55 μm, for example (where minimum attenuations are to be expected), the figures are $\alpha_R = 3.27 \times 10^{-5}$ m^{-1} and 0.14 dB km^{-1}.

Absorption losses. Absorption losses in the visible and near-infrared regions arise mainly from the presence of impurities, particularly traces of transition metal ions

Fig. 8.23 Illustration of the dramatic reduction in minimum fiber absorption loss over the first fifteen years of manufacture.

(e.g. Fe^{3+}, Cu^{2+}) or hydroxyl (OH^-) ions. The latter are responsible for absorption peaks at 0.95 µm, 1.24 µm and 1.39 µm. Most of the dramatic successes in reducing fiber losses to date have come about because of better control of impurity concentrations (see Fig. 8.23).

At wavelengths greater than about 1.6 µm, the main losses are due to transitions between vibrational states of the lattice. Although the actual fundamental absorption peaks occur at wavelengths well into the infrared (in SiO_2, for example, the main peak is at 9 µm), the anharmonic nature of the interatomic forces gives rise to combination and overtone bands – i.e. an incoming photon can simultaneously excite two or more fundamental lattice vibrations (or phonons). Thus, a number of strong absorption

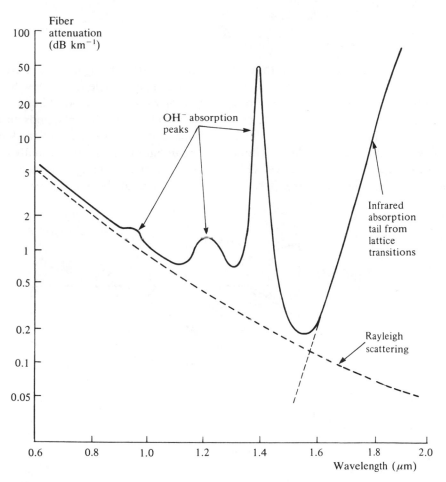

Fig. 8.24 Typical attenuation versus wavelength plot for a silica-based optical fiber. The contribution from Rayleigh scattering is shown, as are the other two main loss mechanisms, namely the infrared absorption tail and the hydroxyl (OH^-) absorption peaks.

bands extend all the way down to about 3 µm with appreciable absorption still occurring below 2 µm. To date, some of the lowest attenuations have been obtained with GeO_2 doped SiO_2 fibers (ref. 8.17), which have shown minimum attenuations of about 0.2 dB km^{-1} at 1.55 µm. This figure is quite close to the limit set by Rayleigh scattering (see Example 8.9). A typical attenuation curve is shown in Fig. 8.24. It is evident that the larger the wavelength at which the lattice vibrational transitions commence, then the lower the minimum attainable attenuation is likely to be. It may be that the 'best' material for optical fibers has not yet been found (see section 9.3.7). A more detailed discussion of absorption losses is contained in ref. 8.18.

8.5 Fiber jointing

8.5.1 Single fiber connectors

The main difficulties associated with joining single fiber strands together arise from the small core diameters that must be accurately aligned. These range from a few hundred microns for multimode fibers down to a few microns for single mode fibers.

The simplest technique for making permanent joins is to butt-join the ends and then glue them together. When done with care, this procedure can result in very low loss joins (less than 0.1 dB per splice). The initial alignment of the fibers is usually carried out using some form of 'V-groove' (Fig. 8.25). Instead of glueing, another jointing method is to fuse the ends together. If the temperature of the joint is raised to the

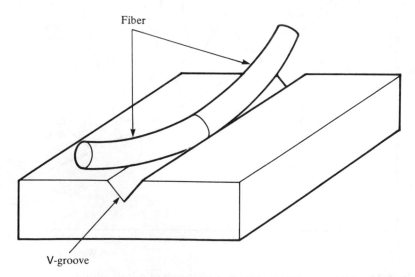

Fig. 8.25 Illustration of the use of a V-groove for the alignment of fiber ends prior to jointing.

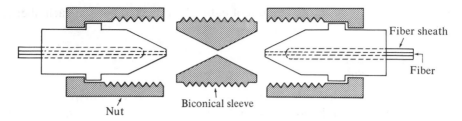

Fig. 8.26 Schematic diagram of a demountable connector for butt-joining optical fibers.

material softening point, then surface tension helps to pull the ends into alignment and the ends then fuse together. Silica fibers require a higher temperature ($\approx 2000°C$) than do glasses ($\approx 700°C$) and an electric arc is usually employed for these.

Often, however, a join needs to be demountable. The commonest method involves holding the butt-jointed ends in mechanical contact. A schematic diagram of a connector to do this is shown in Fig. 8.26. Such connectors, unless very well made, have rather greater losses than those associated with permanent splices. Two main factors contribute to these losses: firstly, Fresnel losses, caused by back reflection at the glass/air interfaces and secondly, end alignment losses.

The Fresnel loss contribution is usually comparatively small and is readily estimated.

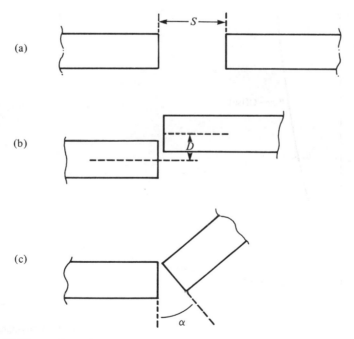

Fig. 8.27 Illustration of the three types of fiber misalignment mentioned in the text: (a) separation; (b) offset; and (c) angular misalignments. The parameters used to describe these (S, D and α) are also shown.

Assuming normal incidence. the amount of light, R_F, reflected back at each fiber end is given by:

$$R_F = \left(\frac{n_1 - n_0}{n_1 + n_0}\right)^2$$

where n_0 is the refractive index of the medium between the fibers and n_1 that of the fiber core. If, for example, we take $n_1 = 1.45$ and $n_0 = 1$ (for air) then we obtain $R_F = 0.034$. This corresponds to a total transmission loss at a join between two fibers of $2 \times 10 \log_{10}\{1/(1 - 0.034)\}$ dB or 0.3 dB.

Fiber end misalignment losses are more difficult to estimate and depend on a knowledge of the power distribution across the fiber faces. Careful control must be maintained over (a) the distance between the fibers along the fiber axis, (b) the offset distance perpendicular to the fiber axis and (c) the angle between the two fiber axes (see Fig. 8.27). Typical coupling losses as a function of the first two of these misalignment parameters are shown in Fig. 8.28 (see also Problem 8.13).

Both the Fresnel and misalignment losses may be reduced by filling the gaps between the fiber ends with a fluid or grease with a refractive index as close as possible to that of the fiber core. However, in practical jointing situations the problem of dirt being picked up in the fluid often outweighs the potential advantages of doing this.

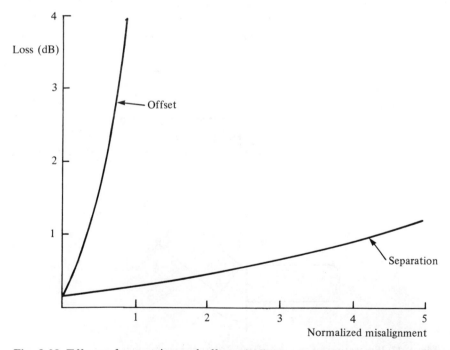

Fig. 8.28 Effects of separation and offset misalignments on fiber connection losses. The misalignments are expressed in terms of the normalized parameters S/a (separation) and D/a (offset), where a is the fiber core radius. The residual loss when the misalignment is reduced to zero is due to Fresnel losses.

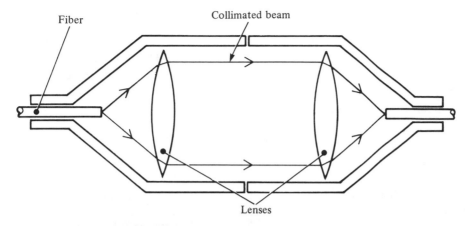

Fig. 8.29 Fiber coupling along a beam expander. The fiber ends are situated at the focal points of the two lenses resulting in an expanded collimated beam between the lenses.

Another type of demountable connector uses a beam expander (Fig. 8.29). Each half of the connector contains a lens with the fiber end at its focal point. Light emerging from one end of a fiber becomes a relatively broad, collimated beam in the region between the lenses and is then focused back onto the other fiber end. Since the expanded beam width can be much greater than the fiber diameter, losses due to offset misalignments are now considerably reduced. Great care is needed, however, in the initial alignment of the fiber with the lens in each half of the connector. An advantage of this type of connector is that further optical components (for example beam splitters) can be inserted into the collimated beam between the lenses.

Most of the jointing techniques described require a square, smooth end to the fibers. A simple and effective way of doing this is to stretch the fiber over a rod of relatively large diameter and then to 'nick' the fiber with a sharp knife edge (Fig. 8.30). A more traditional method is to polish the ends mechanically whilst they are held in a suitable jig.

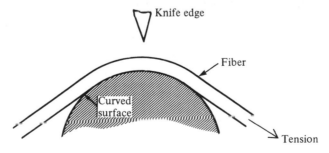

Fig. 8.30 Illustration of one technique used to prepare plane fiber ends normal to the fiber axis. The fiber is held in tension over a curved surface. Then the surface is scored with a tungsten carbide or diamond knife edge. This initiates crack formation and, provided the tension and radius of curvature are set correctly, a smooth break is obtained.

8.5.2 Multi-fiber couplers

A problem arising in some applications of optical fibers is how to share a signal from one or more input fibers amongst several output fibers. Often there are as many input fibers as there are output fibers. In this case, the device is usually symmetrical and may be operated in 'reverse', i.e. with outputs becoming inputs and vice versa. A four port device is shown schematically in Fig. 8.31. Here, an input to either of fibers 1 or 2 will result in an output from other fibers 3 and 4. The way in which the input power in fiber 1 is distributed amongst the output fibers is governed by the *insertion loss L* where *L* is defined by

$$L = -10 \log_{10}\left(\frac{P_j}{P_i}\right) \tag{8.32}$$

Here P_i is the power flowing in the ith input fiber and P_j is the power flowing in the jth output fiber. P_j will always be less than P_i for two reasons; first, because of the splitting of the input power amongst the output fibers, and secondly because there will inevitably be some loss of optical power within the device itself. The total power loss is described by the so-called *excess loss* (EL) where

$$EL = -10 \log_{10}\left(\frac{\sum P_j}{\sum P_i}\right) \tag{8.33}$$

Another term used is the *output ratio*. This describes the nominal insertion loss from one input fiber to each of the output fibers (as in eq. (8.32)) assuming there is no power loss within the device. It thus represents an ideal situation rather than the actual performance of the device.

A simple way of implementing such a coupler is to butt the input fibers up against the end of a 'mixing rod'. This is essentially a short length of large diameter optical fiber. The output fibers are similarly butted up against the other end of the rod (Fig. 8.32(a)). The rod must be long enough to insure that the output from each input fiber becomes sufficiently diffused within the rod to cause an even spread of energy over the output face. To avoid excessive loss, it is best if the fibers can be 'close packed' over the end of the rod, a situation that can only be achieved for certain numbers (e.g. 7, 19, etc.) of input fibers (see Fig. 8.32(b)). A calculation of the insertion loss resulting from this type of coupler is given in Example 8.10.

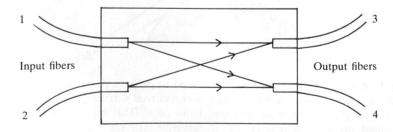

Fig. 8.31 Schematic representation of a four port (2 × 2) fiber coupler. Power flowing in either fiber 1 or 2 will be distributed in some predetermined ratio into fibers 3 and 4.

Fig. 8.32 Design for a 14 port (7 × 7) fiber coupler using a 'mixing rod'. The overall scheme is shown in (a), whilst (b) shows how seven fibers can form a 'close-packed' array across the face of the mixing rod.

Example 8.10 Fiber coupler insertion loss

We consider the rod type coupler illustrated in Fig. 8.32(a). We suppose that seven input and output fibers are used, each having core and core + cladding diameters of d_1 and d_2 respectively. If the mixing rod has a diameter of $3d_2$, the close-packed arrangement of Fig. 8.32(b) can be obtained. Losses will inevitably result because the cores of the output fibers have a combined area that is less than that of the end of the mixing rod. Assuming a uniform spread of light from any one input fiber over the end of the coupling rod, we may write $P_i = C\pi(3d_2)^2/4$ whilst $\Sigma_j P_j = 7C\pi d_1^2/4$, where C is a constant. Thus from eq. (8.33) the insertion loss is given by

$$L_i = -10 \log_{10}\left(\frac{7C\pi d_1^2/4}{c\pi(3d_2)^2/4}\right) \quad \text{or} \quad -10 \log_{10}\left(\frac{7d_1^2}{9d_2^2}\right)$$

Taking $d_1 = 200$ μm and $d_2 = 300$ μm gives $L = 4.6$ dB.

Of course, this figure represents the minimum possible loss; inevitably there will be some Fresnel loss at the ends of the fibers and rod, and also some loss within the rod itself.

Fig. 8.33 Fabrication of a fused biconical taper coupler.

Another technique used to make multi-mode fiber couplers is to twist the fibers around each other and then subject them to tension whilst heating in an oxypropane torch flame (Fig. 8.33). The bundle fuses together and then elongates, the resulting device is known as a *fused biconical taper coupler*. The coupling mechanism is believed to be as follows: when light from one fiber enters the fused region, the narrowing core causes mode conversion into cladding modes. These subsequently spread out across the whole of the cross section in the fused region. Then, when the fibers separate, the cladding modes are converted back into core modes – but with the energy now distributed amongst all the fibers. Such couplers generally exhibit low coupling losses, but the coupling efficiency can vary significantly between supposedly identical devices. This should not be too surprising, perhaps, in view of the mode of manufacture.

Single mode fiber couplers can also be made by this technique. In this case, however, the coupling mechanism is thought to be somewhat different. Thus, when the radiation encounters the narrowing fiber core, the mode energy extends further and further into the cladding. Some will overlap with the other cores, and hence couple energy into them. This is illustrated in Fig. 8.34. The coupling mechanism is wavelength dependent

Fig. 8.34 Coupling mechanism in a single mode biconical taper coupler involving evanescent mode overlap.

and so the intended output ratio will only be obtained at one particular wavelength. One of the most useful single mode couplers is a four port device with a coupling efficiency of 50% (see, for example, section 10.1.3.1). When expressed as a decibel loss, 50% is equivalent to 3 dB, and hence such couplers are known as 3 dB couplers.

8.6 Measurement of fiber characteristics

8.6.1 Introduction

In this section, we describe techniques used to measure those properties such as loss and bandwidth which are important in optical fiber communications. Measurements of purely mechanical properties are not included here for reasons of space (ref. 8.19 may be consulted for a fuller discussion of all types of measurement). Before addressing ourselves to specific techniques, however, it is useful to understand the importance of obtaining the correct launch conditions.

Suppose we illuminate the whole of a fiber end with a uniform highly convergent beam of light. All modes in the fiber will then be excited, including leaky and cladding modes. If we were to examine the mode structure a considerable distance along the fiber, almost all of the lossy modes would have died out. In addition, mode coupling would have thoroughly 'scrambled' the remaining modes to generate an 'equilibrium' mode distribution that would then remain more or less the same throughout the rest of the length of the fiber. In principle, such a distribution would be obtained whatever the launch conditions, provided the initial portion of fiber were sufficiently long. In practical terms, however, this might amount to several kilometers. Most of the properties we are concerned with here do indeed depend of the mode distribution, so that it makes sense to try and measure these for the equilibrium distribution. On the other hand, it may be very inconvenient if many kilometers of fiber have to be used to obtain this distribution. It would obviously be preferable if the launch conditions were such as to set up equilibrium right from the start. Unfortunately this is not at all easy to do, although several procedures can be adopted that go some way towards realizing this goal.

First, to avoid launching cladding and leaky modes we must be careful to illuminate only the core area and to use a light cone whose numerical aperture does not exceed that of the fiber.

One device that aims to produce an equilibrium 'mix' of modes over a short length of fiber is the *mode scrambler*. In this, a large amount of mode coupling is deliberately introduced as a result of bending or microbending distortion (see section 8.4.1). For example, a few turns of the fiber may be wrapped round a post or mandrel a few tens of millimeters in diameter (Fig. 8.35(a)). Alternatively, the fiber may be passed through a series of posts to produce a serpentine distortion (Fig. 8.35(b)). A simple variant of the latter technique is to sandwich the fiber between two layers of coarse sandpaper over a length of some tens of millimeters.

The use of a mode scrambler may itself cause cladding modes to be excited, and these will have to be removed subsequently. One device designed to remove cladding modes

(a)

(b)

Fig. 8.35 Techniques for producing an equilibrium mode distribution. In (a), the 'mandrel wrap' method, a few turns of fiber are taken around a *rod*. In (b), the fiber makes a serpentine path through a series of posts.

is called a *mode stripper*, which may be simply constructed by taking a short length of fiber that has been stripped of its outer protective cladding and pressing it against a felt pad that has been saturated in oil (such as paraffin oil). The oil should have a slightly higher refractive index than the cladding. Any light incident on the cladding–oil interface will then be refracted out of the fiber and absorbed within the felt.

8.6.2 Fiber attenuation measurements

At its simplest, the measurement of attenuation involves focusing a known amount of monochromatic radiation onto the end of a known length of fiber and then measuring the amount emerging from the other end (Fig. 8.36). Use of eq. (8.1) then enables the fiber attenuation to be determined. Radiation at different wavelengths can be obtained by using a white light source, such as a tungsten halogen lamp, in conjunction with either a monochromator or a set of narrow bandpass optical filters. This method ignores the inevitable losses incurred when coupling light into the fiber, although for fibers with high attenuation, this may prove acceptable. With low-loss fibers, however, the cutback technique is generally used. In this, light is launched as in the simple technique and the emerging power measured. Then, without changing the launch conditions in any way, the fiber is cut back to within a few meters of the launch end, and the emerging energy again measured. The fractional transmittance of the length of fiber removed is then given by the ratio of these two measurements, since the launch energy is the same in both cases.

As outlined in the introduction to this section, care must be taken with the launch

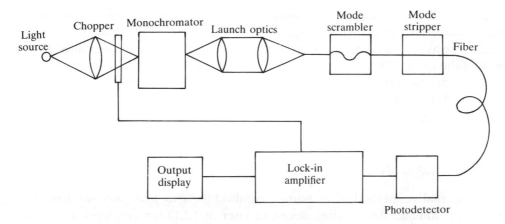

Fig. 8.36 Schematic layout of apparatus for measuring attenuation in optical fibers. To enhance the signal-to-noise ratio, an optical chopper-lock-in amplifier combination is often used.

conditions to obtain an equilibrium mode distribution (i.e. by using a mode scrambler/mode stripper combination).

The above techniques measure the total attenuation within the fiber, which, from a practical point of view, is usually all that is required. However, it may be that the individual contributions from absorption and scattering need to be assessed separately. Absorption losses may be determined by measuring the temperature rise within a given length of fiber, whilst scattering loss measurements depend on determining the amount of light scattered out of the fiber (see problem 8.14 and also ref. 8.20).

8.6.3 Fiber dispersion measurements

When a narrow optical pulse is launched into a fiber, the various dispersion mechanisms discussed in sections 8.3.2 to 8.3.5 will cause the pulse to broaden as it travels down the fiber. In the so-called *time domain* measurement of dispersion, this broadening is measured directly. A narrow optical pulse may conveniently be generated by using a semiconductor laser in conjunction with fast driving electronics. Pulses with widths of a few nanoseconds or less may be readily obtained. If the main source of dispersion is intermodal dispersion then, as in attenuation measurements, an equilibrium mode distribution will have to be created. The detector used must obviously have a fast response and often avalanche photodiodes are used. Since the pulses may be too narrow to be displayed easily on an ordinary oscilloscope, a fast sampling oscilloscope may be required.

Ideally, the input pulse should be narrow compared to the output pulse, so that it can be thought of as a 'spike' of negligible width. The shape of the output pulse is then called the *transfer function* (see Appendix 5). Normally, however, the input pulse has appreciable width and determining the transfer function becomes more difficult. A simplification often adopted is to assume that the input pulse, transfer function and output pulses are all Gaussian in shape (see Fig. A5.1(b)), i.e. that we can describe each curve by the function $A(t) = A(0) \exp(-t^2/2\sigma^2)$, where time is measured from the peak of the curve. The parameter σ essentially determines the r.m.s. pulse width. If σ_i, σ_t and σ_f are the values for the initial, transfer and final pulse widths, then it can be shown (ref. 8.21) that:

$$\sigma_f^2 = \sigma_i^2 + \sigma_t^2 \tag{8.34a}$$

i.e.

$$\sigma_t = (\sigma_f^2 - \sigma_i^2)^{1/2} \tag{8.34b}$$

An alternative method of measuring bandwidths is to take measurements in the frequency domain. An emitting device (a laser or LED for example), is amplitude modulated with a variable frequency sinusoidal waveform, and the output from the fiber is recorded as a function of frequency. This allows the bandwith of the fiber to be measured directly. In fact two bandwidths are commonly used, the optical bandwidth Δf_{op}, and the electrical bandwidth Δf_{el}. The former is defined as the frequency at which the signal falls to half its maximum low frequency value, the latter as the frequency when the signal falls to one over root two of its maximum value. Appendix 5 should be consulted for a discussion of the relationships between pulse dispersion and bandwidths. For a Gaussian pulse, the electrical bandwidth and pulse width σ are related by

$$\Delta f_{el} = \frac{1}{7.56\sigma}$$

8.6.4 Refractive index profile measurement

Several methods are available for measuring the refractive index profile of a fiber; however, none of them is entirely satisfactory. Probably the most accurate is the interferometric slab technique. A thin slice of fiber of thickness d is taken perpendicular to the fiber axis. The thickness must be uniform to within a fraction of the wavelength of the light used. Light passing through the slab will normally 'see' an optical thickness of $n(r)d$, where $n(r)$ is the refractive index a radial distance r from the fiber center. The slab is then placed in the arm of an interferometric system (see section 6.6.1). The nonuniform optical path differences through the slab then show up as distortions in the uniform fringe pattern observed in the absence of the slab. Using this technique, the refractive index can be measured to better than one part in 10^4, the main limitation being the accuracy with which the thickness of the slab can be measured. The disadvantages are that it requires precise and time consuming sample preparation and it is also, of course, destructive to the fiber.

A nondestructive technique capable of yielding reasonably accurate results is the near field scanning method. If all the guided modes are uniformly excited then the variation of light irradiance across the end of the fiber may be written (ref. 8.22)

$$I(r) = I(0)\left[\frac{n_1^2(r) - n_2^2}{n_1^2(0) - n_2^2}\right] \tag{8.35}$$

Here $I(r)$ is the irradiance observed a distance r from the center, and $n_1(r)$ and n_2 are the core and cladding refractive indices respectively. If any leaky modes are present, however, a correction factor has to be applied (ref. 8.23). The apparatus requirements are reasonably straightforward. One end of the fiber is illuminated by focusing a Lambertian source (for example a tungsten filament lamp) onto it. This insures that the guided modes are all excited equally. A magnified image of the other end of the fiber is then formed in the plane of a small-area photodiode, which is then scanned across the image allowing $I(r)$ to be determined.

The two methods described here are by no means the only ones that can be used to determine refractive index profiles; other techniques will be found in ref. 8.24.

8.6.5 Optical time domain reflectometer

All the measurement techniques described hitherto are essentially laboratory based and would be difficult to implement in the field. Furthermore, they often require access to both ends of the fiber, a situation causing not a little difficulty if the fiber ends are several kilometers apart! One instrument capable of yielding considerable information about an optical fiber system, and yet which requires access to one end only of the fiber, is the optical time domain reflectometer (OTDR).

The basis of this instrument is that a narrow pulse of radiation is launched into one end and then the amount of radiation re-emerging from this same end is monitored as a function of time after the initial launch. As the pulse travels down the fiber, it will suffer attenuation so that after a length L of fiber the pulse energy will be given by $E(L)$, where

$$E(L) = E(0)\exp(-\alpha L)$$

where α is the absorption coefficient. At every point along the fiber, a certain portion, S_R say, of the pulse energy will be lost by Rayleigh scattering. Of this a fraction, f_g (see Problem 8.17) will enter guided modes propagating back down the fiber. The amount of energy scattered back down the fiber may therefore be written $S_R f_g E(L)$. This backscattered radiation will itself be subject to attenuation as it travels back to the launch end of the fiber. The energy finally re-emerging that is due to backscattering from a point a distance L into the fiber is given by $E_b(L)$, where

$$E_b(L) = S_R f_g E(L)\exp(-\alpha L)$$

or

$$E_b(L) = S_R f_g E(0)\exp(-2\alpha L)$$

Thus

$$\log_e E_b(L) = \log_e \{S_R f_g E(0)\} - 2\alpha L \tag{8.36}$$

Now the time delay, t, between the initial launch of the pulse and the subsequent return of the backscattered energy from L may be approximated by $t = 2Ln_1/c$, so that we may transform eq. (8.36) to give

$$\log_e E_b(t) = \log_e \{S_R f_g E(0)\} - \frac{\alpha c t}{n_1} \qquad (8.37)$$

Thus a plot of the logarithm of the returning energy as a function of time should yield a straight line with gradient $-\alpha c/n_1$. If n_1 is known, α may be determined.

In this analysis, we have assumed a perfectly uniform fiber as far as α, S_R and f_g are concerned. In practice, α may well vary along the fiber and discontinuities, such as fiber joins, will give local increases in the amount of backscattering. For example, a butt join will give something like a 4% backscatter because of Fresnel reflection. These will show up on the graph as 'spikes'. In addition, the power loss within a join or splice will lead to a corresponding reduction in the signal backscattered from beyond it. The end of the fiber will be marked by a Fresnel spike with the signal falling to zero immediately afterwards. If α does in fact vary along the length of the fiber, the above graph will not be a straight line; however, the local gradient should reflect the local value of α. In commercial instruments, the returning energy is measured in decibels and time is converted into distance along the fiber. Figure 8.37 shows a typical trace.

Fig. 8.37 Typical trace from an OTDR.

One of the main difficulties in realizing this instrument is that the amount of energy returning from the fiber is very small, and even under optimum conditions it is difficult to examine more than 15 km of fiber. The light source is invariably a pulsed semiconductor laser, whilst the returning radiation is measured with an avalanche photodiode. To both launch light and to measure the returning light, some form of beam-splitter arrangement is required. Usually, the instrument sends out a continuous stream of pulses. After each of these, the amount of energy returning after a certain (variable) time delay is measured. The time delay is gradually increased, thus enabling the whole curve to be built up point by point. To increase accuracy, several readings may be taken for each delay and an average taken. Typical pulse widths are the order of 50 ns with launched energies of a few hundred milliwatts per pulse. Features on the curve can be located to within $\pm$ 5 m, or perhaps a little better. Overall, a dynamic range of something like 40 dB ($= 2 \times 20$ dB single-way loss), i.e. 20 dB of signal loss, can be tolerated over the length of fiber under inspection before the returning signal becomes too weak to measure.

8.7 Fiber materials and manufacture

Only two main types of material have been seriously considered to date for use in optical waveguides, these being plastics and glasses. Plastic fibers offer some advantages in terms of cost and ease of manufacture, but their high transmission losses preclude their use in anything other than short distance (i.e. less than a few hundred meters) optical links.

8.7.1 Glass fibers

A broad distinction may be made between glasses based on pure SiO_2 and those derived from low softening point glasses such as the sodium borosilicates, sodium calcium silicates and lead silicates. For convenience, we shall refer to these as silica fibers and glass fibers respectively. An obvious requirement of the material used is that it must be possible to vary the refractive index. Pure silica has a refractive index of 1.45 at 1 μm. B_2O_3 can be used to lower the refractive index, whilst other additives such as GeO_2 raise it. Thus a typical fiber might consist of an $SiO_2 : GeO_2$ core with a pure SiO_2 cladding. Glass fibers can be made with a relatively wide range of refractive indices, but control of the impurity content is more difficult than with silica.

At present there are two main techniques for manufacturing low loss fibers, these being the *double crucible* method and various *vapor phase deposition* techniques.

The apparatus for the former is illustrated in Fig. 8.38. Pure glass, usually in the form of rods, is fed into two platinum crucibles. At the bottom of each crucible is a circular nozzle, that of the inner vessel being concentric with that of the outer and slightly above it. The inner crucible contains the core material, the outer that of the cladding. When the temperature of the apparatus is raised sufficiently by using an external furnace, the core material flows through the inner nozzle into the center of the flow stream from the

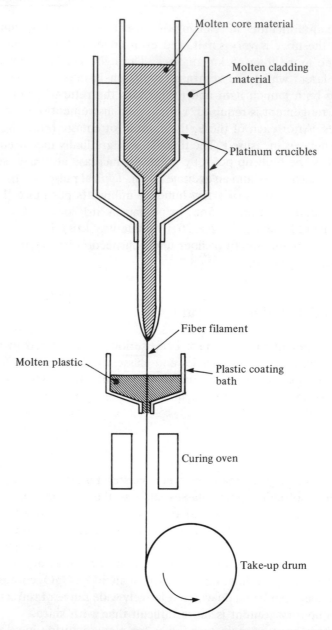

Molten core material

Molten cladding material

Platinum crucibles

Fiber filament

Molten plastic

Plastic coating bath

Curing oven

Take-up drum

Fig. 8.38 Schematic diagram of fiber drawing apparatus using the double crucible technique. Omitted for clarity is the furnace surrounding the double crucibles. Immediately the fiber is formed, it is customary to give it a protective coating of plastic by passing it through a bath of molten plastic and a curing oven.

outer crucible. Below the crucibles is a rotating drum, and the composite glass in the form of a fiber is wound onto it.

If the two types of glass remain separate, then a step index fiber will result. However, by using glasses that interdiffuse (or by having dopants that do so), graded index fibers can be obtained. One problem with this approach is that the index profile will be determined by diffusion processes, and these are usually difficult to control accurately. The resulting fibers, however, will almost certainly have smaller inter-modal dispersion than step index fibers.

A number of vapor phase deposition techniques have been developed that are capable of producing fibers having minimal impurity content. One of the most popular is the *modified chemical vapor deposition* (MCVD) method. In this, a doped silica layer is deposited onto the inner surface of a pure silica tube. The deposition occurs as a result of a chemical reaction taking place between the vapor constituents that are being passed down the tube. Typical vapors used are $SiCl_4$, $GeCl_4$ and O_2 and the reactions that take place may be written as follows:

$$SiCl_4 + O_2 \rightarrow SiO_2 + 2Cl_2$$

and

$$GeCl_4 + O_2 \rightarrow GeO_2 + 2Cl_2$$

The zone where the reaction takes place is moved along the tube by locally heating the tube to a temperature in the range 1200–1600°C with a traversing oxy-hydrogen flame (Fig. 8.39). If the process is repeated with different input concentrations of the dopant vapors, layers of different impurity concentrations may be built up sequentially. This technique thus allows a much greater control over the index profile than does the double crucible method. Once the deposition process is complete, the tube is heated to its softening temperature (≈ 2000°C). Surface tension effects then cause the tube to collapse into a solid rod called a *preform*. A fiber may subsequently be produced by drawing from the heated tip of the preform as it is lowered into a furnace (Fig. 8.40). To exercise tight control over the fiber diameter, a thickness monitoring gage is used before the fiber is drawn onto the take-up drum, and feedback is applied to the drum take-up speed. In addition, a protective plastic coating is often applied to the outside of the fiber (see Fig. 8.38 and section 8.8) by passing it through a bath of the plastic material; the resulting coating is then cured by passing it through a further furnace.

In other vapor deposition techniques, the preform may be made by growing doped silica layers on the outside of a silica rod. The reactant gasses, such as $SiCl_4$ and $GeCl_4$, are fed into H_2/O_2 flames where a stream of fine doped silica soot particles is formed. In the *outside vapor deposition* (OVD) technique, these are deposited on the outside of a rotating mandrel, the growth taking place radially. Alternatively, as in the *vapor axial deposition* (VAD) technique, deposition may take place on the end surface of a silica 'seed' rod. Both of these deposition techniques produce a porous material, and so a further production stage is required involving heating the rod to between 1000°C and

(a)

Core material formed by chemical reaction

$\approx 1600\,°C$

$SiCl_4 + GeCl_4 + O_2$

Silica tube

Traversing heat source

(b)

$\approx 2000\,°C$

Collapsed preform

Traversing heat source

Fig. 8.39 Production of fiber preform by modified chemical vapor deposition. In the first stage (a) the reactants are introduced into one end of a silica tube and the core material deposited on the inside of the tube in the reaction zone, where the temperature is maintained at about 1600°C. Several traverses of the heating assembly may be necessary to build up sufficient thickness of core material. In the second stage (b) the tube is collapsed into a solid preform rod by heating to the silica softening temperature (about 2000°C).

1500°C in the presence of chlorine. The solid preform thus formed may be drawn into a fiber in exactly the same way as described above for the MCVD technique.

8.7.2 Plastic fibers

Other types of fiber are possible using plastics. For example, fibers are available with silica cores and plastic claddings. These *plastic coated silica* (PCS) fibers are easy to manufacture; the fiber core may simply be drawn through a bath of a suitable polymer which is subsequently cured by heating to a higher temperature to provide a solid cladding. This process readily lends itself to the production of step index fibers with large core diameters where very little of the energy is carried in the cladding. Such fibers are attractive for medium distance, moderate bandwidth communication systems, where cost is a major consideration. Typical losses are of the order of 10 dB km^{-1}.

Fibers can also be made entirely from plastics, but these suffer from very high attenuations mainly because of a large Rayleigh scattering contribution. Figure 8.41

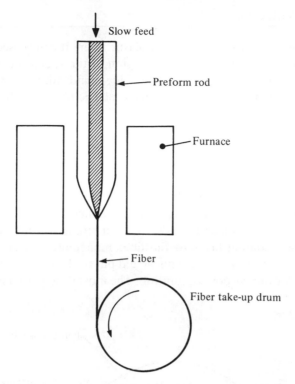

Fig. 8.40 Conversion of a preform rod into a fiber. The end of the rod is heated to its softening point and the fiber drawn off from the tip and wound onto a take-up drum. The fiber is usually coated with a layer of plastic for protection before being wound on the drum, but this process is omitted here (see Fig. 8.38).

Fig. 8.41 Typical attenuation versus wavelength plot for a wholly plastic fiber.

shows a typical attenuation spectrum for a plastic fiber. It can be seen that such fibers are only of any practical use in the visible region of the spectrum, preferably around 600–700 nm, and then only for short-distance, low-bandwidth systems. Since plastic is an inherently more flexible material than glass, plastic fibers can be made with larger diameters (up to a millimeter or so).

8.8 Fiber cables

When freshly made, the tensile strength of glass and silica fibers is very high and indeed compares favorably with steel. However, surface damage caused by handling or even atmospheric attack rapidly leads to a decrease in strength. Surface damage may be reduced by adding a coating layer to the fiber immediately after manufacture. The material used must provide a good chemical and physical barrier and yet must be fairly readily removable for the purpose of jointing. The most widely used material is kynar, a

Fig. 8.42 Typical cable designs. In (a) the coated fibers are wrapped helically round a central strengthening member, while (b) shows the BICC 'PSB' design where the bare fiber lies loosely within a cavity in the cable.

vinylidene fluoride polymer commonly used as an electrical insulator. Although this primary coating helps preserve the intrinsically high tensile strength of the fiber, it does little to protect it from external mechanical damage. One major aim is to avoid the occurrence of microbends. As we saw in section 8.4.1, the presence of microbends can lead to significant additional optical losses. Typical increases observed in practice have been of the order of 1 or 2 dB km^{-1}.

In many cable designs, the fiber is given a further coating or contained within a tight-fitting tube to make the fibers easy to handle during cable manufacture. Great care must be taken to ensure that differential thermal expansion or aging processes do not again give rise to microbends or place undue stress on the fiber. The most successful materials for this have been relatively hard ones such as nylon, polypropylene or polyurethane.

When incorporated into the main cable structure, the fibers are often wrapped helically round a central strengthening member and contained within a polymer sheath (Fig. 8.42(a)). Designs have also been proposed where the second fiber coating is dispensed with, and the fiber lies loosely in a cavity within the cable structure, an example of this being the BICC 'PSB' design shown in Fig. 8.42(b). Further details on fiber cable design are contained in ref. 8.25.

Problems

8.1   An optical power of 1 mW is launched into an optical fiber of length 100 m. If the power emerging from the other end is 0.3 mW, calculate the fiber attenuation (ignore reflection losses).

8.2   Calculate the critical angle for a water-air interface. The refractive index of water is 1.33.

8.3   Verify that the phase change δ on reflection at an interface for radiation polarized with its electric vector parallel to the plane of incidence is given by eq. (8.7b).

8.4   Calculate the phase shifts introduced on reflection inside a planar waveguide where $n_1 = 1.53$, $n_2 = 1.50$ and the angle of incidence is $85°$.

8.5   Light of wavelength 0.85 μm traveling in a medium of refractive index 1.53 is incident upon an interface with another medium of refractive index 1.50 at an angle of $80°$. Calculate the relative amounts of power that penetrate distances (a) 5 μm and (b) 50 μm into the second medium.

8.6   Show how graphical solutions to eq. (8.10) may be obtained for TE modes. (Hint: plot a graph of $y = \phi(\theta) + \pi m$ between θ_c and $90°$. For TE modes, $\phi(\theta)$ is given by 2ψ where ψ is given by eq. (8.6b). Also plot a graph of $y = 2dn_1 \cos \theta/\lambda_0$; the solution to eq. (8.10) is then given by the point of intersection of these two curves).

8.7   Estimate the attenuation of a 'nonguided' ray down a planar dielectric waveguide where $n_1 = 1.53$, $n_2 = 1.5$ and $d = 200$ μm. Assume that the ray has an internal angle of $0.1°$ less than the critical angle and that the electric field is perpendicular to the plane of incidence.

8.8   A step index fiber has a numerical aperture of 0.16, a core refractive index of 1.450 and a core diameter of 90 μm. Calculate (a) the acceptance angle of the fiber, (b) the refractive index of the cladding and (c) the approximate maximum number of modes with a wavelength of 0.9 µm that the fiber can carry.

If the fiber is immersed in water (refractive index 1.33) will the acceptance angle change?

8.9   Estimate the inter-modal dispersion expected for the fiber in Problem 8.8.

8.10   Calculate the maximum core diameter needed for a circular dielectric waveguide (refractive index 1.5) to support a single mode of radiation with wavelength 1 µm when (a) the waveguide has no cladding and (b) the waveguide is of the step index type with $\Delta = 0.01$.

Comment on the values obtained from the point of view of manufacturing single mode waveguides.

8.11   A laser operating at 0.9 µm with a linewidth of 2 nm is used in conjunction with a single mode silica-based fiber. Estimate the material dispersion expected.

8.12   Consider a waveguide bent into an arc of a circle of radius R. Determine the point in the cladding at which the phase velocity of a guided mode equals the phase velocity of a plane wave in the cladding. Assume that the mode phase velocity is given by c/n_1. Hence show that the rate of energy loss is proportional to $\exp(-R/R_0)$ where R_0 is a constant. Assume that the mode electric field intensity declines exponentially with distance into the cladding.

8.13   Show that when two identical fibers of radius a are separated by a longitudinal displacement S in air, the fractional coupling loss can be written as $a^2/(a + S\tan\alpha_{max})^2$. Ignore Fresnel losses and assume that radiation emerges from the end of the fiber with a uniform angular distribution up to the fiber acceptance angle, α_{max}.

Also obtain an expression for the energy coupled between the two fibers when they are in contact but offset by an amount D. Assume that the energy is uniformly distributed across the area of the fiber core.

Sketch your results in the form of dB loss versus normalized misalignment in both cases (i.e. S/a and D/a) and compare with the practical results shown in Fig. 8.28 (you will have to decide on a suitable value for α_{max}).

8.14   In an experiment to determine fiber scattering loss, a fiber is passed diagonally through a cube whose sides are lined with Si photodetectors. Light from an He–Ne laser ($\lambda = 633$ nm) is coupled into one end and 4 mW of power is found to be emerging from the other (a short length is used). If the cube side length is 30 mm and the short-circuit current obtained from all the cells connected in parallel is 100 nA, estimate the scattering loss in the fiber (in dB/km^{-1}). You will need to assume a value for the quantum efficiency of the detector.

8.15   A Gaussian pulse with $\sigma = 2$ ns is launched into a fiber 10 km long. At the other end, the pulse is still Gaussian but with $\sigma = 10$ ns. Estimate the (electrical) bandwidth of the fiber.

8.16   The refractive index profile of a graded index fiber is to be examined using the near-field scanning technique. Draw a schematic diagram of the apparatus required and show explicitly how the profile parameter may be obtained from the results.

8.17   When scattering takes place within a fiber, some of the scattered radiation will enter guided modes and travel back down the fiber. Assuming isotropic scattering within a step index fiber of numerical aperture NA and core refractive index n_1, show that the fraction of the scattered radiation so captured is approximately given by $(NA)^2/4n_2^2$.

8.18 Deduce all you can about the fiber whose OTDR trace is shown in Fig. 8.37.

8.19 It has been suggested that if a material could be found where the onset of the infrared lattice absorption bands is at a higher wavelength than in glass/silica fibers, then much lower ultimate fiber losses could be obtained. Estimate from Fig. 8.24 what the minimum absorption is likely to be at 4 μm, assuming Rayleigh scattering is the dominant loss mechanism.

References

8.1 J. Tyndall, *Royal Institution of Great Britain Proceedings*, **6**, 1870–1872, 189.

8.2 K. C. Kao and G. A. Hockham, 'Dielectric–fibre surface waveguides for optical frequencies', *Proc. IEEE*, **113**, 1966, 1151.

8.3 A. C. S. van Heel, 'A new method of transporting optical images without aberrations', *Nature*, **173**, 1954, 39.

8.4 M. Born and E. Wolf, *Principles of Optics* (6th edn), Pergamon, Oxford, 1980, section 1.5.2.

8.5 A. H. Cherin, *An Introduction to Optical Fibers*, McGraw-Hill, 1983, Chapter 4.

8.6 J. E. Midwinter, *Optical Fibers for Transmission*, Wiley–Interscience, New York, 1979, Chapter 5.

8.7 A. H. Cherin, *op. cit.*, Chapter 5.

8.8 H.-G. Unger, *Planar Optical Waveguides and Fibres*, Oxford University Press, Oxford, 1977, Chapter 5.

8.9 D. Gloge and E. A. J. Marcatili, 'Multimode theory of graded-core fibres', *Bell Syst. Tech. J.*, **52**, 1973, 1563.

8.10 R. Olshansky and D. B. Keck, 'Pulse broadening in graded index optical fibres', *Appl. Opt.*, **15**, 1976, 483.

8.11 W. A. Gambling, H. Matsumara and C. M. Ragdale, 'Zero total dispersion in graded index single mode fibres', *Electron. Lett.*, **15**, 1979, 474.

8.12 A. Kawana, M. Kawachi, T. Miyashita, M. Saruwatari, K. Asatani, J. Yamada, and K. Oe, 'Pulse broadening in long span single-mode fibres around a material-dispersion-free wavelength', *Opt. Lett.*, **2**, 1978, 106.

8.13 M. K. Barnoski (ed.), *Fundamentals of Optical Fiber Communications*, Academic Press, New York, 1976, section 2.3.1.

8.14 J. E. Midwinter, *op. cit.*, Appendices 1, 2 and 3.

8.15 W. A. Gardener, 'Microbending loss in optical fiber', *Bell Syst. Tech. J.*, **54**, 1975, 457.

8.16 R. Olshansky, 'Propagation in glass optical waveguides', *Rev. Mod. Phys.*, **51**(2), 1979, 341.

8.17 T. Miya, Y. Terunuma, T. Hosaka and T. Miyashita. 'Ultimate low-loss single-mode fibre at 1.55 μm', *Electron Lett.*, **15**, 1979, 106.

8.18 J. E. Midwinter, *op cit.*, section 8.3.

8.19 J. E. Midwinter, *op. cit.*, Chapters 10, 11.

8.20 J. Senior, *Optical Fiber Communications, Principles and Practice*, Prentice Hall International, Hemel Hempstead, 1985, section 5.2.3.

8.21 J. Gowar, *Optical Communication Systems*, Prentice Hall, Hemel Hempstead, 1984, section 2.4.3.

8.22 A. H. Cherin, *op. cit.*, section 6.8.

8.23 M. J. Adams, D. N. Payne and F. M. E. Sladen, 'Correction factors for the determination of optical fibre refractive index profiles by near field scanning techniques', *Electron. Lett.*, **12**, 1976, 281.

8.24   G. Keiser, *Optical Fibre Communications*, McGraw-Hill, London, 1983, section 9.4.

8.25   G. Mahlke and P. Gösling, *Fibre Optic Cables, Fundamentals, Cable Engineering, Systems Planning*, Wiley, Chichester, 1987.

Chapter 9

Optical communication systems

It has long been realized that light, with its carrier frequency of some 10^{14} Hz, has the potential to be modulated at much higher frequencies than either radio or microwaves and thus opens up the possibility of a single communication channel with extremely high information content. The main difficulties in implementing this have been twofold. First, until recently it was impossible to modulate and demodulate light at anything other than very low frequencies. Secondly, without some sort of guiding medium, atmospheric transmission was limited to line-of-sight communications and was subject to the vagaries of the weather. It was the development of the laser that initially aroused interest in optical communications, but early systems still had to employ atmospheric transmission. The cladded dielectric waveguide soon became a practical proposition, however, and the problem of a suitable guiding medium was solved. Not all lasers are equally suitable for launching adequate power into these waveguides, and the semiconductor laser has proved the best in this respect. Interestingly, an incoherent source – the LED – is in many ways equally useful, which brings home the point that at present the coherence properties of lasers are little used in optical communications. This is in spite of the fact that the techniques of heterodyne and homodyne mixing do offer prospects of improved signal-to-noise ratios as discussed in section 9.1.1. It must be realized that at present we are still a long way from utilizing the full potential of possible bandwidths. The highest modulation frequencies so far achieved are some 10^5 times smaller than those which the carrier frequency could, in theory, allow.

In the present chapter, we review current optical communication systems and give a brief indication of the directions in which they may develop. One such area of interest in this respect is that of integrated optics. Here, the aim is to miniaturize optical components such as sources, detectors, modulators, filters, etc. and to fabricate complete optical processing systems containing these items onto a single semiconductor chip. First of all, however, we discuss the types of modulation that are most applicable to optical communications.

9.1 Modulation schemes

The term modulation describes the process of varying one of the parameters associated with a carrier wave to enable it to carry information. Variations in the amplitude, irradiance, frequency, phase and polarization can all be used for this purpose. When the carrier wave oscillates at optical frequencies, however, not all of these possibilities are equally suitable. For example, most optical detectors respond only to the irradiance of the light, and so modulation of anything other than the irradiance is not very useful unless special techniques are employed, such as those discussed in the next section.

There are many different techniques for transferring information into variations of the appropriate wave parameter, and we may divide these into three categories which we term *analog, pulse* and *digital*. With analog modulation, the primary information signal, which we take to be a time varying electrical voltage, continuously varies the appropriate wave parameter. Thus, at any one time there is a one-to-one relationship between the original signal amplitude and the magnitude of the wave parameter. In both of the other two methods, the signal amplitude is only sampled at regular intervals and this information is then conveyed by means of a series of 'pulses'. In pulse modulation, the pulse width may be varied in proportion to the required signal; alternatively, a fixed width pulse may be used and its time of occurrence within a fixed time slot used for the same purpose. There are also several other possibilities. In digital modulation, the information is provided by using a string of pulses whose timing and widths are both fixed but whose amplitudes are restricted to certain 'quantized' values. Fig. 9.1(a)–(c) illustrate these techniques.

Of these three categories, the most common are analog and digital and it is to these that we now direct our attention.

9.1.1 Analog modulation

As we indicated above, not all of the wave parameters: amplitude, irradiance, frequency and phase, are equally suitable for modulation purposes at optical frequencies. The main difficulties arise in signal demodulation, since detectors respond only to the irradiance of the radiation falling on them (see, for example, eq. (7.27)).

We suppose that the magnitude electric field of the carrier wave can be written

$$\mathscr{E}_c(t) = A_c \cos(\omega_c t + \phi_c)$$

where A_c, ω_c and ϕ_c represent the carrier amplitude, angular frequency and phase respectively, all of which may be modulated and hence be time dependent. If this signal is allowed to fall directly onto a detector (in so-called *direct* detection), then the output of a detector O_d which responds only to irradiance is given by

$$O_d = R A_c^2 \langle \cos^2(\omega_c t + \phi_c) \rangle$$
$$O_d = R A_c^2/2 \tag{9.1}$$

Here R is the detector responsivity and the angle brackets indicate an average taken over a complete period of the function inside. Thus eq. (9.1) indicates that in direct

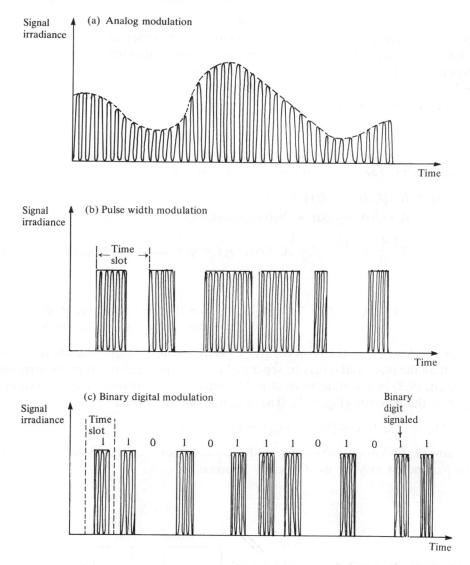

Fig. 9.1 Three possible schemes for modulating an optical light beam. In (a) the amplitude (and hence the irradiance) is continuously modulated in time. The envelope of the waveform gives the required signal. In (b) the signal is sampled at the beginning of each 'time slot' and a pulse is emitted during that time slot whose length is proportional to the signal amplitude. (c) Shows a digital signal. Here, as in (b), the signal is sampled at regular intervals and then a series of pulses is emitted which indicates (in binary notation) the signal amplitude. The presence of a pulse during a time slot indicates a 'one', while the absence of a pulse indicates a 'zero'. It should be noted that for illustrative purposes the carrier frequency has been made much smaller than would occur in practice.

detection only the signal irradiance (which is proportional to the square of the amplitude) is recoverable, and hence modulation of the other wave parameters is not possible. There are, however, other techniques available that which do enable further information to be extracted from the received signal.

In *heterodyne* detection, the incoming signal is mixed with one from a local oscillator (Fig. 9.2) which we describe by

$$\mathscr{E}_0(t) = A_0 \cos (\omega_0 t + \phi_0)$$

where ω_0 is chosen to be very close to ω_c, in contrast to A_c, A_0 will not, in general, be time dependent. The output of the detector now becomes

$$\begin{aligned} O_d &= R \langle [\mathscr{E}_c(t) + \mathscr{E}_0(t)]^2 \rangle \\ &= R \langle \mathscr{E}_c^2(t) + \mathscr{E}_0^2(t) + 2\mathscr{E}_c(t) \cdot \mathscr{E}_0(t) \rangle \\ &= R \left[\frac{A_c^2}{2} + \frac{A_0^2}{2} + 2A_c A_0 \langle \cos (\omega_c t + \phi_c) \cdot \cos (\omega_0 t + \phi_0) \rangle \right] \end{aligned} \qquad (9.2)$$

Now

$$\langle \cos (\omega_c t + \phi_c) \cdot \cos (\omega_0 t + \phi_0) \rangle = \tfrac{1}{2} \langle \cos [(\omega_c + \omega_0)t + (\phi_c + \phi_0)] \\ + \cos [(\omega_c - \omega_0)t + (\phi_c - \phi_0)] \rangle$$

Since ω_0 is close to ω_c, the term $\cos [(\omega_c - \omega_0)t + [(\phi_c - \phi_0)]$ oscillates much more slowly than the other and it may be separated from all the other terms on the right-hand side of eq. (9.2) by inserting an electrical bandpass filter centered on $(\omega_c - \omega_0)$ in the output of the detector (Fig. 9.2). The resulting output is then

$$O_d = RA_c A_0 \cos [(\omega_c - \omega_0)t + (\phi_c - \phi_0)] \qquad (9.3)$$

O_d is now dependent on carrier amplitude, angular frequency and phase; hence each of these parameters may be used to carry information.

Fig. 9.2 Schematic diagram of a heterodyne detection receiver. The incoming signal (frequency ω_c) is mixed with one from a local oscillator (frequency ω_0) using a beam-splitter arrangement. The combined signals fall onto a photodetector and the electrical output of the latter is passed through a bandpass filter centered on frequency $\omega_c - \omega_0$.

In *homodyne* detection, the local oscillator is set to the same frequency as the carrier. In this situation, eq. (9.2) becomes

$$O_d = R\left[\frac{A_c^2}{2} + \frac{A_0^2}{2} + A_cA_0\cos(\phi_c - \phi_0) + A_cA_0\langle\cos(2\omega_ct + (\phi_c + \phi_0))\rangle\right]$$

By inserting a low frequency (i.e. $\ll \omega_c$) bandpass electrical filter after the detector, we may block all the terms except $A_c^2/2$ and $A_cA_0\cos(\phi_c + \phi_0)$. If, in addition, $A_0 \gg A_c$ then we have

$$O_d = RA_cA_0\cos(\phi_c - \phi_0) \tag{9.4}$$

Thus, in homodyne detection both amplitude and phase modulation are possible.

We also see that for both heterodyne and homodyne detection the signal amplitude is a factor A_0/A_c larger than for direct detection. When $A_0 \gg A_c$ we have signal amplification, which enables higher sensitivity to be obtained. Thus A_0 can be increased to the level where the receiver noise is dominated by the shot noise contribution from the local oscillator. In practice, improvements of some 10 to 20/dB in signal-to-noise ratios have been observed over direct detection techniques (ref. 9.1). At present, however, such systems are not in common use mainly because of the difficulty of achieving the required frequency, phase and polarization stability when using semiconductor laser/ optical fiber systems. For most of the rest of this chapter, we shall assume the use of direct detection, although coherent detection will be re-examined in section 9.3.7.

9.1.2 Digital modulation

In digital modulation, the information is coded into a series of pulses whose temporal positions are fixed but whose amplitudes are quantized. The simplest scheme is called *two level binary*; in this, the pulse amplitude has only one of two levels conveniently referred to a 'zero' and 'one'. The 'one state may be represented by the presence of a pulse of greater than a predetermined amplitude and the 'zero' state by the absence of a pulse. The pulses are taken to occur within predetermined time slots. If the pulse width is less than that of the time slot it occupies, then we refer to a *return-to-zero* (RZ) signal. Conversely, if the pulse fills the time slot, we refer to a *non-return-to-zero* (NRZ) signal. Figure 9.3 illustrates these ideas. The complete process whereby an analog signal is converted into a digital pulse coded one is shown in Fig. 9.4. The amplitude of the incoming signal is measured at discrete intervals and the result converted into a binary number. For example, in Fig. 9.4(a) the signal amplitude at $t = 2$ ms is 3 V, the number 3 in binary notation is 011 (we employ a three digit binary number for simplicity; in practice, eight or more binary digits or *bits* are required). During the time interval between this sampling event and the next, the information is transmitted as a series of pulses. Thus, the signal heights at the five times indicated in Fig. 9.4(a) become converted into the digital signal shown in Fig. 9.4(b).

At first sight, this scheme seems somewhat wasteful since it requires a much higher system frequency bandwidth than would be needed for the corresponding analog signal. To reproduce a given signal, the *sampling theorem* (see ref. 9.2) tells us that we must

Fig. 9.3 In a nonreturn-to-zero (NRZ) signal (a), each pulse exactly fills the time slot. In a return-to-zero (RZ) signal (b), the pulse width is less than that of the time slot.

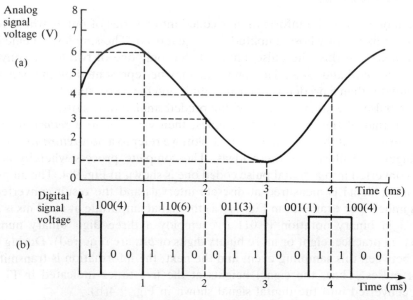

Fig. 9.4 Transformation of an analog signal into a digital one. The analog voltage (a) is sampled every millisecond and each measurement converted into a three-bit two-level binary NRZ signal (b).

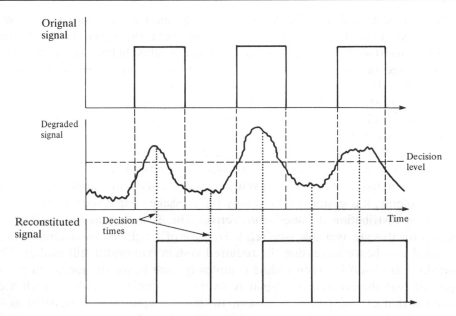

Fig. 9.5 Showing how a severely degraded digital signal can still be restored to its original condition.

sample at a rate that is at least twice that of the highest frequency component in the signal. Since an eight-bit number is commonly used for each amplitude measurement the bit rate required will then be sixteen times the highest signal frequency. Furthermore, to reproduce each bit with reasonable accuracy, the system frequency response must be at least equal to the bit rate. In telephone communication systems, for example, the highest signal frequency is usually 4 kHz; a digital system would therefore require a bit rate of some 4×16 kb s^{-1} or 64 kb s^{-1} and the frequency bandwidth would have to be greater than this to enable the shape of the pulses to be reasonably accurately reproduced.

The great advantage of a digital system over an analog one is its relative freedom from noise or distortion. Inevitably in any communication system the transmitted and received signals will not be identical. Noise may be introduced and nonlinearities in component response will distort the signal. In analog systems, there is no means of rectifying this (at least when using direct detection), but a digital signal can suffer severe distortion and still be capable of accurately imparting the original information. In Fig. 9.5 we show a severely degraded signal but, provided that during each time slot we can always make the correct decision as to whether it contains a 'one' pulse or a 'zero', the original signal can be exactly reproduced. The way in which this decision is usually reached is to set up a *decision level*. If the signal exceeds this level at a particular time (the *decision time*) within each time slot, then a 'one' is recorded; if it does not do so, then a 'zero' is recorded.

We have assumed in this discussion that the detector 'knows' the position and spacing of the time slots and also the height of the decision level. Usually, however, the detector can only determine this information from the signal itself. A standard sequence inserted before the message proper begins helps, but both the pulse amplitude and the timing of the pulses may very well 'wander' somewhat during the message, and the detector circuitry must be able to maintain updated values of these parameters. Fortuitous groupings of pulses can make this task difficult. For example, the decision level is usually determined by taking the average amplitude of a sequence of received bits. If there are equal numbers of 'ones' and 'zeros', this procedure will be adequate. If, however, a fairly long sequence of the same bit occurs, which is not as unlikely as it seems at first, then the decision level may change to such an extent that errors become likely when a more even sequence subsequently appears. To overcome these problems, it is customary to modify the message so that sequences do contain a fairly even distribution of 'ones' and 'zeros'. This is known as *line coding*. Several schemes for this are available (see ref. 9.3), most of which involve adding extra bits to the signal and hence increasing the required system bandwidth still further. There is generally a trade-off between coding complexity (and hence the increased bandwidth required) and the accuracy of signal recovery. A simple example of such a coding scheme is that of *coded mark inversion*. In this, an input of '0' is signaled as 01 and sequential '1s' are signaled alternatively as 00 and 11. Long runs of the same bit are then avoided and some measure of error detection is also possible since the sequence 10, when referring to an original bit, should not occur. However, the number of bits required has now been increased by a factor two. Most line coding schemes adopted in practice increase the required bit rate by some 20%.

9.2 Free space communications

Although fiber optical communication is becoming increasingly important, free space communication over distances of up to a few kilometers is a practical possibility if only fairly low bandwidths are required. Systems are easily set up and costs are considerably less than if using fibers. The low beam divergence of a well-collimated optical beam can have considerable security advantages over radio communications. A typical application would be a voice or low data rate link between nearby buildings in an urban environment.

An obvious disadvantage is that adverse atmospheric conditions, such as rain or snow, may introduce severe distortion or even render the system completely inoperable. Even when conditions are favorable, the signal will be attenuated by both absorption and scattering processes in the atmosphere. The former arises from the presence of molecular constituents such as water vapor, carbon dioxide and ozone, whose exact concentrations depend on many variables such as temperature, pressure, geographical location, altitude and weather conditions. Strong absorption occurs around wavelengths of 0.94 μm, 1.13 μm, 1.38 μm, 1.90 μm, 2.7 μm, 4.3 μm and 6.0 μm. Between these values lie the so-called *atmospheric windows*, where losses are mainly determined by scattering. Scattering is primarily due to particles that are large

Fig. 9.6 Typical atmospheric absorption spectrum. The curve shows the percentage transmission through 1 km of atmosphere at sea level. The eight atmospheric 'windows' are indicated.

compared to the wavelength, such as smoke or fog, and this is known as *Mie* scattering. The effective absorption coefficient in Mie scattering varies relatively little with wavelength. A typical atmospheric absorption spectrum is shown in Fig. 9.6.

Another problem is that of atmospheric turbulence; heating of the air in contact with the earth's surface creates convection currents in the atmosphere. Since the refractive index of air is temperature dependent, we experience refractive index variations through the atmosphere causing beam deviation and spreading. This can give rise to sudden and pronounced fading of the signal. Under ideal conditions, ranges of up to 150 km are possible; but to achieve reasonably large bandwidths, distances should be less than 2 km. Fluctuations in absorption and scattering tend to modulate the signal strength in the range 1 to 200 Hz so that amplitude modulation is not usually very satisfactory.

The simplest systems make use of LED emitters and silicon photodiode detectors. These may be coupled by a suitable lens or mirror system such as those shown in Fig. 9.7. More complex systems have been proposed using laser sources such as Nd: YAG and CO_2 lasers where the inherently small laser beam divergence greatly simplifies beam collimation. Modulation may be achieved in a variety of ways. For example, a CO_2 laser operating in a single mode may be frequency modulated by introducing an electro-optic crystal into the resonant cavity. A signal voltage applied to the crystal changes its length and hence the optical length of the cavity. This in turn causes the single mode resonance frequency to change. Alternatively, lasers such as the Nd: YAG can be made to emit a train of pulses by the technique of mode locking (see section 6.3). A signal can then be impressed on this pulse train using an external modulator. Such

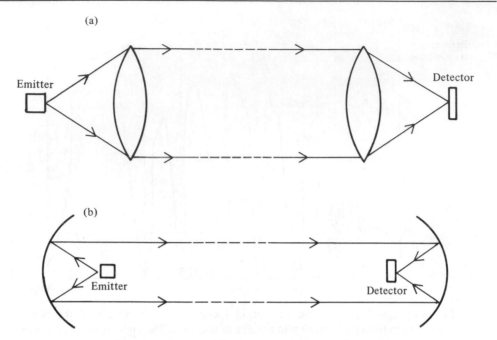

Fig. 9.7 Two simple types of optical antenna using (a) lenses and (b) concave mirrors.

systems as these, however, have been rarely used, primarily because of their complexity and high cost. A possible future use is in deep space communications, where the low beam divergences are useful in view of the large distances involved.

9.3 Fiber optical communication systems

The advent of optical fibers with losses less than 1 dB km^{-1} and with high information carrying capacity (e.g. bandwidths > 100 MHz) has meant that they have become very attractive alternatives to twisted wire or coaxial cables in communication links. Whilst it is true that at present optical fiber cable costs are appreciably greater than copper-based cables, this position is likely to change in the future as increased demand leads to increasing manufacturing efficiency. In contrast with copper-based cables, the raw material costs of fiber cable are a relatively small fraction of the total cost.

A schematic diagram of an optical transmission system is shown in Fig. 9.8. The emitter is usually an LED or semiconductor laser, whilst the detector may be a *pin* or avalanche photodiode. Repeater units may be necessary over relatively long links to counter the effects of fiber transmission losses and dispersion. In these, the signal is detected, amplified and then re-emitted. For digital signals it is possible to restore a distorted digital signal pulse to its original shape, whilst with analog signals any signal distortion or noise present is, of necessity, passed on. A separate power supply line

Fig. 9.8 Schematic diagram of the main components of a fiber optical communication system.

must be provided for these units. The presence of repeaters greatly adds to the cost of a link.

The main advantages of optical fiber links include relatively low attenuations (at optimum wavelengths), high bandwidths (up to several gigahertz), small physical size and weight, the elimination of ground loop problems and immunity from electrical interference. This latter advantage is especially useful in electrically noisy environments such as densely populated urban areas and power stations. There is, of course, no spark hazard – an additional advantage in explosive environments such as chemical plants. Fiber cables may also survive better than copper cables in certain corrosive environments (for example sea water).

So far, the largest application has been for telephone 'trunk' links, i.e. links capable of carrying a large number of simultaneous telephone conversations between telephone buildings. These may be from a few kilometers up to several hundred kilometers apart. We saw in section 9.1.2 that a single channel telephone link required a bandwidth of 4 kHz or, for a digital signal, 64 kb s^{-1}. Telephone networks require links capable of carrying many single channels simultaneously. Digital transmission is able to cope with this easily since the separate bit streams required may be interleaved. In both the United States and Europe, standard rates have been agreed upon for various 'levels' of transmission and these are shown in Table 9.1. Fiber optical links offer obvious economic advantages for medium to long haul links (10 km and upwards) with large capacities (100 Mb s^{-1} and upwards).

There are numerous other potential applications including, for example, undersea links, video transmission, computer links and, in the military sphere, missile guidance. The concept of an optical local area network (LAN) will be discussed further in 9.3.6. Several pilot experiments have been run to examine the feasibility of providing the whole of a community's communication/information needs (such as telephone, TV, radio, etc.) using fiber optics. It is obviously impossible in a text of this size to cover these and other areas of interest adequately, and the reader is referred to ref. 9.4 for further information.

The individual components involved in a fiber optical communication link (emitters, fibers and detectors) have all been covered in general terms in previous chapters, and we now review the possibilities open to us which meet the requirements of a communication system. We consider first the choice of operating wavelength.

Table 9.1. Digital rates used in telecommunications in Europe and the United States

Digital rates	Approximate number of telephone channels (1 channel = 64 kb s^{-1})
Europe	
2.048 Mb s^{-1}	32
8.448 Mb s^{-1}	120 (= 4 × 30)
34.368 Mb s^{-1}	480 (= 4 × 120)
139.364 Mb s^{-1}	1920 (= 4 × 480)
565 Mb s^{-1}	7680 (= 4 × 1920)
United States	
1.544 Mb s^{-1} T1	24
6.312 Mb s^{-1} T2	96 (= 4 × T1)
46.304 Mb s^{-1} T3	672 (= 7 × T2)
281 Mb s^{-1} T4	4032 (= 6 × T3)

9.3.1 Operating wavelength

The two crucial characteristics of optical fibers that depend on wavelength are attenuation and material dispersion. A typical attenuation versus wavelength curve for a silica-based fiber was shown in Fig. 8.24. Material dispersion, i.e. the variation in group velocity with wavelength, was covered in section 8.3.4. There we saw (eq. (8.29)) that the pulse spread $\Delta\tau$ over a fiber of length L could be written as

$$\Delta\tau = \left| \frac{-L\lambda}{c}\left(\frac{\mathrm{d}^2 n}{\mathrm{d}\lambda^2}\right)\Delta\lambda \right|$$

The pulse spread per unit length per unit wavelength interval $\Delta\tau/L\Delta\lambda$ for silica is shown in Fig. 9.9. At the present time a number of glass/silica fiber systems use GaAs or GaAlAs devices which emit in the range 0.82 µm to 0.9 µm. This is by no means an ideal region; at 0.85 µm, silica fibers have a minimum attenuation of about 2 dB km^{-1} whilst their material dispersion is some 80 ps nm^{-1} km^{-1}. Thus, a typical LED with a linewidth of 50 nm would suffer a dispersion of 4 ns km^{-1}. Over a 10 km fiber length, the system bandwidth would then be restricted to some 25 MHz by material dispersion alone. Lasers, of course, have a considerable advantage here; by virtue of their narrow linewidths, they enable much higher bandwidths to be achieved (see Example 8.11).

Fiber attenuation may be considerably reduced by working at longer wavelengths, the minimum (about 0.15 dB km^{-1}) occurring at a wavelength of about 1.55 µm. Material dispersion also decreases at longer wavelengths, becoming very small indeed at around 1.3 µm; values of a few picoseconds per nanometer per kilometer having been achieved experimentally. In fact, many high capacity telecommunication links now operate at 1.3 µm or 1.55 µm. Suitable detectors for these longer wavelengths (mainly based on GaInAs) have become increasingly available over the last few years. However, they are still considerably more expensive than those based on silicon.

Plastic fibers have extremely high attenuations outside the visible region (Fig. 8.41) and are usually used in conjunction with red emitting LEDs.

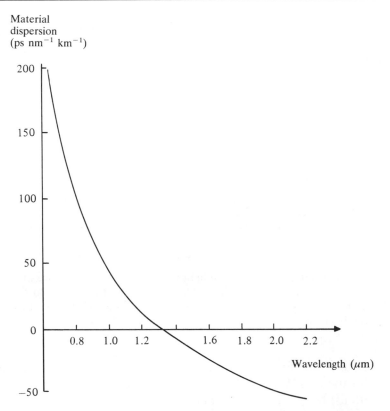

Fig. 9.9 Material dispersion for pure silica as derived from the data shown in Fig. 8.19. The quantity plotted here is $-(\lambda/c)(d^2n/d\lambda^2)$ rather than the quantity designated $\Delta\tau$ in the text.

9.3.2 Emitter design

The main requirements for an emitter in an optical communication system are that it must be able to couple a useful amount of power into the fiber and that it must be capable of operating with the required bandwidth. The relatively small diameters of optical fiber cores (that is, from several hundred microns down to a few microns) imply correspondingly small source sizes for efficient coupling of radiation into the fiber. Consequently, the most commonly used sources are LEDs and semiconductor lasers which were discussed in Chapters 4 and 5 respectively. Other sources have been used. for example miniature Nd: YAG lasers with rod diameters similar to those of fiber cores, but none has the operating convenience of the semiconductor devices.

Compared with semiconductor lasers, LED sources are easy to drive, have long lifetimes and are inexpensive. Their principal disadvantage is that they are much less efficient at launching power into fibers than are lasers. This is mainly because of their larger emitting area and greater beam divergence. To illustrate this, we calculate the maximum power that can be coupled into a step index fiber from a source in contact

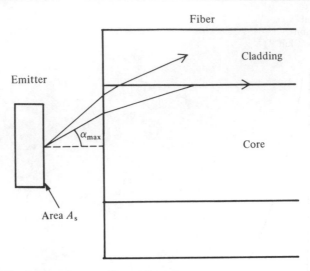

Fig. 9.10 Illustrating the coupling of light from a flat emitting surface into a fiber optical waveguide. Rays making an angle less than α_{max} with the normal to the fiber end will become trapped in the fiber core. The gap between the emitter and the fiber has been deliberately enlarged to show the angles involved.

with it. We assume that the source area A_s is less than or equal to the fiber core area A_c. Figure 9.10 illustrates the basic geometry. If the fiber core diameter is much larger than the wavelength used, a large number of modes can be carried by the fiber and a simple ray treatment of light propagation within the fiber is adequate. For each point on the emitting surface, only light emitted up to an angle α_{max} with the normal to the surface will be propagated down the fiber, where $\alpha_{max} = \sin^{-1}(\mathrm{NA}/n_0)$ (see eq. (8.17)). If the source brightness as a function of angle is written $B(\alpha)$, the total energy Φ_F coupled into the fiber is given by

$$\Phi_F = A_s \int_\Omega B(\alpha)\, d\Omega \tag{9.5}$$

Here $d\Omega$ is the solid angle subtended by rays between the angles α and $\alpha + d\alpha$, and the integral is carried out for all rays that remain trapped in the fiber. The relation between $d\Omega$ and $d\alpha$ can readily be derived with the aid of Fig. 9.11 and is

$$d\Omega = 2\pi \sin \alpha\, d\alpha$$

Hence, inserting this value for $d\Omega$ into eq. (9.5) yields

$$\Phi_F = 2\pi A_s \int_0^{\alpha_{max}} B(\alpha) \sin \alpha\, d\alpha \tag{9.6}$$

LED sources are usually approximately Lambertian so that we may put $B(\alpha) = B(0) \cos \alpha$ (see Problem 4.6). Equation (9.6) then becomes

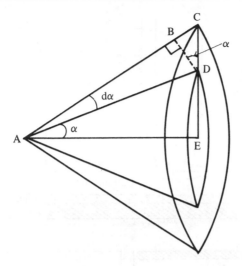

Fig. 9.11 Relation between solid angle $d\Omega$ and $d\alpha$ may be derived by noting that: $d\Omega = (BD \cdot 2\pi \cdot DE)/AB^2$. Since $BD = AB\, d\alpha$ and $DE/AB \approx \sin \alpha$, we have $d\Omega = 2\pi \sin \alpha\, d\alpha$.

$$\Phi_F = 2\pi A_s B(0) \int_0^{\alpha_{max}} \cos \alpha \sin \alpha\, d\alpha$$

$$= \pi A_s B(0) \sin^2\alpha_{max}$$

or

$$\Phi_F = \pi A_s B(0) \frac{(NA)^2}{n_0^2}$$

We know from eq. (4.21b) that the total output of the LED is $\pi A_s B(0)$ and hence the coupling efficiency η_c may be written

$$\eta_c = \frac{\text{Energy coupled into fiber}}{\text{Energy emitted by LED}}$$

$$= \frac{\pi A_s B(0)(NA)^2}{\pi A_s B(0) n_0^2}$$

$$\eta_c = \frac{(NA)^2}{n_0^2} \tag{9.7}$$

A typical value for NA would be 0.3, and for this and an air interface only some 9% at most of the total radiation emitted by the surface in contact with the core area will enter the fiber. (Fresnel losses have been ignored in this derivation since they amount to only a few per cent of the energy coupled into the fiber.) If A_s is larger than A_c, the coupling efficiency will be further reduced by the factor A_c/A_s.

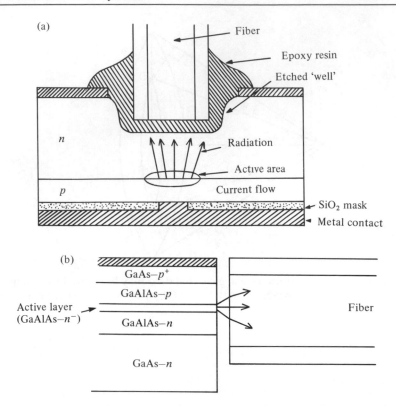

Fig. 9.12 Two types of emitter designed for more efficient coupling into optical fibers. (a) The etched well or 'Burrus' type and (b) an edge emitter. In (a), the active light emitting area is restricted to a small region just below the end of the fiber by the use of an SiO_2 mask. The fiber is held in position by the use of a transparent epoxy resin which also helps to reduce Fresnel losses. In the edge emitter, the radiation is confined to a narrow light guiding layer with a structure very similar to that of the double heterostructure laser.

For a graded index fiber the calculation is more complicated. As a rough guide, however, we note from eq. (8.24) that a parabolically graded fiber (i.e. $\gamma = 2$) with the same diameter as a step index fiber is only able to support about half the number of modes. Hence the coupled energy will be about a factor of two smaller.

To maximize the coupling efficiency, we therefore require a fiber with as large an NA as possible and a source with an area no larger than the fiber core area. Increasing the NA, however, incurs the penalty of decreasing the signal bandwidth because of increasing mode dispersion (see eq. (8.21b)). In addition, reducing the emitting area of an LED whilst maintaining total output tends to reduce the operating lifetime.

The two main types of LED most often used in fiber optical systems are the surface etched well emitter ('Burrus' type) and the edge emitter, both of which are illustrated in Fig. 9.12. In the former, a well is etched into the top of a planar LED structure to enable the fiber end to be as close as possible to the light emitting region (that is to the

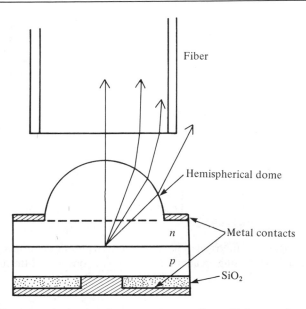

Fig. 9.13 One way devised to increase the coupling efficiency of emitters to fibers is to incorporate some type of lens system; here, the effect of a hemispherical lens is illustrated.

p–n junction). If the emitting area is less than that of the fiber core, then some form of optical coupling between the source and the fiber may be advantageous. For example, a hemispherical lens on top of the emitting surface will magnify the effective surface area, but demagnify the solid angular distribution of the radiation, so that the resulting radiation pattern may be a better match to the fiber acceptance pattern (Fig. 9.13). The construction of the edge emitting LED is very similar to that of the semiconductor laser (see section. 5.10.2), the radiation being confined to a narrow channel. On emerging, the beam divergences tend to be narrower than those of the surface types with typical half-power divergences of 50° and 30° perpendicular and parallel to the junction plane respectively (a Lambertian source has a half-power beam divergence of 60°). To couple this distribution into a fiber efficiently, an anamorphic optical system is sometimes used, such as that shown in Fig. 9.14.

Total optical powers from edge emitters are typically several times smaller than from surface emitters, but the narrower beam divergence can give rise to more coupled power. Edge emitters are usually preferred for use with small NA fibers (i.e. NA ≤ 0.4), whereas surface emitters, because of their greater total power output, are better for large NA fibers.

As far as output characteristics are concerned, LEDs have an almost linear relationship between drive current and light output. This makes the LED more suitable for amplitude than pulse modulation, since for the latter the drive current must be switched from high to low values to obtain a wide ratio between 'on' and 'off' outputs (say from 300 mA to below 50 mA). The factors that ultimately affect the modulation

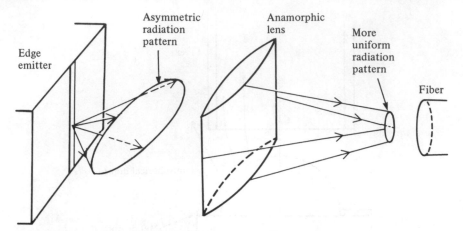

Fig. 9.14 Strongly asymmetrical radiation patterns from edge emitters can more efficiently be coupled into a fiber using an anamorphic lens (i.e. one with differing curvatures in two orthogonal directions).

bandwidths attainable are discussed in section 4.6.5. At low current levels, the limiting factor is usually the junction capacitance, whilst at high current levels it is the lifetime of carriers injected into the recombination region. It is shown in Appendix 3 that when the response is lifetime limited, we may write

$$R(f) = \frac{R(0)}{(1 + 4\pi^2 f^2 \tau^2)^{\frac{1}{2}}}$$

where $R(f)$ is the relative response at frequency f and τ is the minority carrier lifetime. In GaAs a typical value of τ would be 10^{-9}s, giving a half-power bandwidth of some 300 MHz.

We turn now to the consideration of laser sources. In comparison with LEDs, they exhibit much higher output power, narrower beam divergence and a smaller emitting area. All of these factors enable lasers to couple significantly more power into fibers than LEDs (see Problem 9.5). They are especially useful for small diameter low numerical aperture fibers, in particular single mode fibers having core diameters of only a few microns. A laser with a 10 μm wide stripe can couple something like 1 mW into a single mode fiber using simple butt jointing techniques.

Lasers, of course, have very different output characteristics from LEDs. From Fig. 5.21 we see that very little radiation is emitted until the current reaches a threshold value, after which the output rises very sharply. Such a characteristic is well-suited to generating digital signals, since the drive current needs only a small swing to provide a high 'on' to 'off' contrast. Analog signals can also be transmitted, since the characteristic above threshold is reasonably linear. (Sometimes, however, 'kinks' can develop as discussed in section 5.10.2.3.) One problem which severely affects both types of signal is that the threshold current is temperature sensitive and can show long-term drift with age. Good temperature stabilization is therefore required together

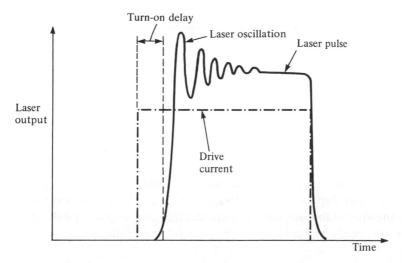

Fig. 9.15 Typical semiconductor laser output pulse in response to a square drive current pulse.

with some form of feedback mechanism to counter any change in the characteristic. The most common technique with digital signals is to maintain a constant power output. For this purpose, a photodetector may be mounted next to the rear facet of the laser. The drive current may then be adjusted until the photodetector (which need not have a very fast response) indicates some predetermined average power level.

Laser modulation bandwidths are significantly larger than those of LEDs because the carrier recombination times are appreciably shortened by the action of stimulated emission. However, the output response to a short current pulse contains two rather undesirable features. First, the optical output is delayed relative to the start of the current pulse; secondly, the output tends to show a rapid oscillation superimposed on the overall square pulse. Figure 9.15 illustrates these points. The turn-on delay may be reduced by operating the laser with a permanent d.c. bias current so that the total current never falls much below threshold. The rapid oscillations are the laser 'spiking' oscillations discussed in section 5.10.1. It is found that the magnitude of the spikes decreases and their frequency increases as the drive current is increased above threshold. Laser construction has a significant effect on the magnitude of the spikes; some single transverse mode narrow stripe lasers show little spiking.

The spectral linewidth also depends on the drive current and can be quite wide if the current is close to threshold. Consequently, if the narrowest spectral width is required it is necessary to bias the laser somewhat above threshold. Unfortunately this will rather degrade the noise performance since the signal will not fall to such a low value when indicating a 'zero', and there will be an increased amount of shot noise. The much narrower bandwidths of lasers, however, means that material dispersion is much less than when using LEDs (see section 8.3.5).

In conclusion, it is evident that lasers have considerable advantages over LEDs from

Fig. 9.16 Assumed equivalent circuit for noise analysis in a *pin* diode. A current source i_s is shunted by the diode junction capacitance C_j and the load resistor R_L.

the point of view of coupling the maximum amount of power into fibers and giving minimum material dispersion. The disadvantages are that they are more expensive, require somewhat elaborate temperature control and output stabilization, and consideration also needs to be given to the possibility of using a d.c. bias.

9.3.3 Detector design

Most detectors have large sensitive areas compared to fiber core areas and, Fresnel losses apart, can easily collect most of the radiation being carried by the fiber. On the other hand, too big a detector area is often a positive disadvantage since this implies an unnecessarily high dark current noise and a high detector capacitance. For this reason, photomultipliers are rarely used. Thermal detectors are, in general, much too slow and insensitive and the almost universal choice has been for some kind of junction device. Of these, the main choice has been between *pin* and avalanche photodiodes. The latter provide a substantial amount of gain which is useful in increasing the system sensitivity but only when this is limited by the noise in the amplifier following the detector. To illustrate this point we now carry out a simplified noise analysis for both a *pin* and an avalanche (APD) detector.

9.3.3.1 NOISE ANALYSIS FOR A *pin* DETECTOR
We suppose that an optical power P at wavelength λ falls on the detector; then, if the detector quantum efficiency is η, the resulting signal current i_s may be written (see eq. (7.22)) as

$$i_s = \frac{\eta Pe\lambda}{hc} \tag{9.8}$$

Inevitably, the signal levels are such that amplification is required and the noise generated in this amplifier is of crucial importance. The equivalent circuit of the detector and amplifier is shown in Fig. 9.16. It is conventional to represent this amplifier noise as if it were Johnson noise originating in the load resistor which is assumed to be at an effective temperature of T_{eff}. From eq. 7.11 the r.m.s. noise voltage will then be

$$\Delta V_J = (4kT_{eff}R_L\Delta f)^{\frac{1}{2}}$$

the r.m.s. noise current Δi_J is then

$$\Delta i_{\mathrm{J}} = \frac{(4kT_{\mathrm{eff}}R_{\mathrm{L}}\Delta f)^{\frac{1}{2}}}{R_{\mathrm{L}}} \tag{9.9}$$

The noise originating in the detector has two main sources; shot noise in the signal current $\Delta i_{\mathrm{s}}^{\mathrm{shot}}$ and shot noise in the dark current $\Delta i_{\mathrm{D}}^{\mathrm{shot}}$. Using eq. (7.9) we may write

$$\Delta i_{\mathrm{s}}^{\mathrm{shot}} = (2i_{\mathrm{s}}e\Delta f)^{\frac{1}{2}}$$

and

$$\Delta i_{\mathrm{D}}^{\mathrm{shot}} = (2i_{\mathrm{D}}e\Delta f)^{\frac{1}{2}} \tag{9.10}$$

where i_{D} is the dark current. For simplicity, we assume that the sensitivity is limited by amplifier noise alone. That is we ignore both $\Delta i_{\mathrm{s}}^{\mathrm{shot}}$ and $\Delta i_{\mathrm{D}}^{\mathrm{shot}}$ in comparison with Δi_{J}. The signal-to-noise power ratio S/N can then be written as

$$\frac{S}{N} = \frac{i_{\mathrm{s}}^2 R_{\mathrm{L}}}{\Delta i_{\mathrm{J}}^2 R_{\mathrm{L}}} = \frac{i_{\mathrm{s}}^2 R_{\mathrm{L}}}{4kT_{\mathrm{eff}}\Delta f} \tag{9.11}$$

At first glance, eq. (9.11) would seem to imply that S/N may be increased indefinitely by increasing R_{L}. However, we must remember that the detector capacitance is in parallel with the load resistor and unless some form of frequency equalization is introduced the bandwidth will be limited to $(2\pi R_{\mathrm{L}}C_{\mathrm{j}})^{-1}$ (see eq. (7.32)). Accordingly we put $\Delta f = (2\pi R_{\mathrm{l}}C_{\mathrm{j}})^{-1}$, and hence eliminating R_{L} from eq. (9.11) we obtain

$$\frac{S}{N} = \frac{i_{\mathrm{s}}^2}{8\pi kT_{\mathrm{eff}}\Delta f^2 C_{\mathrm{j}}}$$

Assuming the minimum detectable signal $(i_{\mathrm{s}})_{\mathrm{min}}$ is given when S/N = 1, we have

$$(i_{\mathrm{s}})_{\mathrm{min}} = 2\Delta f(2\pi kT_{\mathrm{eff}}C_{\mathrm{j}})^{\frac{1}{2}} \tag{9.12}$$

Using eq. (9.8), the minimum detectable optical power P_{min} then becomes

$$P_{\mathrm{min}} = \frac{2hc}{\eta e\lambda} \Delta f(2\pi kT_{\mathrm{eff}}C_{\mathrm{j}})^{\frac{1}{2}} \tag{9.13}$$

Figure 9.17 shows a plot of P_{min} versus Δf for the case where $\lambda = 0.85\ \mu\mathrm{m}$, $C_{\mathrm{j}} = 1\ \mathrm{pF}$, $\eta = 1$ and $T_{\mathrm{eff}} = 1000\ \mathrm{K}$.

9.3.3.2 NOISE ANALYSIS FOR AN AVALANCHE PHOTODIODE

If the gain of the APD is M, the signal current will be Mi_{s}. We might then expect that the shot noise terms would be given by replacing i_{s} and i_{D} in eqs. (9.10) by Mi_{s} and Mi_{D} respectively. However, we must remember that the avalanche process itself introduces excess noise, which is allowed for by introducing an excess noise factor $F(M)$ (see eq. (7.33a)). Consequently, we must write

$$\Delta i_{\mathrm{s}}^{\mathrm{shot}} = M[2i_{\mathrm{s}}e\Delta f F(M)]^{\frac{1}{2}}$$

and

$$\Delta i_{\mathrm{D}}^{\mathrm{shot}} = M[2i_{\mathrm{D}}e\Delta f F(M)]^{\frac{1}{2}}$$

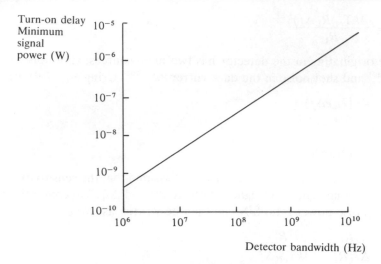

Turn-on delay
Minimum
signal
power (W)

Detector bandwidth (Hz)

Fig. 9.17 Minimum signal power required to give a signal-to-noise ratio of unity in a *pin* detector when $\lambda = 0.85$ μm, $C_j = 1$ pF, $\eta = 1$ and $T_{eff} = 1000$ K. All sources of noise except amplifier noise have been ignored.

The signal-to-noise ratio can be written:

$$\frac{S}{N} = \frac{i_s^2 M^2}{M^2 F(M) 2e\Delta f(i_s + i_D) + 8\pi k T_{eff}\Delta f^2 C_j} \tag{9.14}$$

If the factor $F(M)$ were unity, the signal-to-noise ratio would increase with increasing M and would attain a maximum asymptotic value of $i_s^2/2e\Delta f(i_s + i_D)$. This is much larger than the corresponding value for a *pin* diode, since the shot noise term was taken to be much smaller than the amplifier noise term. However, because the factor $F(M)$ increases with increasing M, the optimum S/N value is somewhat less than this. It may be shown (see Problem 9.7) that the value of M that gives the largest S/N value is given by solving the equation

$$M_{opt}\left(\frac{M_{opt}^2}{r} - \frac{1}{r} + 1\right) = \frac{2 \times \text{Amplifier noise}}{\text{Shot noise}} = \frac{8\pi k T_{eff}\Delta f C_j}{(i_s + i_D)e} \tag{9.15}$$

where r is the ratio of the electron-to-hole ionization probabilities. The optimum signal-to-noise ratio $(S/N)_{opt}$ may then be obtained by substituting for the value of M_{opt} obtained from eq. (9.15) into eq. (9.14).

We may obtain a useful approximate expression for $(S/N)_{opt}$ by eliminating the quantity $(i_s + i_D)e$ from eqs. (9.14) and (9.15). We then have

$$\left(\frac{S}{N}\right)_{opt} = \frac{i_s^2 M_{opt}^2}{\left[\dfrac{2 M_{opt} F(M)}{\left(\dfrac{M_{opt}^2}{r} - \dfrac{1}{r} + 1\right)} + 1\right]} 8\pi k T_{eff}\Delta f^2 C_j$$

Since, in the absence of any gain (i.e. $M = 1$), the amplifier noise is taken to be much larger than the shot noise, we may put

$$\frac{i_s^2}{8\pi k T_{\text{eff}} \Delta f^2 C_j} = \left(\frac{S}{N}\right)_{M=1}$$

We note that if $M \gg 1$ then $F(M) \simeq M/r$ and if $M^2 \gg r$ then $(M^2/r) - (1/r) + 1 \simeq M^2/r$. Consequently, if we assume that $M_{\text{opt}}^2 \gg 1$ and $M_{\text{opt}}^2 \gg r$, eq. (9.16) finally reduces to

$$\left(\frac{S}{N}\right)_{\text{opt}} \simeq \left(\frac{S}{N}\right)_{M=1} \frac{M_{\text{opt}}^2}{3} \tag{9.17}$$

Example 9.1 Calculation of optimum avalanche gain of a silicon APD

We consider a silicon APD where $r = 50$, $T_{\text{eff}} = 1000$ K, $C_j = 1$ pF and where the signal strengths and noise signals are such that when the gain is unity a signal-to-noise ratio of 1 is obtained. (A *pin* detector would then also give S/N = 1.) To use eq. (9.17) we require the value of M_{opt} obtained by solving eq. (9.15). If we assume that $i_D \ll i_s$ we may substitute for i_s from eq. (9.12) into eq. (9.15) and obtain

$$M_{\text{opt}}\left(\frac{M_{\text{opt}}^2}{r} - \frac{1}{r} + 1\right) = \frac{2}{e}(2\pi k T_{\text{eff}} C_j)^{\frac{1}{2}}$$

Insertion of the numerical values yields $M_{\text{opt}} = 57$. From eq. (9.17), we then have

$$\left(\frac{S}{N}\right)_{\text{opt}} \simeq \frac{(57)^2}{3} \simeq 10^3$$

We must remember that this value refers to the electrical power delivered by the detector; in terms of the optical power falling onto the detector, the improvement factor is only some $\sqrt{10^3}$, i.e. 33 (since $P \propto i_s$). This implies that when the gain is optimized, this particular APD would yield the same signal-to-noise ratio (i.e. unity) as that of a *pin* detector with an optical power some 33 times smaller (i.e. 15 dB less).

With relatively large signal powers, then, there may be little or no advantage to be gained in using an APD rather than a *pin* photodiode. In fact an APD may worsen the signal-to-noise ratio. Thus if the dominant noise term is shot noise in the signal current, then from eq. (9.14) (S/N) = $i_s/(F(M)2e\Delta f)$. Since the corresponding signal-to-noise ratio in a *pin* detector may be obtained by putting $F(M) = 1$, we see that the signal-to-noise ratio in the APD is now a factor $F(M)$ smaller than in a *pin* photodiode.

9.3.3.3 FUNDAMENTAL LIMITATIONS ON SIGNAL SIZE

Although in practical situations amplifier noise may often be the limiting factor, it is of interest to examine the fundamental limitations on the pulse size in a digitally coded optical signal. The smallest conceivable size corresponds to the arrival of a single photon. On this scale, however, the statistical nature of the photon emission processes become evident. For example, although the average power transmitted may correspond to 20 photons per pulse, pulses will arrive at the detector containing perhaps 18 or 24 or indeed any other number of photons including zero. It can be shown (ref. 9.5) that the process is governed by Poisson statistics and that the probability, $p(n, n_m)$ of detecting n photons per unit time interval when the mean arrival rate is n_m is given by

$$p(n, n_m) = \frac{(n_m)^n}{n!} \exp(-n_m) \tag{9.18}$$

For example, with a mean arrival rate of 20 photons per pulse, the probability that a given pulse contains no photons at all is given by

$$p(0, 20) = \frac{(20)^0}{0!} \exp(-20) = 2 \times 10^{-9}$$

Thus if we have a signal consisting of equal numbers of 'ones' and 'zeros', where the 'ones' correspond to pulses containing on average 20 photons and the 'zeros' to the arrival of no photons, then there is a probability of 2×10^{-9} that a 'one' pulse will be mistaken for a 'zero'. No error can be made in any of the 'zeros' since there can be no fluctuations in zero photons. Hence, including both 'ones' and 'zeros', a signal containing an average of 10 photons per bit will give rise to a *bit error rate* (BER) of 10^{-9}. With less than an average of 10 photons per bit the BER will be worse (i.e. larger than 10^{-9}), conversely with more than 10 photons per bit the BER will be better (i.e. smaller than 10^{-9}). A BER of 10^{-9} has, in fact become fairly standard for optical communication systems. Remarkably perhaps, practical systems can approach to within an order of magnitude or so of this theoretical limit. That is, they can operate with an average of 100 photons per bit and achieve a BER of 10^{-9}. Figure 9.18 shows examples of typical detector performances that have been achieved in practice.

It is also of interest at this stage to examine the comparable signal limitations for analog transmission. The limiting factor on signal-to-noise ratio will be signal shot noise, which is given by eq. (9.10). The maximum signal-to-noise ratio obtainable, $(S/N)_{max}$, is then given by

$$\left(\frac{S}{N}\right)_{max} = \frac{i_s^2}{(\Delta i_s^{shot})^2} = \frac{i_s}{2e\Delta f}$$

Substituting for i_s from eq. (9.8) yields

$$\left(\frac{S}{N}\right)_{max} = \frac{\eta P \lambda}{2hc\Delta f} \tag{9.19}$$

A reasonable value for $(S/N)_{max}$, comparable to a bit error rate of 10^{-9} in a digital system, is 50 dB. Thus, to achieve this we require a minimum signal level, P_{min}, given in watts by

$$P_{min} = \frac{2 \times 10^5 hc\Delta f}{\eta \lambda} \tag{9.20}$$

If we take $\lambda = 0.85$ μm. $\eta = 1$ and $\Delta f = 6.25$ MHz (which is comparable to a digital signal of 6.25×16 Mb s^{-1} or 100 Mb s^{-1}), we obtain from eq. (9.20) that $P_{min} = 0.3$

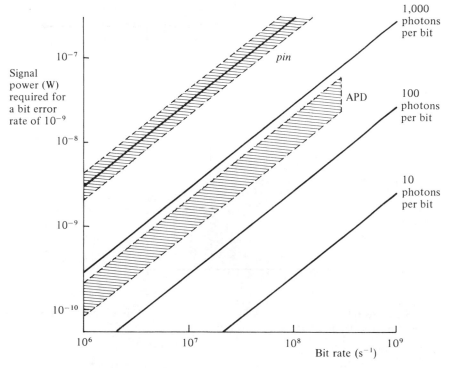

Fig. 9.18 Shaded regions indicate typical minimum signal powers required for *pin* and APD receivers to achieve a bit error rate of 10^{-9} as a function of bit rate at a wavelength of 0.85 μm. Also shown are the powers corresponding to a constant number of photons per bit. As demonstrated in the text, the 10-photon-per-bit line represents the fundamental limit for this error rate.

μW. A glance at Fig. 9.18 shows that a 100 Mb s^{-1} digital system requires substantially less power than this to achieve a bit error rate of 10^{-9}, thus illustrating the clear superiority of digital systems for low noise long distance optical transmission.

9.3.3.4 AMPLIFIER FRONT END DESIGN

We have seen in the previous section that the characteristics of the first stage of the amplifier immediately following the detector can have a decisive effect on the ultimate performance of the detector/amplifier combination. In this section, we consider in more detail several ways in which this first stage (the amplifier 'front-end') may be implemented.

We start with the so-called 'low impedance' front end. Referring to Fig. 9.16, the current source (i.e. the photogenerated current) 'sees' the diode capacitance (C_j) and load resistor (R_L) acting in parallel. With an amplifier connected across R_L, the bias resistor (R_a) associated with the amplifier first stage will also be in parallel with R_L. The effective load resistance (R_e) is then given by $R_e = R_L R_a/(R_L + R_a)$.

To obtain maximum signal bandwidth (see the arguments leading to eq. (7.32)) both R_L and R_a should be as small as possible. The major penalty to be paid with such a

Fig. 9.19 Possible design for a low impedance front end circuit using a bipolar junction transistor. The value of R_L is relatively low.

design is that the system is then likely to be dominated by Johnson noise (see eq. (9.11)), and the signal-to-noise ratio correspondingly poor. If the noise limitations are not a problem, however, then this design has the advantage of relative simplicity. The circuit is usually implemented using a bipolar junction transistor (see, for example, Fig. 9.19, which shows a grounded emitter configuration).

If a low noise level is of paramount importance, then a 'high impedance' front end may be considered. That is R_a is made relatively large. The disadvantage is then that the electrical bandwidth is limited to $(2\pi R_a C_j)^{-1}$. At high frequencies the response falls off as $1/f$, the capacitor and parallel resistor combination acting to give an output proportional to the integral of the signal current. Thus it is possible to extend the frequency response by including a differentiating circuit (i.e. one whose output is proportional to f) later on in the amplifier chain. One problem that arises with this approach is that, although it allows an increased bandwidth, it also leads to a reduction in the available dynamic range, i.e. in the ratio of maximum to minimum usable input signals. A possible circuit for a high impedance front-end is shown in Fig. 9.20. Here, a silicon field effect transistor (FET) is used in a common source configuration. The silicon FET exhibits both very high input impedance and low noise performance. One difficulty is that the gain of the device is rather limited, above 25 MHz the gain drops to values close to unity, so that above this frequency a bipolar transistor is more often used.

Recently, a considerable effort has been put into the development of GaAs FETs, since they operate up to several gigahertz and thus offer an alternative to bipolar devices. There has been particular interest in hybrid integrated circuits utilizing *pin*

Fig. 9.20 Possible design for a high impedance front end circuit using an Si FET. The value of R_a is high.

photodiodes with GaAs FET front-ends. The integration technique enables stray capacitances to be reduced to very low values (i.e. ~ 0.5 pF).

A modification of the high impedance front end is the 'transimpedance' design. Here, a high impedance amplifier is used in conjunction with negative feedback; i.e. a portion of the output signal is fed back into the input with its phase reversed. The advantage of this is that the frequency response is extended thereby reducing or even eliminating the need for any differentiating circuitry. At the same time, the amplifier noise is still much less than would be the case in a low impedance circuit. A simple and popular transimpedance circuit suitable for fairly modest bandwidth requirements is shown in Fig. 9.21. Here, an FET operational amplifier is used in conjunction with a high value feedback resistor R_f. The voltage output of this circuit is given by $i_\lambda R_f$, where i_λ is the current output from the photodiode.

9.3.3.5 DETECTOR MATERIALS

Most systems operating at present use silicon detectors, which limits the operating wavelength to a maximum of 1.1 μm. As we have seen in section 9.3.1, it is advantageous from the point of view of both fiber loss and material dispersion to use longer wavelengths, either around 1.3 μm or 1.55 μm. Germanium detectors, which respond out to 1.9 μm, are readily available. Unfortunately they suffer from much higher dark currents than do the corresponding silicon devices. In addition, germanium has a carrier ionization ratio r of about unity. This implies that the optimum gains in APDs are relatively small, and hence so are any improvements in signal-to-noise ratio. For

Fig. 9.21 Design for a transimpedance front end using an FET operational amplifier. The voltage output is given by $V_{out} = i_\lambda R_f$, where R_f is the feedback resistor and i_λ the photogenerated current.

example, taking $r = 1$ in eq. (9.15), together with the other values assumed for the calculation on the silicon APD gives an optimum gain of 15 and hence an improvement in the optical power signal-to-noise ratio by a factor of only 8.9 (i.e. 9.5 dB).

Other possible candidates include GaInAs and GAInAsP (ref. 9.6). As in the case of Ge, however, detectors made from these materials also suffer from relatively high dark currents and low carrier ionization ratios. A considerable amount of research is being carried out to determine the optimum material and type of detector, although it is by no means obvious that a 'clear winner' will emerge.

9.3.4 Fiber choice

As we have seen in Chapter 8, there are three main types of fiber available, namely step index, graded index and single mode fibers. Step index fibers are relatively inexpensive, have large NA values, but suffer from high inter-modal dispersion. Graded index fibers show greatly reduced intermodal dispersion but have relatively small NA values; only about half of the energy can be carried by a graded index fiber as can be carried by a step index fiber with the same core diameter. Single mode fibers are subject only to material dispersion, but have such small diameters that usually only laser sources can couple sufficient amounts of power into them; in addition, much higher accuracy is required in alignment at splices to avoid excessive jointing loss. Because of the low amounts of power carried by such fibers, it is almost essential to work at wavelengths where transmission losses are very low.

For many applications, a key factor is the maximum length possible between transmitter and receiver (or between repeaters). Two factors influence this, namely fiber attenuation and fiber dispersion. In low bandwidth systems, the former is usually the limiting factor, whereas in high bandwidth systems it is the latter. For example,

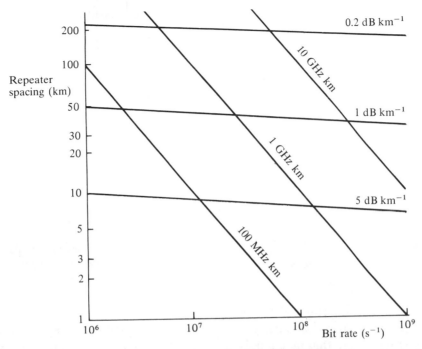

Fig. 9.22 Maximum repeater spacing allowing for differing fiber attenuations and dispersions as a function of bit rate. A total loss of 50 dB at 1 Gb s^{-1} has been assumed. The attenuation limited lines are slightly inclined due to reduced receiver sensitivity with increasing bit rate.

suppose we consider a fiber with an attenuation of 5 dB km^{-1}; if a signal loss of some 40 dB can be tolerated, then a cable length of 40/5 or 8 km can be used. If the cable were of the step index type, however, with an inter-modal dispersion of some 50 ns km^{-1} (or 20 MHz km) and a signal bandwidth of 10 MHz were required, then the maximum fiber length would be restricted to 20/10, i.e. 2 km. Figure 9.22 illustrates the effects of these factors on the maximum fiber lengths achievable for several fiber loss and dispersion values.

One additional source of noise that has arisen in practical systems is the so-called *modal noise*. This manifests itself as a signal amplitude modulation appearing when a cable containing misaligned joints is flexed. It is essentially an interference effect. Across any fiber cross-section the modes traveling down the fiber will produce an interference pattern. The number of dark and bright areas in this pattern will be roughly proportional to the number of modes propagating down the fiber. Physical movement of the fiber will slightly change the relative irradiance and phase of the modes and hence also the appearance of the interference pattern (see Fig. 9.23). At a misaligned fiber join, any change in the power distribution across the fiber cross section may then lead to a change in the power coupled across the join. The interference pattern can only form, of course, if the modes remain coherent. Over long lengths of

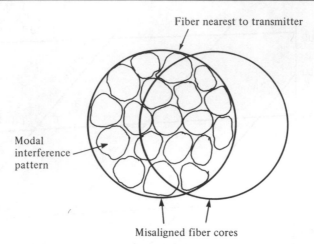

Fiber nearest to transmitter

Modal interference pattern

Misaligned fiber cores

Fig. 9.23 Mode interference pattern at a fiber join. Any slight change in the pattern will result in a change in the coupled power if there is any misalignment between the fibers.

fiber, the differences between the corresponding ray paths may exceed the coherence length of the radiation and modal noise will then become less important. In this respect, it is interesting to note that modal noise is much more of a problem with laser sources than with LEDs. One way of diminishing its effect when using lasers is to operate the laser well below threshold, thus increasing the spectral width and hence decreasing the coherence length. This procedure will, of course, give rise to increased material dispersion, and a compromise may have to be sought.

9.3.5 System design considerations

We now briefly consider typical component combinations used in practical systems. Three main factors are paramount in influencing system choice, namely bandwidth, maximum transmission distance and, of course, total system cost.

For short distances (i.e. up to a few hundred meters) and for bandwidths of up to a few megahertz, a very popular combination is a red emitting LED, a high-NA all-plastic fiber and a Si photodiode detector. More exacting distance and bandwidth requirements may be met with a move to silica (or glass-) based fibers with an LED (or laser) source operating at about 850 nm. The detector may be either a silicon photodiode or a silicon avalanche photodiode.

To determine whether or not a chosen system will perform satisfactorily, a number of checks must be made. Obviously both detector and emitter must be capable of handling the required signal bandwidth. The fiber dispersion over the length required must not degrade the signal excessively. For a given bit error rate (for a digital signal) there will be a minimum average signal power that must reach the detector. If the power launched into the fiber is known, together with the fiber attenuation, the maximum length of fiber that can be used may be calculated. Allowance must be made for any splices or joins, and a safety margin of say 5 dB is also usually included. Such a calculation is called a *flux budget* (see Example 9.2). It is customary to use a logarithmic power unit called the *dBm*, where the power is referred to 1 mW; thus a power of 10 mW becomes $10 \log_{10}(10/1)$ or 10 dBm.

Example 9.2 Flux budget calculation

Suppose we have to construct a link of length 15 km and bandwidth 100 Mb s^{-1}. Components are chosen with the following characteristics: receiver sensitivity -50 dBm (at 100 Mb s^{-1}), fiber loss 2 dB km^{-1} and transmitter launch power (into fiber) 0 dBm. It is anticipated that in addition ten splices will be required, each involving a loss of 0.5 dB. The following table itemizes each loss term and allows a ready determination of whether the system should operate with sufficient power margin:

Transmitter output	0 dBm
Receiver sensitivity	-50 dBm
Margin	50 dBm

System loss:

Fiber loss 2 dB km^{-1}, 15 km	30 dB
Detector coupling loss	1 dB
Total splicing loss (0.5 dB × 10)	5 dB
Headroom for temperature range, aging effects and future splices	5 dB
Total attenuation	41 dB

Hence, the excess power margin is 9 dB, which should be sufficient for the operation of the link.

In Fig. 9.24 we show typical launch powers for both LED and laser sources and also the received powers required by state-of-the-art receivers to achieve a 10^{-9} bit error rate as a function of bit rate. It is obvious from Fig. 9.24 that as the required bit rate increases the relative advantage of the laser/APD combination also increases.

When the transmission rate exceeds 400 Mb s^{-1} or so, the usable fiber length is determined primarily by fiber dispersion rather than by fiber attenuation (see Fig. 9.22). It then becomes necessary to move to single mode fibers using operating wavelengths close to the material dispersion minimum (i.e. near 1.3 μm in SiO$_2$ based fibers). Transmission rates of several gigabits per second over fibers of several tens of kilometers in length have already been demonstrated and at a more modest bit rate of a few hundred Mbs^{-1} unrepeated transmissions over distances of 100 km or so have been achieved, albeit in a 'laboratory' environment.

Fig. 9.24 Simple illustration of the system flux budget for representative systems. The upper shaded areas show typical launch powers from LED and laser sources, whilst the lower curves show typical receiver sensitivities for a 10^{-9} bit error rate at 0.85 μm wavelength. The fundamental detection limit corresponding to 10 photons per bit is also shown.

9.3.6 Local area networks

Although the initial drive to develop optical fibers has come mainly from the telecommunications area, i.e. from the need to develop high-bandwidth long-distance repeaterless point-to-point links, there are other types of link, for example the local area network (LAN), to which optical fibers may make significant contributions. The LAN concept is difficult to specify exactly, but aims to provide a communication network between a group of users within a restricted geographical area (for example within a factory or group of offices). Typical maximum distances involved might be a few kilometers. Local telephone networks are nothing new, of course, but the availability of wider bandwidth links enables several other types of communication to be contemplated – for example data communication and even video links. Situations

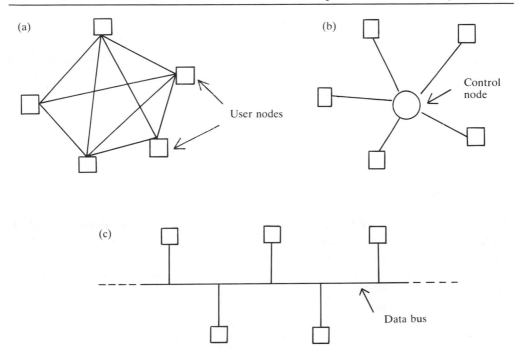

Fig. 9.25 Three possible topologies for a local area network: (a) fully connected, (b) star, and (c) data bus.

can easily be envisaged where networks linking up to say one hundred users with data rates of up to several megabits per second may be required.

One of the main problems with realizing such networks using optical fibers is that the most loss the system can readily tolerate between launch and detection is about 40–50 dB (i.e. see Fig. 9.24). With coaxial cable, the corresponding figure is nearly 90 dB. In long haul communication links, the advantage this gives to coaxial links is more than offset by their relatively high transmission losses. For the comparatively short distances involved in typical LANs, however, coaxial links do not have this built-in disadvantage. Furthermore, any such system will inevitably involve a large number of 'taps' and switches. For an optical system these components, at present, involve appreciable inherent losses, whilst the the corresponding losses in copper wire technology are negligible. Nevertheless, fiber systems do offer the advantages of high bandwidth, immunity from local interference, a higher level of information security and small size and weight of the cable.

There are many ways in which the interconnections in LANs may be made with trade-off occurring between such factors as system complexity (and hence cost) and flexibility. Figure 9.25 illustrates several possible topologies for such systems. In Fig. 9.25(a), the 'fully connected' network, every potential user (or 'node') is connected to every other. This arrangement is complex, expensive and needs considerable rewiring if more nodes are added. Some of the links may never be used. It is, however, the most reliable topology, since if one link is damaged then only this one linkage is affected, and rerouting may be possible to restore the connection.

Figure 9.25(b) illustrates the 'star' network. Here, each node is connected to a central 'control' node. If this is 'active', each signal can be switched to its required destination. If 'passive', a particular node will direct a signal to all nodes, each of which will then have to decide themselves whether or not to accept the signal. The number of links is reduced to a minimum, but the reliability of the whole system hinges on that of the central node. If this fails in some way, the entire system is likely to break down. Figure 9.25(c) illustrates the 'data bus', this offers great flexibility and is efficient in its use of fiber.

Some of the optical components likely to be useful in LANs have been touched upon in section 8.5.2. For example, the 'mixing rod' type of fiber coupler could form the basis of the central (passive) node of a star network. Similarly, the fused coupler could be designed to provide a small amount of coupling from one fiber into another, thus providing a 'tap' as required in a data bus configuration.

The diversity of possible systems and their realization makes any comprehensive discussion here impossible, and the interested reader may consult the texts of ref. 9.8 for further information.

9.3.7 Future developments

Most of this chapter has been concerned with direct detection, but, as was mentioned in section 9.1.1, there are considerable advantages in adopting a coherent detection system. The main difficulty with such schemes is the necessity for providing semiconductor lasers with the required frequency and phase stability. A normal semiconductor laser might have a linewidth of about 500 GHz (i.e. 3 nm in wavelength terms). For coherent detection, on the other hand, linewidths of 15 MHz or less are required. As seen in section 6.2, semiconductor lasers with narrow stable bandwidths can be obtained by using DFB structures and external cavities. However, the use of such lasers adds considerably to overall system complexity.

A further problem is that normal fibers are not very good at maintaining polarization (see section 10.1.3.3), and for coherent detection both signal and local oscillator beams must have the same plane of polarization if they are to 'mix' properly at the detector. Polarization-maintaining fiber is available, but it is both more expensive and more lossy than conventional fiber. Thus, a coherent link would also probably have to monitor the state of polarization of the incoming signal and allow for any variation in the polarization direction. This again would also add to the system complexity (and hence cost). Nevertheless, for long haul high bandwidth links, the ability to operate without repeaters may make coherent systems attractive.

On way of increasing the bandwidth of an existing optical link with no modification to the fiber itself is to employ the technique of *wavelength multiplexing*. In this, several different signals are transmitted simultaneously down the fiber on carriers of different wavelength. This scheme has been demonstrated as a practical proposition. The main

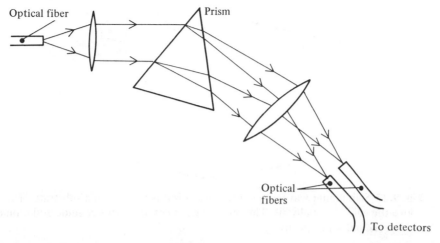

Fig. 9.26 Illustrating one method that has been used to separate a wavelength multiplexed signal into its two (or more) constituent wavelength signals. The light is passed through a prism. Dispersion within the prism causes the two signals to become spatially separated.

difficulty is in separating the signals at the detector end; inevitably, both appreciable signal loss and a certain amount of crosstalk between the signals are entailed. Several possible methods of signal separation have been proposed (see ref. 9.9) and in Fig. 9.26 we show one of these. A miniature prism spatially separates the transmitted wavelengths, thus enabling separate fibers to pick up the signals and carry them to their respective detectors. This approach also enables two-way transmission to take place over a single fiber, the different directions of propagation corresponding to different wavelengths.

The performance of silica-based fibers has been dramatically improved over the last few years to the stage where, over a wide wavelength range, the theoretical attenuation limit given by Rayleigh scattering has been reached. Since Rayleigh scattering gives an attenuation that varies as λ^{-4}, it is desirable to work at as high a wavelength as possible. In silica fibers, the highest usable wavelength is limited by the onset of strong lattice absorption bands at about 1.6 μm (see Fig. 8.24). Consequently, materials are being sought where this type of absorption occurs at higher wavelengths. Several promising materials are being investigated, such as the alkali halides and non-oxide glasses (ref. 9.9) and minimum attenuations of the order of 10^{-3} dB km^{-1} have been predicted between 5 and 7 μm. Such low attenuations hold out the highly attractive possibility of repeaterless links some hundreds or even thousands of kilometers long. Development of fibers of these materials is, however, at a fairly preliminary stage and they are unlikely to be readily available for several years. In addition, of course, new emitters and detectors will have to be developed for use at these longer wavelengths.

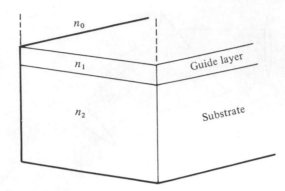

Fig. 9.27 Slab planar waveguide. The guide itself is formed on a substrate. The medium above the guide is usually air. The refractive indices of substrate, guide and topmost layer are n_2, n_1 and n_0 respectively.

9.4 Integrated optics

Although signal transmission using light waves is now well established, if any processing is required the signal usually has to be converted into an electrical one before this can be done. The aim of integrated optics (IO) is to be able to do as much signal processing as possible directly on the optical signal itself. It is envisaged that a family of optical and electro-optical elements in thin film planar form will be used, allowing the assembly of a large number of such devices on a single substrate. Most device elements are expected to be based on single mode planar optical waveguides. Similar advantages are expected to those accruing when the idea of the integrated circuit was adopted in electronics.

The basic concept of IO was first proposed by Anderson in 1965 (ref. 9.10) and considerable progress has been made since then. For a fairly comprehensive coverage of the field, the reader is referred to ref. 9.11. It must be admitted at the outset, however, that not a great deal of *integration* has in fact been achieved. Although several types of device of the kind that will undoubtedly be needed have been successfully demonstrated, they require a wide range of different substates. We now briefly consider possible structures for these devices.

9.4.1 Slab and stripe waveguides

It is generally assumed that in IO the signal will be carried within planar waveguides in either slab or stripe form and formed by modifying the surface of a substrate. Planar waveguides were discussed in section 8.2, but there we confined our attention to symmetrical guides. In the present instance we are usually dealing with asymmetrical guides, i.e. guides where the layers above and below the guiding layer have different refractive indices (Fig. 9.27). The topmost layer (of refractive index n_0) is often air and consequently has a much lower refractive index than the other two. If the refractive

Fig. 9.28 Two basic geometries used for making stripe waveguides.

indices of the guide and substrate layers are n_1 and n_2 respectively, and the thickness of the guide is d, the guide can support a mode of order m provided that (ref. 9.12)

$$d \geqslant \frac{\lambda(m + \frac{1}{2})}{2\,(n_1^2 - n_2^2)^{\frac{1}{2}}} \quad (m \geqslant 0) \tag{9.21}$$

Here λ is the wavelength within the guide, and it is assumed that $n_1 > n_2 > n_0$. We see that in contrast to the symmetrical guide case, there is a minimum thickness below which it is not possible for the guide to support any modes at all. Assuming typical values for n_1 and n_2 of 1.50 and 1.49, we see from eq. (9.21) that for a guide to support only one (i.e. the $m = 0$) mode, its thickness must lie between 1.4λ and 4.3λ. A typical waveguide thickness would therefore be approximately equal to 2λ.

A wide variety of different techniques have been used to make slab waveguides; for example, sputtering one type of glass onto another, in-diffusion of a layer of titanium deposited on a substrate of lithium niobate, and liquid phase epitaxy. This last technique can be used with semiconducting materials such as GaAs and GaAlAs.

The two basic geometries for stripe waveguides are shown in Fig. 9.28. They may be formed by exploiting one of the methods for making slab waveguides and delineating the stripe pattern using photo- or electron-beam lithography. One problem which may hinder miniaturization is that any bends in the waveguide must have comparatively large radii (i.e. of the order of millimeters) otherwise losses can become prohibitively large (see the discussion in section 8.4.1.) Losses in IO waveguides are usually much higher than in optical fibers, being of the order of 0.1 dB mm^{-1}. One reason for this is the large scattering from the upper waveguide surface (i.e. the waveguide/air interface) which is often relatively rough.

9.4.2 Devices

We consider first a phase shifter that can be made the basis for several other more complex devices. In its simplest form, it is similar to the Pockels cell discussed in section 3.4. A stripe waveguide is formed within a suitable optically active material such as lithium niobate and electrodes are formed on the substrate surface on either side of the waveguide (Fig. 9.29). If the electrodes are a distance D apart and extend for a length

Fig. 9.29 Integrated optical version of the Pockels cell that may be used as a phase shifter.

L, then the additional phase shift $\Delta\phi$ produced when a voltage V is applied across the electrodes is, from eq. (3.14).

$$\Delta\phi = \frac{\pi}{\lambda} rn_1^3 V \frac{L}{D} \tag{9.22}$$

where n_1 is the guide refractive index and r is the guide material electro-optic coefficient (see Table 3.1 for representative values). A big advantage over bulk Pockels effect devices is that the ratio L/D may be made relatively large, say about 1000, and a phase change of π may then be achieved with voltages as low as 1 V or so.

A high-speed switch may be made by incorporating the phase shifter into one arm of the interferometric arrangement shown in Fig. 9.30. Here, the guide splits into two – both paths rejoining after an identical pathlength. With no voltage across the phase shifter, the radiation in the two arms will have the same phase when recombining, and hence the device will have no effect on the radiation flowing along the guide. However, if the phase shifter is activated to give a phase shift of π, then, on recombining, the radiation in the two arms will destructively interfere and no radiation will proceed down the guide. (The initial energy in the beam will be radiated out into the substrate.) Devices of this type have been made with switching times of the order of several hundred picoseconds.

Fig. 9.30 Interferometric modulator.

(a)

(b)

Electrodes

Guides

Substrate

Electric field lines

Fig. 9.31 Two waveguides (a) are shown in close proximity for a coupling distance z. The amount of coupling may be modified by depositing electrodes on top of the guides and applying a voltage between them as shown in (b). This arrangement forms the basis for an IO switch.

When two stripe waveguides are brought into close proximity, there can be a transfer of energy from one to the other (Fig. 9.31(a)). This is because the mode energy penetrates to a small extent into the region round the guiding layer (see section 8.1). The amount of energy coupled depends on the factor $\{C^2 \sin^2 [z(C^2 + \delta^2)^{\frac{1}{2}}]\}/(C^2 + \delta^2)$ (ref. 9.13 and Problem 9.10). Here, C characterizes the coupling between the guides and z is the length over which the guides are coupled. The factor δ is given by

$$\delta = \frac{\pi}{vc}(n_\mathrm{I} - n_\mathrm{II})$$

where n_I and n_II are the refractive indices of the two guide materials and v is the frequency of the radiation. If $\delta = 0$, then after a coupling length $z = L_c$, where $L_c = \pi/(2C)$, all the power in one guide will be transferred to the other (and after a further distance L_c all the power will have been transferred back). However, if $\delta \neq 0$ then not only will less energy be exchanged but also the energy change will take place more rapidly with distance. Indeed, if δ is such that $L_c (C^2 + \delta^2) = \pi$, then after a coupling length L_c no energy will be exchanged. By substituting $L_c = \pi(2C)$, this requirement reduces to $\delta = \sqrt{3}C$. Thus if we can 'switch' δ from zero to the value $\sqrt{3}C$, we can switch energy from one waveguide to the other. Several configurations have been built to implement this, of which the simplest is shown in Fig. 9.31(b). Electrodes are

Fig. 9.32 Waveguide with a corrugation of period D etched upon it.

deposited above each waveguide and a potential is applied between them. The opposing vertical fields in the two waveguides can, if the material axes have been chosen correctly, induce opposite changes in the guide refractive indices and hence change the value of δ. The main problem with this arrangement is that it is not easy to achieve total energy transfer because of the difficulty of ensuring that the coupling length is exactly L_c. More complicated electrode configurations have been proposed to overcome these difficulties (ref. 9.13).

Devices such as filters and resonators may be realized in IO by incorporating periodic structures into optical waveguides. Consider, for example, a waveguide with a 'corrugation' etched upon its surface perpendicular to the direction of beam propagation (Fig. 9.32). This structure is encountered in the distributed feedback laser (section 6.2); it acts as a wavelength dependent mirror, strong reflection occurs when $2D = m\lambda_0/n_1$, where D is the grating period, λ_0 the vacuum wavelength, n_1 the guide material refractive index and m an integer.

Reflection bandwidths are usually narrow, but may be increased by 'chirping' the grating (Fig. 9.33).

Although both devices discussed above use stripe waveguides, there are also useful devices based on slab waveguides. One of these is a beam deflector based on diffraction from an acoustic wave. An interdigital electrode structure deposited on a suitable acousto-optical material (see Fig. 9.34) can generate a beam of surface acoustic waves

$\longleftarrow$ Decreasing grating spacing

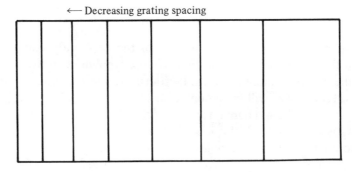

Fig. 9.33 'Chirped' diffraction grating structure. A plan view is shown with the vertical lines representing the grating peaks (or troughs).

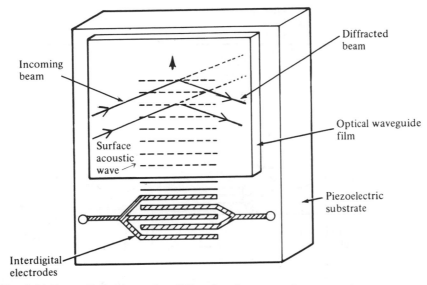

Incoming beam

Diffracted beam

Surface acoustic wave

Optical waveguide film

Piezoelectric substrate

Interdigital electrodes

Fig. 9.34 Beam deflection using diffraction from a surface acoustic wave generated by applying an alternating voltage to an interdigital structure evaporated onto the surface of a piezoelectric substrate.

which can then serve to diffract light traveling along the guide. The angle through which the beam is diffracted may be changed be varying the frequency (and hence the wavelength) of the acoustic wave. This device is of course a version of the acousto-optic deflector discussed in Chapter 3.

Such a beam deflector forms the heart of an integrated optical spectrum analyzer, designed to display the frequency spectrum of a radio frequency signal (ref. 9.14). The layout is illustrated in Fig. 9.35. Light from a semiconductor laser is launched into a waveguide and the beam subsequently rendered parallel by the use of a 'geodesic' lens. This is a circular indentation made in the substrate layer with the guide layer thickness being unchanged across the indentation (Fig. 9.36). Such a structure behaves like a 'one-dimensional' lens (ref. 9.15). The parallel beam then passes through the beam deflector. If a single RF frequency is present, the amount of deflection will depend on the instantaneous value of the frequency. A second lens subsequently focuses the light onto a particular photodetector in a photodetector array. Each detector element corresponds to a particular frequency (or, more acurately, a narrow range of frequencies). If, in fact, more than one frequency is present, the light will be divided into different components that are then focused onto different elements, thus enabling the frequency spectrum of the RF signal to be obtained.

9.4.3 Emitters and detectors

The most obvious choice for an emitter would seem to be a semiconductor laser, since here the radiation is generated within a channel of similar dimensions to those of stripe

Fig. 9.35 Integrated optical spectrum analyzer based on an acousto–optic deflector.

Fig. 9.36 Cross section of a geodesic lens in a waveguide.

Fig. 9.37 IO semiconductor laser based on GaAs/GaAlAs using Bragg reflectors instead of cleaved end mirrors. Light from the active layer is coupled into the layer beneath, which then acts as a waveguide.

Fig. 9.38 IO detector using a Si p–n junction as a photodetector. Light is carried to the detector via a doped SiO$_2$ waveguide grown on the surface of the silicon.

waveguides. Because of the difficulties of obtaining cleaved end mirrors, a distributed feedback structure (section 6.2) is usually employed. One problem with the laser source is that when not being pumped the lasing region is absorbing; consequently, arrangements are made for coupling the radiation in the active layer into a non-lossy guiding layer, which is usually beneath it (see Fig. 9.37).

A similar difficulty arises with detectors: the detecting region must of necessity be lossy and again some form of coupling from a non-lossy guide layer is required. Figure 9.38 shows a detector based on silicon using SiO$_2$ as the waveguide material.

Problems

9.1 How accurately may the amplitude of a signal waveform be specified using an eight bit number?

9.2 The analog modulation bandwidth required for television is about 4 MHz; what bit rate is required for a digital transmission?

9.3 Estimate the maximum range for a line-of-sight optical communication system using mirror collimators of 0.1 m diameter and operated at 0.85 μm. The minimum power required by the detector is 10^{-9} W and the emitter launches 1 mW of radiation. Take into account losses arising from both (a) beam diffraction and (b) atmospheric attenuation (which you may assume to be 10 dB km^{-1}).

9.4 An LED source emits 1 mW from an area 100 μm in diameter. It is butt joined to a step index fiber with a core diameter of 50 μm and core refractive index of 1.5. The cladding has a refractive index of 1.48. Assuming the LED source is a Lambertian emitter, estimate the energy coupled into the fiber.

9.5 The much narrower beam divergence of lasers enables them to couple substantially more power into a fiber than LEDs. A convenient expression for their surface luminance as a function of angle is given by $B(\theta) = B(0) \cos^n\theta$. A Lambertian source is characterized by $n = 1$, whereas lasers may have $n \approx 20$. Using this expression, calculate the coupling efficiency into a fiber as a function of n assuming NA $\ll 1$.

9.6 An emitter can couple some 10^{-3} W of optical radiation at 900 nm into a fiber which has an attenuation of 5 dB km^{-1}. A receiver is used which requires an average of 500 photons per bit for an acceptable bit error rate. Plot a graph of the maximum fiber length possible for bit rates of between 10^5 and 10^9 bits s^{-1}. Modify this diagram to include the effects of fiber dispersion (5 ns km^{-1}) if the maximum allowable pulse spreading must not exceed $1/(2 \times$ bit rate).

9.7 Show that the optimum gain of an avalanche photodiode may be obtained by solving the equation

$$M^3 \frac{dF(M)}{dM} = \frac{2 \times \text{Amplifier noise}}{\text{Shot noise}}$$

where $F(M)$ is the excess noise factor. Hence, by substituting for $F(M)$ from eq. (7.33a) verify eq. (9.15).

9.8 Determine the optimum gain for an avalanche photodiode made from GaAs where the electron-to-hole-ionization ratio, r, is 5. Assume the ratio of amplifier-to-shot noise is 1000. Estimate the improvement in signal-to-noise ratio obtained in this instance by using an APD rather than a *pin* photodiode.

9.9 A simple communication link 8 m long is to be set up using the components detailed below. Check that the system will operate satisfactorily and, if not, suggest possible modifications. You may assume that the bandwidth requirements are met.

> Fiber (plastic): attenuation = 0.6 dB m^{-1}
> Emitter (LED): a maximum power of 800 μW can be launched into the fiber at a drive current of 60 mA
> Detector (photodiode): the minimum and maximum optical powers required are −20 dBm and −8 dBm respectively

9.10 Show that the power coupled from one waveguide to another when their propagation constants are equal is proportional to the factor $\sin^2(Cz)$, where C is the coupling factor.
[Hint: assume that the equations governing the behavior of the electric field amplitudes (A_1, A_2) in the guides can be written

$$\frac{dA_1}{dz} = ik_1A_1 + iCA_2$$

$$\frac{dA_2}{dz} = ik_2A_2 + iCA_1$$

Here, the second term on the right-hand side of these equations represents the coupling between the waveguides; k_1, k_2 are the propagation constants $(= 2\pi/\lambda)$; and i $= \sqrt{-1}$. By taking $k_1 = k_2$ and using the boundary conditions $A_1(z = 0) = A(0)$ and $A_2(z = 0) = 0$, the required result follows. It is also possible to derive the general result when $k_1 \neq k_2$ from the above equations, but the going is rather harder!]

References

9.1 T. G. Hodgkinson, D. W. Smith, R. Wyatt and D. J. Mayon, 'Coherent optical fiber transmission systems', *Br. Telecom. Technol. J.*, **3**, 1985, 5.

9.2   H. Taub and D. L. Schilling. *Principles of Communication Systems*. McGraw-Hill. New York, 1971, Chapter 5.

9.3   *Ibid.*, Chapter 13.

9.4   H. F. Wolf (ed.), *Handbook of Fibre Optics*. Granada. St Albans. 1979, Chapters 8 and 11.

9.5   R. M. Gagliardi and S. Karp. *Optical Communications*. Wiley-Interscience. New York, 1976, Chapter 2.

9.6   M. J. Howes and D. V. Morgan (eds). *Optical Fiber Communications*, Wiley, New York. 1980, Chapter 3.

9.7   (a) M. J. Howes and D. V. Morgan (eds). *op. cit.*, section 6.8.2.
         (b) H. Ishio, J. Minowa and N. Kiyoshi, 'Review and status of wavelength division multiplexing technology and its applications', *J. Lightwave Technol.*, **LT2**, 1984, 448.

9.8   (a) S. D. Personick, *Fiber Optics: Technology and Applications*, Plenum Press, New York, 1985, Chapter 9.
         (b) S. Matsushita, K. Kawai, H. Uchida, 'Fibre-optic devices for LAN applications', *J. Lightwave Technol.*, **LT-3**, 1985, 554.

9.9   G. H. Sigal, Jr, and D. C. Tran, 'Ultra-low loss optical fibres: an overview', *Proc. SPIE Int. Soc. Opt. Eng.*, **484**, 1984, 2.

9.10 D. B. Anderson. *Optical and Electro-optical Information Processing*. MIT Press. Cambridge, Mass., 1965, 221.

9.11 (a) T. Tamir (ed.). *Integrated Optics* (2nd edn.). Springer Verlag, Berlin. 1979.
         (b) R. G. Hunsperger, *Integrated Optics: Theory and Technology*, Springer-Verlag, Berlin, 1982.

9.12 L. Levi. *Applied Optics*. Vol. 2. Wiley. New York, 1980, Chapter 13.

9.13 M. J. Howes and D. V. Morgan (eds). *op. cit.*, Chapter 4.

9.14 T. J. Joseph, T. R. Ranganath, J. Y. Lee and M. Pedicott, 'Performance of the integrated spectrum analyser', *Proc. SPIE Int. Soc. Opt. Eng.*, **321**, 1982, 134.

9.15 B. Chen and O. S. Ramer, 'Diffraction limited lens for integrated optical circuits', *I.E.E.E. J. Quant. Electron.*, **15**, 1979, 853.

Chapter 10

Noncommunications applications of fibers

Although the main driving force in the development of optical fibers has been their use for telecommunications purposes, other perhaps more mundane uses promise to be of equal, if not greater, importance. In this chapter we examine some of these.

10.1 Optical fiber sensors

In many industrial processes there is a need to monitor such quantities as displacement, pressure, temperature, flow rate, liquid level and chemical composition. Ideally, the measurement technique should be reliable, robust, corrosion resistant, intrinsically safe and free from external interference. Obviously, optical fibers (and indeed optical methods in general) have the potential for making a significant contribution in this area. In many cases, however, well developed alternative technologies already exist and it may be some time before optical fibers can make any significant penetration into the market.

Optical fiber sensors themselves can be divided into two main categories, namely 'active' and 'passive'. In the latter, the modulation takes place outside the optical fiber, which acts merely as a convenient transmission channel for the radiation. In active sensors, on the other hand, the quantity to be measured acts directly on the fiber itself to modify the radiation passing down the fiber.

As far as active sensors are concerned, it is possible to modulate either the amplitude, the phase or the state of polarization of the radiation in the fiber. In multimode fibers, however, mode coupling and the usually random relationships between the phases and polarization states of the propagating modes generally preclude the use of either phase or polarization modulation. Thus, active multimode fiber sensors almost invariably involve amplitude modulation. On the other hand, with single mode fibers both phase and polarization modulation becomes possible. Phase modulation opens up the possibility of fiber based interferometric sensors that can offer exceptionally high sesitivities. It is convenient therefore to treat multimode and single mode active fiber sensors separately.

10.1.1 *Multimode passive optical fiber sensors*

Some of the simplest passive fiber-based sensors are concerned with the measurement of movement or position. For example, when two fiber ends are moved out of alignment, the coupling loss depends on the displacement. The actual relationships between loss and displacement were considered in section 8.5 (see also Problem 8.13). A similar type of sensor uses a shutter moving between two fiber ends that are laterally displaced (Fig. 10.1(a)). Improvements in sensitivity are possible by placing a pair of gratings within the gap, one fixed, the other movable (Fig. 10.1(b)). Here, however, although the sensitivity has increased, the range has decreased, since the output will be periodic in the spacing of the grating. The range of movement possible for the single shutter sensor is obviously limited by the fiber core diameter. If a beam expander is employed between the fibers (Fig. 10.1(c)), then the range can be greatly expanded. The displacement–transmission characteristic of the expanded beam system is considered in Problem 10.1.

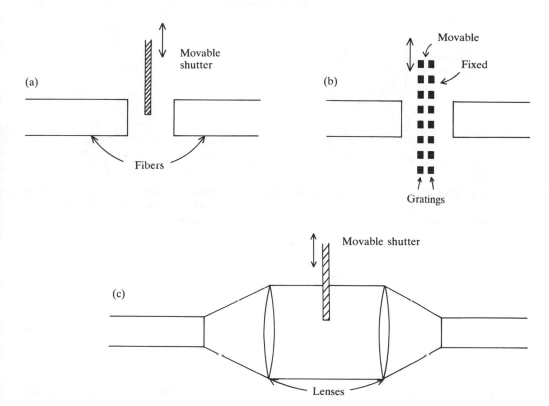

Fig. 10.1 Simple displacement sensors. In (a) a movable shutter varies the light coupled between two longitudinally displaced fibers. In (b), the use of two gratings increases the sensitivity. In (c), a beam expansion system enables an increase in the range of measurable displacements to be increased.

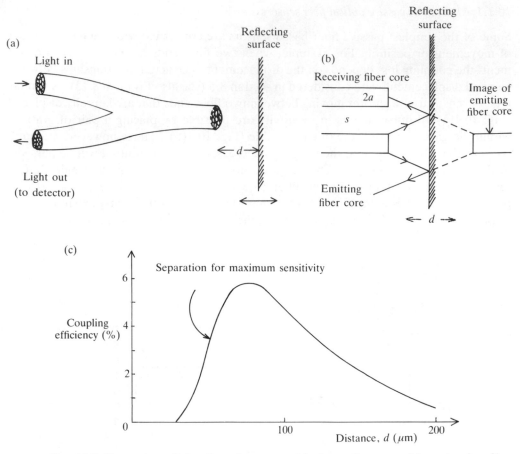

Fig. 10.2 Illustration of the Fotonic sensor. (a) shows the general layout using fiber bundles. A two fiber version is shown in (b), which can be used to derive the form of the output–distance (d) relationship. The typical result of such a calculation taking $a = 100$ μm, $s = 100$ μm and NA = 0.4 is shown in (c).

One of the first commercially available displacement sensors was the 'Fotonic sensor'. This uses a bundle of fibers, half of which are connected to a source of radiation, the other half to a detector (Fig. 10.2(a)). If the bundle is placed in close proximity to a reflecting surface then light will be reflected back from the illuminating fibers into the detecting fibers. The amount detected will depend on the distance from the fiber ends to the surface. To analyze this dependence, we consider the somewhat simpler situation where there are just two fibers. If we regard the reflecting surface as a mirror, the problem then reduces to that of the coupling between two fibers that are diplaced both laterally and longitudinally (Fig. 10.2(b)). The form of the relationship between displacement and light output may be determined by considering the overlap between the sensing fiber core area and the cross section of the light cone emitted by the image of the emitting fiber. We can readily appreciate that at very small fiber-surface distances, no light will be coupled between the two fibers. Then, beyond a certain critical distance, there will be an increasing overlap between the above areas and the coupled radiation will increase rapidly. Once the detecting fiber area is completely filled, however, the output will fall with increasing distance. At large distances, an inverse square law will then be obtained. The displacement output characteristic is considered in more detail in Problem 10.2, and a typical result of this analysis is shown in Fig. 10.2(c).

In practice, when a fiber bundle is used instead of just two fibers the displacement–output characteristic will be somewhat different and will depend on how the emitting and receiving fibers are distributed (usually randomly), but the overall shape remains similar to that of Fig. 10.3(b). Because of the very nonlinear nature of the curve, the sensor is not very suitable for the measurement of large displacements, although it is possible to increase the range by using a lens system. In fact, the sensor was developed originally for noncontact vibration analysis.

Any displacement measurement technique is readily adapted to the measurement of pressure, and those mentioned above are no exception. For example, the Fotonic sensor may be placed close to a reflecting diaphragm with a constant pressure maintained on the sensor side. Any change in external pressure will cause flexing of the diaphragm and a consequent change in the instrument's output. It should be remembered, however, that none of the instruments described above are linear except over

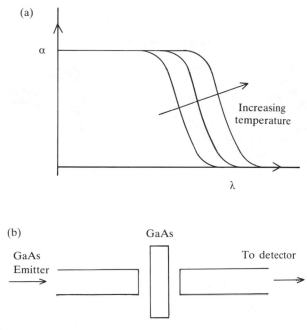

Fig. 10.3 Variation of the absorption coefficient (α) of GaAs with both wavelength (λ) and temperature (a). Temperature sensor utilizing variations in the transmittion of GaAs with temperature (b).

very small displacements. Accurate calibration over the whole range is therefore required.

As well as displacement/pressure sensors, a number of extrinsic fiber temperature sensors have been proposed. For example, the position of the bandgap of semiconductors such as GaAs is temperature dependent (Fig. 10.3(a)), and a simple sensor can be made in which a piece of the semiconductor is placed in the gap between the ends of two fibers (Fig. 10.3(b)). Light with a wavelength corresponding approximately to the semiconductor bandgap is sent down one of the fibers and the power emerging from the other is measured and can be related to the temperature.

Another temperature sensor, the Fluoroptic sensor, is available commercially and is claimed to have a sensitivity of 0.1°C over the range −50 to 250°C. The instrument relies on the temperature variation of the fluorescence in europium doped lanthanum oxysulfide (Eu : La$_2$O$_2$S). A small amount of this material is placed on the object whose temperature is to be measured, and the fluorescence is excited by illuminating it with ultraviolet light transmitted down a fairly large diameter (400 μm) plastic coated silica fiber. The source of radiation is a quartz-halogen lamp whose output has been filtered to remove any unwanted higher wavelengths. Another, similar, fiber picks up some of the emitted fluorescence and carries it back to the detector system (Fig. 10.(4)). In

Fig. 10.4 Schematic layout of the Fluoroptic temperature sensor. The fluorescent radiation generated in the phosphor is separated into its two main constituent wavelengths (λ_1 and λ_2) and the relative optical power of these wavelengths is determined by using a beam splitter followed by two optical bandpass filters to isolate the two wavelengths.

fact, the phosphor emits at more than one wavelength and it is the intensity ratio of two of the lines which is measured. Because a ratio is measured, any fluctuations in the irradiance of the source are not important.

At the detector end of the fiber, the radiation is split into two using a beam splitter. Each beam falls onto a Si photodiode having an appropriate filter in front to isolate the particular wavelength required. The ratio of the two signals then provides the temperature information, which is usually contained in a 'look-up' table. Because the ultraviolet output from a quartz–halogen lamp is small and the fiber absorption relatively large at short wavelengths, the output of the phosphor is quite small. Efficient detectors and low noise preamplifiers are required, and the maximum fiber length is restricted to about 15 m. Nevertheless, the device provides a performance superior to thermocouples, and allows point temperature measurements in semi-remote hostile environments.

10.1.2 Multimode active optical fiber sensors

One of the ways in which we can influence the amount of radiation flowing down a fiber is by means of microbending loss (section 8.4.1), and this can therefore be made the basis of a displacement or pressure transducer. In a typical device, the fiber passes between a pair of ridged plates which impart a periodic perturbation to the fiber. In fact, we have met such an arrangement before in the guise of a mode scrambler (section 8.6). If step index fiber is used, a particular periodicity of wavelength will only couple together a few modes (see eq. (8.30)). However, it may be shown (ref. 10.1) that, with a graded index fiber where the profile parameter, γ is equal to 2, all modes are coupled together when

$$\Lambda = \frac{2\pi a}{(2\Delta)^{\frac{1}{2}}} \tag{10.1}$$

Typically $\Lambda \sim 1$ mm (see Example 10.1), with this value for Λ we expect very high microbending loss and hence high sensitivity for the sensor.

Example 10.1 Microbending sensor periodicity

We consider the periodicity required for maximum microbending loss in a graded index fiber. We assume the fiber has a 50 μm core diameter and maximum core and cladding refractive indices of 1.53 and 1.50 respectively. We have therefore that $a = 25$ μm and that $\Delta = (1.53 - 1.50)/1.53$ or 0.0196. Hence from eq. (10.1):

$$\Lambda = \frac{2\pi \times 25 \times 10^{-6}}{(2 \times 0.0196)^{\frac{1}{2}}}$$

$$= 0.79 \text{ mm}$$

When the modes in a fiber are excited by a coherent source, they are capable of interfering with each other and thus of producing an interference pattern across the end of the fiber (see the description of modal noise in section 9.3.4). The pattern obtained will depend on the phase differences developed between the modes as they travel along the fiber, which is impossible to predict. Provided there are no perturbations acting on the system, however, the pattern should remain unchanged. If now the fiber is slightly flexed in any way, mode coupling will change the distribution of energy amongst the modes, and hence produce a change in the interference pattern across the fiber end. Of course, unless there is a significant amount of coupling into lossy modes, there will not be any great change in the total amount of energy emerging from the fiber. If, however, we consider only a small portion of the whole area of the fiber end, any change in the interference pattern as a whole is almost certain to produce quite significant changes in the emerging energy. Thus, if a detector is so placed as to intercept only a small portion of the total light emerging from the fiber, its output should vary when there is any deformation of the fiber. By its very nature, such a detector will be very nonlinear, though in some circumstances this may not be a great disadvantage. For example, by laying the fiber just below ground level it may be possible to detect the presence of intruders, since their footsteps will cause deformation of the fiber. All that is required is for the output to trigger an alarm when the signal exceeds some predetermined level.

10.1.3 Single mode fiber sensors

In single mode fiber sensors, we are mainly dealing with the effect of the external quantities to be measured on the phase (or mode velocity) of the light within the fiber. Some sensors also involve polarization modulation, but these will be discussed later.

10.1.3.1 PHASE MODULATED SENSORS

One obvious way in which we may detect phase changes is to construct an interferometric system where the phase of the beam through a sensing fiber is compared with that of a reference beam. The Mach–Zehnder configuration is easily realized using optical fibers; a typical laboratory version is shown in Fig. 10.5(a). A beam splitter is used to divide a laser beam into two parts that are then launched into the sensing and reference fibers respectively. The outputs from these fibers meet at another beam splitter and the two resulting beams are then allowed to fall onto suitable photodetectors. It is not strictly necessary to use both output beams, but there are advantages to be gained from so doing, as will be explained later. A more compact 'practical' version may be constructed (Fig. 10.5(b)), where the He–Ne laser is replaced by a semiconductor laser, and the beam splitters by '3 dB couplers' (section. 8.5.2). However, the following discussion will be in terms of the former configuration.

Let us consider the form of the output expected from the interferometer. We write the electric fields of the two beams at one of the detectors as $A_s \exp i(\omega t + \Delta\phi)$ and $A_r \exp i\omega t$, where $\Delta\phi$ represents the phase difference between the two. In general, $\Delta\phi$ will be made up of a 'static' differential phase term $\Delta\phi_d$ and a signal term, ϕ_s, where $\Delta\phi = \Delta\phi_d + \phi_s$. The total signal amplitude is the sum of the two fields and

(a)

(b)

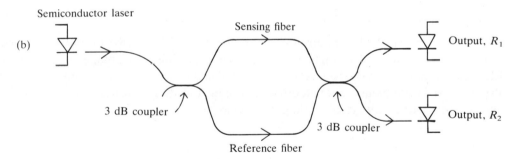

Fig. 10.5 Optical fiber Mach–Zehnder interferometer configurations. (a) shows a 'laboratory' version based on an He–Ne laser, whilst (b) shows an 'all fiber' version based on a semiconductor laser. In both cases, two outputs are available.

the detector output (R_1) is proportional to the product of this sum and its own complex conjugate, i.e.

$$R_1 \propto (A_s \exp \mathrm{i}(\omega t + \Delta\phi) + A_r \exp \mathrm{i}\omega t)(A_s \exp\{-\mathrm{i}(\omega t + \Delta\phi)\} + A_r \exp\{-\mathrm{i}\omega t\})$$

i.e.

$$R_1 \propto A_r^2 + A_s^2 + A_r A_s (\exp(\mathrm{i}\Delta\phi) + \exp(-\mathrm{i}\Delta\phi))$$
$$R_1 \propto A_r^2 + A_s^2 + 2A_r A_s \cos \Delta\phi$$

Assuming for simplicity that $A_r = A_s = A$, then

$$R_1 \propto 2A^2(1 + \cos(\Delta\phi)) \tag{10.2}$$

This response as a function of $\Delta\phi$ is shown in Fig. 10.6, illustrating the basic problem with any interferometric sensor, namely that its output is periodic in the phase difference. It is also evident that the greatest sensitivity to small variations in signal are obtained when the system is operating at points half way between the maximum and minimum values of R. These are known as the 'quadrature' points. Midway between these maximum sensitivity points the sensitivity falls off to zero. Thus, even if we could ensure that we started off operating at a quadrature point it is quite likely that changes in ambient conditions around the reference arm would cause $\Delta\phi$ to change and hence cause a drift away from the quadrature point.

So far, we have considered the output from only one of the signal detectors (R_1). It may be shown (see, for example, ref. 10.2) that there is a phase difference between the beams falling on the two detectors of π. Thus, the response of the second detector (R_2) may be written as

$$R_2 \propto 2A^2\{1 + \cos(\Delta\phi + \pi)\}$$

or

$$R_2 \propto 2A^2\{1 - \cos(\Delta\phi)\}$$

Thus

$$(R_1 - R_2) \propto 4A^2\cos(\Delta\phi) \tag{10.3}$$

In effect, therefore, by taking the difference between the two output signals we double the sensitivity of the interferometer. There are also advantages to be gained when the amplitudes of the beams from the signal and reference arms are not equal (see Problem 10.4). Problems associated with phase drift in the reference arm still remain, and we now examine ways of dealing with them.

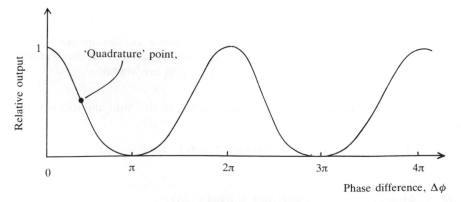

Fig. 10.6 Relative output of a Mach–Zehnder interferometer as a function of the phase difference between the two arms ($\Delta\phi$).

The most straightforward technique is known as the 'active phase tracking homo-dyne' scheme. This requires that we control the phase of the reference arm. One of the simplest ways of doing this is to wrap the fiber around a cylinder of a piezoelectric material. When a voltage is applied to the cylinder it expands radially, thus stretching the fiber and hence inducing a phase change. If this induced phase change is written as $\Delta\phi_m$, we may hold the system at the quadrature point by requiring that $\Delta\phi_m + \Delta\phi_d = (2m + 1)\pi/2$. In the absence of any modulation of the sensing arm, this may be achieved by making the differential signal part of a feedback loop connected to the reference arm modulator, and then ensuring that the differential signal remains close to zero (Fig. 10.7).

In the presence of a signal, we cannot use the differential signal directly in the feedback loop in this way, or the signal information will also be canceled out. However, it is likely that any ambient modulation of the reference arm will take place at a lower frequency than that of the signal, so that the difference signal $(R_1 - R_2)$ integrated over a time long compared to the signal modulation but short compared to the ambient modulation can often provide a suitable feedback signal.

Fig. 10.7 Active phase tracking homodyne detection using either the low gain-bandwidth–product (GBP) mode or the high GBP mode. in the latter case, an integrating circuit is not present in the feedback electronics.

The signal obtained before the feedback electronics will now be the same as in a 'perfect' system operating at the quadrature point and with no reference arm phase drift. Such a mode of operation is known as a low gain–bandwidth-product (GBP) mode. One problem with this mode of operation is that the output is only linear in $\Delta\phi$ when $\Delta\phi \ll \pi$. In addition, phase excursions greater than 2π can lead to ambiguities because of the periodicity of the output. One way to avoid these difficulties is to operate the above system in a high GBP mode. That is, we remove the integrating circuit and operate the feedback circuit at frequencies high enough to include the signal frequency. At first sight, this would seem to obliterate the signal entirely; however, it is still present in the signal applied to the phase modulator in the reference arm. Thus we may now write for operation at the quadrature point

$$\Delta\phi_d + \phi_s + \Delta\phi_m = (2m + 1)\frac{\pi}{2}$$

or

$$\Delta\phi_m = (2m + 1)\frac{\pi}{2} - \Delta\phi_d - \phi_s$$

Assuming therefore that $\Delta\phi_m$ is proportional to the feedback voltage applied to the modulator, the feedback signal is linear in $(\Delta\phi_d + \phi_s)$. This linearity will extend over many 2π phase changes, the only limitation being the phase range that can be covered by the modulator itself. The required phase signal ϕ_s may be separated out using electrical filters since ϕ_s usually oscillates at a higher frequency than $\Delta\phi_d$.

The active phase tracking homodyne scheme outlined above is easy to implement and usually extremely linear in operation. There are a number of drawbacks, however. For example the scheme uses an electrically active element which may require several hundred volts for the piezoelectric element. This may well reduce the sensor's usefulness as an electrically 'safe' device. Furthermore, although the phase range is much in excess of 2π it is still limited, and large phase excursions may necessitate resetting the system with consequent loss of information.

Many other schemes have been implemented to overcome these two basic difficulties. Nearly all involve a considerable increase in complexity and none is really ideal. (There is insufficient space to consider them here, the interested reader may consult ref. 10.3 for more details.)

We turn now to the effects which various quantities to be measured have on the phase of the radiation in a single mode fiber. For simplicity, we assume that the mode behaves like a ray traveling down the center of the fiber with velocity c/n_1. Thus the phase (in radians) associated with a length L of fiber may be written as

$$\phi = \frac{2\pi L n_1}{\lambda_0}$$

If X represents the quantity to be measured, then the change in ϕ caused by a change ΔX in X may be written as

$$\Delta\phi = \frac{2\pi L}{\lambda_0}\frac{dn_1}{dX}\Delta X + \frac{2\pi n_1}{\lambda_0}\frac{dL}{dX}\Delta X \tag{10.4}$$

Because we have assumed that the internal ray angles (θ) are 90°, any changes in the fiber core radius (a) do not affect ϕ. However, even if $\theta < 90°$, it can be shown (see Problem 10.5) that ϕ is much less sensitive to changes in a than it is to changes in L and n. Consequently, we may safely neglect the dependence of ϕ on a. We now estimate the sensitivity of ϕ to variations in axial force, pressure and temperature.

An axially applied force, F, will extend the fiber thus changing L (it also changes n, but we ignore this). If Young's modulus for the fiber is E, then we have

$$\frac{\Delta L}{L} = \frac{F}{AE}$$

where A is the fiber cross-sectional area. Thus

$$\Delta\phi_F = \frac{2\pi n_1 LF}{\lambda_0 AE} \tag{10.5}$$

For silica, $E = 2 \times 10^{11}$ Pa and, taking an external fiber diameter of 100 μm, a core refractive index of 1.45 and light of wavelength 1 μm eq. (10.5) gives

$$\Delta\phi_F = \frac{2\pi \times 1.45}{10^{-6} \times \pi \times \dfrac{(10^{-4})^2}{4} \times 2 \times 10^{11}}$$

$$= 6 \times 10^3 \text{ rad N}^{-1} \text{ m}^{-1}$$

If the fiber is subject to a uniform radial pressure P, the fiber radius will decrease and its length will increase. We neglect the effects of any radius change, but the length change is then given by:

$$\frac{\Delta L}{L} = \frac{2\nu P}{E}$$

where ν is Poisson's ratio, and hence

$$\Delta\phi_P = \frac{4\pi n_1 L\nu P}{\lambda_0 E} \tag{10.6}$$

For silica, $\nu \sim 0.2$ and hence taking the same fiber parameters as above we obtain:

$$\Delta\phi_P = \frac{4\pi \times 1.45 \times 0.2}{10^{-6} \times 2 \times 10^{11}}$$

$$- 1.8 \times 10^{-5} \text{ rad Pa}^{-1} \text{ m}^{-1}$$

Both refractive index and fiber length are affected by temperature T, and so we may write

$$\Delta\phi_T = \frac{2\pi L}{\lambda_0}\left(\frac{dn_1}{dT} + n_1\alpha\right)\Delta T \tag{10.7}$$

where α is the linear expansivity.

For a typical fiber

$$\frac{\mathrm{d}n}{\mathrm{d}T} = 7 \times 10^{-6} \quad \text{and} \quad \alpha = 5 \times 10^{-7}\,\text{K}^{-1}$$

Again taking a wavelength of 1 μm and a core refractive index of 1.45 we have

$$\Delta\phi_T = \frac{2\pi}{10^{-6}}\,(7 \times 10^{-6} + 1.45 \times 5 \times 10^{-7})$$

$$= 49\,\text{rad}\,\text{K}^{-1}\,\text{m}^{-1}$$

In this case, the thermally induced change in the refractive index has a much larger effect than does the thermal expansion term. We may also note that in practical terms the temperature sensitivity is extremely high when compared to the effects of pressure and axial force. Assuming that the interferometer can be operated with a resolution of 10^{-6} radians, then the corresponding temperature sensitivity is 2.3×10^{-8} K. It would appear that we have the potential for a very accurate thermometer. Unfortunately, every part of the interferometer including the reference arm has the same high sensitivity. To maintain the whole of the reference arm at the same temperature to within 10^{-8} K is a formidable task. In fact, the high sensitivity to temperature remains a difficulty when measuring other parameters and often limits the sensitivities that can be achieved.

Quantities other than force, pressure and temperature can be measured, but usually by indirect means (rotation sensing is an exception which will be dealt with separately). For example, attaching the sensing fiber to a magnetostrictive element (whose dimensions change in the presence of a magnetic field) enables magnetic fields to be measured. One problem with this technique is that such materials do not usually respond linearly to a magnetic field. Typical sensitivities are about 10^{-7} T m^{-1}.

The system limitations on sensitivity are rather too complex to go into in any detail here. The ultimate limits will be determined by shot noise at the detectors (see section 7.2.4). Generally speaking, in fiber interferometers using low powered lasers (i.e. 1 mW), shot noise considerations limit the minimum detectable phase shift to 10^{-6}–10^{-7} rad Hz^{-1}. Other factors that may be important are intensity fluctuations in the laser output and laser frequency jitter. The latter will only give rise to a noise signal if the optical pathlengths in the two arms of the interferometer are unequal (if they were exactly equal, of course, the phase difference would always be zero). Although it may not always be practicable to have nearly equal arm lengths, to achieve 10^{-6} rad sensitivity requires that the path differences be less than ~ 1 mm.

10.1.3.2 FIBER OPTIC GYROSCOPE

The fiber optic gyroscope relies on the Sagnac effect, i.e. the phase shift induced between two light beams traveling in opposite directions round a fiber coil when the coil is rotating about an axis perpendicular to the plane of the coil. To calculate the phase shift expected, we assume we have a single circular turn of fiber of radius R. It turns out that we must also assume that both light beams are effectively traveling in a vacuum. If there is no rotation, then both beams will return to their starting points in a time t, where $t = 2R/c$. If, however, the ring is rotating in a clockwise direction at a rate of Ω rad s^{-1}, then the counterclockwise beam will arrive at its starting point sooner since the effective velocity of the beam will be $c + R\Omega$ and the time taken, t', will given by

$$t' = \frac{2\pi R}{c + R\Omega}$$

Similar arguments for the clockwise beam give a transit time of t'', where

$$t'' = \frac{2\pi R}{c - R\Omega}$$

The difference between the two transit times Δt_s ($= t'' - t'$) is then

$$\Delta t_s = 2\pi R \left(\frac{1}{c - R\Omega} - \frac{1}{c + R\Omega} \right)$$

$$= \frac{4\pi \Omega R^2}{c^2 - R^2 \Omega^2}$$

Since $c^2 \gg R^2 \Omega^2$,

$$\Delta t_s = \frac{4\Omega A}{c^2}$$

where $A = \pi R^2$ is the ring area.

In terms of a phase shift:

$$\phi_s = 2\pi \Delta t_s f$$

where f is the frequency of the light, or

$$\phi_s = \frac{8\pi \Omega A}{\lambda_0 c}$$

This expression is independent of the medium in which the light is traveling. Most optical fiber Sagnac interferometers are constructed with many turns (N say) wrapped round a circular former. We then have

$$\phi_s = \frac{8\pi \Omega A N}{\lambda_0 c} \tag{10.8}$$

Example 10.2 Phase shift in a Sagnac gyroscope.

Suppose we have 1000 turns of fiber on a 0.1 m radius former; we may calculate the phase shift induced between two counter-rotating beams by the earth's rotation. The earth's rotation rate is about 15° per hour or $15\pi/(180 \times 3600)$ rad s^{-1}.

Assuming that $\lambda = 1$ μm, eq. (10.8) then gives a phase shift of

$$\phi_s = \frac{8\pi \times 15\pi \times 10^3 \times \pi (0.1)^2}{180 \times 3600 \times 10^{-6} \times 3 \times 10^8}$$
$$= 1.9 \times 10^{-4} \text{ rad}$$

The basic arrangement of the Sagnac fiber interferometer is shown is Fig. 10.8. The form of the output signal from the detector as a function of the rotation rate (and hence the phase change) will again be as shown in Fig. 10.6. At low rotation rates, the sensitivity will be very low and will reach maximum values at the quadrature points (i.e. when $\phi_s = (2n + 1)\pi/2$). Another problem to be surmounted involves what is known as 'reciprocity'. This concerns the necessity to ensure that the two counter-propagating beams travel absolutely identical paths, since any path difference is a potential source of phase noise. In fact, the simple arrangement shown in Fig. 10.8 is not reciprocal, the clockwise beam is reflected twice at the beam splitter whereas the anticlockwise beam is transmitted twice through it. The reciprocity requirements can be very stringent. For example, in a system capable of measuring rotation rates as low as 10^{-4} deg^{-1}h the paths must be reciprocal to 1 part in 10^{16}! Problems also arise with circularly cored fibers since they are not very good at maintaining polarization. Thus, although the two

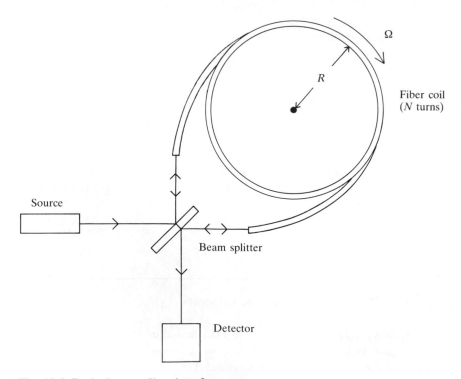

Fig. 10.8 Basic Sagnac fiber interferometer.

counter-propagating beams with identical polarizations may be launched into the fibers, energy may be coupled from one polarization mode into the orthogonal one as the beams traverse the fiber. The energy in the orthogonal modes will not be able to interfere with the original beams and a reduction in output will result which may be interpreted as a phase signal.

Although it might be expected that a highly coherent source would be needed in any interferometric system, this is not always the case. For example, in the present system any scattering within the fiber (for example Rayleigh scattering) will give rise to spurious beams, which will be able to interfere with the two main beams at the detector if the coherence length of the radiation is the order of the fiber length. With a low coherence source, on the other hand, only those beams which have nearly identical pathlengths (such as the two main beams) can interfere. A 'superluminescent' diode is often used as a source in fiber gyros. This is basically a semiconductor laser into which is introduced sufficient loss to prevent full laser action. It acts as a high radiance LED. However, because of its laser like structure it is able to couple significantly more radiation into single mode fibers than can an ordinary LED.

It can be appreciated that, with all these many problems to be overcome, the fiber gyro may never be capable of the highest levels of performance. Its place may well lie in lower cost systems of moderate resolution.

10.1.3.3 POLARIZATION MAINTAINING SINGLE MODE FIBERS

One interesting type of interferometric single mode fiber sensor is the 'polarimetric' sensor, where the two beams are two orthogonally polarized modes within the same fiber. Before we can discuss such sensors, we need to examine the behavior of polarized modes and the possible structures for a polarization maintaining fiber.

It is important to realize from the outset that although we frequently talk of single mode fibers, in fact two orthogonally polarized modes can usually propagate equally well down ordinary single mode fiber. In a circularly symmetrical fiber, these two modes will have identical propagation velocities. However, real fibers will not be perfectly symmetrical; there are bound to be some slight anisotropies in both shape and refractive index. The result of this is that the two modes will inevitably have slightly differing velocities. In other words, the fiber is to some extent birefringent (section 3.2). The birefringence may result either from the intrinsic properties of the fiber (intrinsic birefringence) or from external perturbations such as bending or squeezing (extrinsic birefringence).

Let us suppose that we launch equal amounts of energy into the two orthogonally polarized modes of a single mode fiber. As the modes travel along the fiber their phase relationship will change (Fig. 10.9), but after a certain length (called the beat length) the phase difference reaches 2π and the original input polarization state recurs.

Suppose the velocities of the two modes can be written as c/n_x and c/n_y, where n_x and n_y are the effective refractive indices for the two modes. After a distance L, the phase difference between the modes $\Delta\phi_p$ is given by

$$\Delta\phi_p = (n_x - n_y)\frac{2\pi L}{\lambda_0}$$

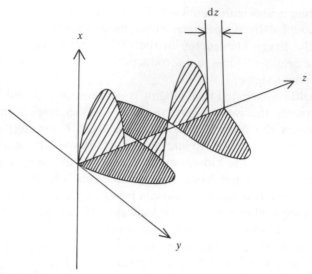

Fig. 10.9 Fiber birefingence causes a phase shift ($\Delta\phi = (2\pi/\lambda)\mathrm{d}z$) to develop between the mode polarized along the y direction (heavy shading) and that polarized along the x direction (light shading) as the modes travel along the fiber in the z direction.

The beat length L_{p} is then given by the condition that $\Delta\phi_{\mathrm{p}} = 2\pi$; hence:

$$L_{\mathrm{p}} = \frac{\lambda_0}{(n_x - n_y)} \tag{10.9}$$

Typical 'circular' cored single mode fibers have beat lengths in the range 100 mm < L_{p} < 5 m.

Now, modes that have very similar propagation constants can be coupled together by fiber perturbations which have relatively long periodicities (see section 8.4.1). Such perturbations are difficult to avoid and hence coupling between the two orthogonally polarized modes becomes a likely possibility. Thus, if one polarization state is launched into one end of a fiber, mode coupling will ensure that some of this mode energy will be coupled into the orthogonal mode. However, if the fiber has a large birefringence, the two polarization modes can be coupled only by short period perturbations, which are much less likely to be present. Such fibers are known as 'Hi-Bi' fibers and are capable of maintaining the initial linear polarization state over quite large distances.

The quality of a polarization maintaining fiber may be judged by the value of the polarization holding parameter, h. This is defined as the amount of polarization mode coupling per unit length. Thus if $P_x(z)$ and $P_y(z)$ represent the mode powers at a point z, then

$$\frac{\mathrm{d}P_x(z)}{\mathrm{d}z} = hP_y(z)$$

Assuming that h is small, $P_y(z)$ will change little with z, and so $P_x(z) = hzP_y(z)$.

Fibers have been fabricated with h values as small as $5 \times 10^{-5}\ \mathrm{m}^{-1}$. This implies, for example, that after 1 km and assuming $P_x(0) = 0$, $P_x/P_y = 10^3 \times 5 \times 10^{-5}$, which is 5×10^{-2} or -13 dB in optical power terms.

Most polarization maintaining fibers are intrinsically birefringent and several different techniques have been used to produce them. All rely on producing a fiber cross section which is in some way asymmetric about two perpendicular axes. Usually, this asymmetry is built into the fiber preform before the fiber is drawn. Some examples are shown in Fig. 10.10. Elliptically cored fibers (Fig. 10.10(a)) have been made with beat lengths of about 1 m, but they are difficult to manufacture. Most Hi-Bi fibers are therefore made by incorporating a localized stress region into the fiber. Thus an elliptical cladding (Fig. 10.10(b)) produces an asymmetric stress across the core which leads to birefringence. Other examples are the 'bow tie' fiber (Fig. 10.10(c)), which is fabricated using a gas phase etching process during manufacture of the preform, to produce stress regions round the core, and the 'side pit' or 'panda' fiber (Fig. 10.10(d)) which incorporates borosilicate rods in the cladding.

A recent development has been fibers in which only one of the orthogonal modes can

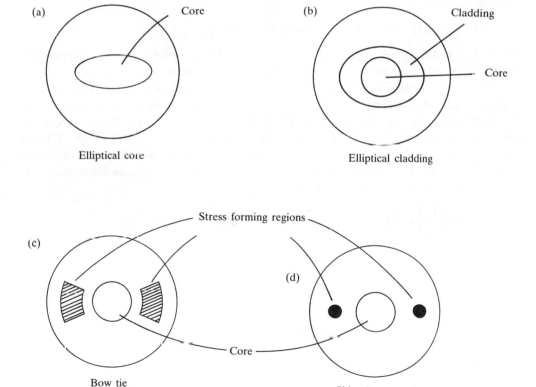

Fig. 10.10 Various types of highly birefringent fiber. (a) shows an elliptically cored fiber. The others all induce stress within a circular core. (b) has an elliptical cladding whilst (c) and (d) have stress forming regions within the cladding.

propagate at the design wavelength. Thus the propagation constants are such that one of the modes is a guided mode whilst the other can only propagate as a leaky mode, and is accordingly highly attenuated.

10.1.3.4 POLARIMETRIC SENSORS

In polarimetric sensors, the two beams in the interferometer are provided by the two orthogonally polarized modes in a Hi-Bi fiber. Because of the nature of such fibers, there are two fixed directions relative to the fiber along which the polarization vectors must lie. These are referred to as the 'fast' and the 'slow' axes, depending on the relative mode velocities. If we illuminate with light polarized at 45° to these directions (Fig. 10.11), then both modes will be excited equally. An increasing phase difference will then develop as the modes travel down the fiber. The polarimetric sensor depends on the effect on external parameters of this phase difference. One problem is how to get the two modes to interfere, since they are orthogonally polarized. This is usually achieved with a combination of a Soleil-Babinet compensator and a Wollaston prism (Fig. 10.11). The former is a variable phase plate and it is set to introduce a phase difference of $\pi/2$ between the slow and fast modes. The amplitudes of the two beams can now be written as

$$A_f \exp (i\omega t) \quad \text{and} \quad A_s \exp \{i(\omega t + \pi/2 + \Delta\phi_p)\}$$

where $\Delta\phi_p$ is the phase difference between the beams before passing through the compensator. For simplicity, we write these two amplitudes as A_f' and A_s'.

A Wollaston prism is a device that splits a beam into two physically separate components having orthogonal polarization. In the present instance, the prism is set to resolve the emergent light into directions at 45° to the fast and slow directions. Thus the amplitudes of the two beams from the prism are given by $(A_s' + A_f')/\sqrt{2} (= A_1')$ and $(A_s' - A_f')/\sqrt{2} (= A_2')$ (Fig. 10.11). These signals are then allowed to fall onto two detectors d_1 and d_2. Expanding the expressions for A_s' and A_f', and assuming for simplicity that $A_s = A_f = A$, we obtain

$$A_1' = \frac{A}{\sqrt{2}} \left\{ \exp (i\omega t) + \exp i\left(\omega t + \frac{\pi}{2} + \Delta\phi_p \right) \right\}$$

The beam irradiance $I_1 (= A_1 \times A_1^*)$ is then

$$I_1 = \frac{A^2}{2} \left\{ \exp (i\omega t) + \exp i\left(\omega t + \frac{\pi}{2} + \Delta\phi_p \right) \right\} \left[\exp (-i\omega t) + \right.$$
$$\left. \exp \left\{ -i\left(\omega t + \frac{\pi}{2} + \Delta\phi_p \right) \right\} \right]$$

When multiplied out this simplifies to

$$I_1 = A^2 (1 + \sin \Delta\phi_p)$$

Similarly, we may show that for the other beam the irradiance is given by

$$I_2 = A^2 (1 - \sin \Delta\phi_p)$$

(a)

(c)

(b)

Fig. 10.11 In Polarimetric sensors, light entering the sensing fiber has its polarization set at 45° to the 'fast' and 'slow' axes, as in (a). The emerging light is processed as shown in (b) by passing it through a Soleil–Babinet compensator (SBC) and then through a Wollaston prism (WP). The latter separates out the two orthogonally polarized beams which then fall onto detectors d_1 and d_2. The amplitudes of the two output signals may be obtained by resolving the amplitudes as in (c).

The interferometer output is then most conveniently processed as

$$\frac{I_1 - I_2}{I_1 + I_2} = \sin \Delta \phi_p$$

For small values of $\Delta \phi_p$, the output is then proportional to $\Delta \phi_p$. The interferometer will of course only respond to the effect of external forces on the intrinsic birefringence of the fiber. If we put $B = n_x - n_y$, then we have

$$\Delta \phi_p = \frac{2\pi L B}{\lambda_0}$$

If X represents the quantity to be measured then we may write

$$\frac{d\Delta \phi_p}{dX} = \frac{2\pi}{\lambda_0} \left(B\frac{dL}{dX} + L\frac{dB}{dX} \right) \tag{10.10}$$

Here, the first term on the right-hand side represents the effect of length change, whilst the second represents the change in birefringence. By comparing eq. (10.10) with eq. (10.4), we can see that the former term is much smaller than the corresponding term in a conventional two fiber interferometer; the ratio being in fact B/n_1 or $(n_x - n_y)/n_1$. For example, taking a fiber with a beat length of 2 mm, $n_x - n_y \sim 5 \times 10^{-4}$ and hence $(n_x - n_y)/n_1 \sim 3 \times 10^{-4}$. For fibers with a longer beat length, the factor will be correspondingly smaller. Thus we are justified in neglecting the first term on the right-hand side of eq. (10.10) in comparison to the second.

We saw earlier that there are basically two types of high birefringence fiber, elliptically cored and internally stressed. The former tend to be relatively insensitive to pressure, axial stress and temperature, since none of these parameters can change the core ellipsisity to any extent. Internally stressed fibers are produced by deliberately incorporating elements into the cladding which have different thermal expansion coefficients, thus producing an internal stress when they cool down during manufacture. Thus the internal stress, and hence B, is expected to be temperature dependent. Equally, axial strain affects the anisotropic stress within the fiber, again affecting B. External pressure, on the other hand, being isotropic with respect to the anisotropic strain, has little effect on B. Obviously the sensitivities to external forces are very dependent on the exact type of fiber used. However, for typical internally stressed fibers these are as follows: temperature, 3 rad K^{-1} m^{-1}; pressure, 10^{-8} rad Pa^{-1} m^{-1}; and strain, 10^6 rad.

It may seem perverse to develop sensors with deliberately reduced sensitivities. However, we have seen earlier that the high sensitivity of the conventional two fiber interferometer, particularly with regard to temperature, can be something of an embarrassment. One interesting possibility is that of a remote polarimetric sensor. This consists of two parts: (a) a 'down' lead fiber and (b) a sensing fiber. Both fibers are of identical types except that they are fusion spliced together so that their fast and slow directions are at 45° (Fig. 10.12). The end of the sensing fiber is silvered to reflect light back down the fiber. In operation, linearly polarized light is launched into only one of the two fiber modes (the 'input' mode). When the light reaches the splice, energy will be launched into both polarization modes of the sensing fiber. After traversing the sensing fiber twice, the light will arrive back at the splice. Light will then in general be coupled into the 'unused' mode of the input fiber (the 'output' mode). Back at the launch end, the light in this mode can be separated out and monitored. Any change in the differential path length in the sensing fiber will give rise to a change in the energy launched into the output mode.

10.1.3.5 POLARIZATION ROTATION SENSORS

All the single mode fiber sensors considered so far have relied on the quantity to be measured affecting the phase of a mode within a fiber. Here, we consider the possibility of altering the angle of polarization of a mode. In section 3.7.1 we saw that such a rotation can take place in a medium in the presence of a magnetic field (Faraday rotation). Obviously, in the light of the discussion in section 10.1.3.3 we require a fiber that is as isotropic as possible, since any anisotropies can cause coupling between the orthogonally polarized modes. Unfortunately the Faraday effect is comparatively weak, and it is only possible to measure relatively large magnetic fields. One area

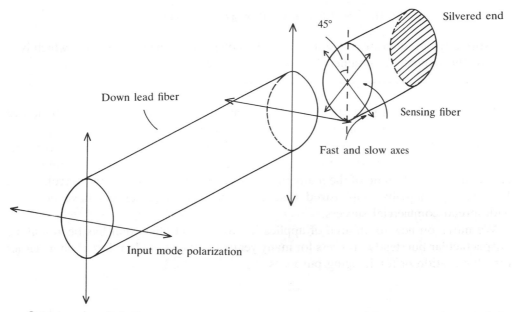

45°
Silvered end

Down lead fiber

Sensing fiber

Fast and slow axes

Input mode polarization

Output mode polarization

Fig. 10.12 Remote polarimetric sensor. The separation shown here between the down lead fiber and the sensing fiber is for pictorial convenience; normally the two are fused together.

where such a sensor has proved valuable is in the monitoring of large currents in electricity generating stations.

Suppose we consider a single turn of fiber of radius r which passes round a wire carrying a current I A (Fig. 10.13). Ampere's circuital theorem gives $H \times 2\pi r = I$, and hence $B = \mu_0\mu_r\, I/2\pi r$. The amount of polarization rotation is, from eq. (3.18), given by $\theta_r = VB2\pi r$; therefore, we have $\theta_r = \mu_0\mu_r VI$, where μ_r is the relative permeability of the fiber and V is its Verdet constant. For n turns of fiber, the amount of rotation will be n times greater, thus:

$$\theta_r = \mu_0\mu_r nVI \tag{10.11}$$

It will be noticed that this result is independent of r, and indeed in general θ_r is independent of the size or shape of the loop and the position of the conductor within the loop. This is a useful result, since it indicates that the device will be insensitive to any vibrations. As stated above the sensitivity of this device is low, and, as Example 10.3 shows, resolutions of some tens of Amperes are all that can be expected.

Example 10.3 Faraday rotation sensor sensitivity

For silica, $V = 4$ rad m^{-1} T^{-1} and $\mu_r = 1$; hence for a 10 turn fiber coil and with $I = 30$ A the polarizatin rotation given by eq. (10.11) is

$$\theta_r = 4\pi \times 10^{-7} \times 10 \times 4 \times 30 \times \pi/180 = 0.09°$$

Normally, a resolution of about 0.1° in the polarization angle is possible, which is just the magnitude obtained here with a 30 A current.

In this section on optical fiber sensors, we have attempted to give an overall view of the many different types of sensor that have been proposed and of the basic physical principles behind them. Inevitably, in view of the limited space available, a comprehensive survey has not been possible. For a more detailed discussion, the interested reader may consult some of the many review articles that have been written (ref. 10.4). It should perhaps be emphasized that as yet few of these sensors have achieved widespread commercial success.

We move on now to an area of application for optical fibers that has been making unspectacular but steady progress for many years, i.e. the use of bundles of fibers either for illumination or for imaging purposes.

10.2 Light guiding fibers

Apart from their use in fiber optic communications, transparent optical fibers have been used for many years in a wide range of light guiding applications in science, medicine

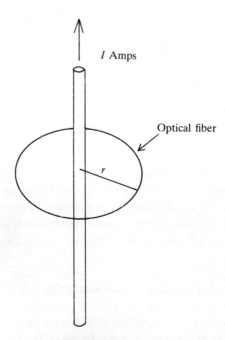

Fig. 10.13 Geometry for investigating the Faraday rotation induced in an optical fiber passing around a current carrying conductor.

and industry (ref. 10.5). Polished acrylic rods were first used for illumination purposes in the late 1920s. Most current applications, however, have arisen following the dramatic reduction in fiber losses resulting from Van Heel's suggestion that the fiber be provided with a cladding to reduce surface losses at each internal reflection within the fiber (ref. 10.6). Light propagates along these fibers in a zigzag manner in precisely the same way as we have described in section 8.3.

The commonest light guiding fibers consist of a core and cladding as in communications fibers. The transmission distances involved are quite short, however, so there is not the same importance attached to having ultra-low loss fibers. Usually, an all glass (i.e. borosilicate) type of fiber is used with losses of the order of 500 dB km^{-1}. There is a sufficiently wide range of compatible glasses to enable fibers with a wide range of numerical apertures to be fabricated. The numerical aperture (NA) is given by eq. (8.18), although in the present case the amount of light which can be accepted by the fiber through the generation of skew rays is often important. With a Lambertian source, the light gathering efficiency may be about 50% higher than that predicted by eq. (8.18), which, of course, applies only to meridional rays.

There are very few applications of single optical fibers (i.e. *monofibers*), but a brief consideration of these reveals the properties of fiber light guides. In general, the fiber can be regarded as an optical system which uses the light entering it to produce an output with the same solid angle but with reduced irradiance. Thus, a simple application is the facility of having a light source remote from, for example, a hazardous environment. Similarly, a fiber may be used to collect light from only part of a source in terms of angular or spatial dimensions. Monofibers can be used in this type of application subject to limitations caused by the dimensions of the fiber and the requirements for flexibility. The allowed bending radius at which a fiber breaks depends on the fiber diameter, and as a rough guide a 50 μm diameter fiber will break at a bending radius of about 1 mm, while a 2 mm diameter fiber may be taken as being nonflexible. If the transfer of light can be achieved without a requirement for flexibility, then larger diameter fibers can be used. Fibers up to 10 mm diameter can be bent, if required, by softening them in a flame, with negligible effect on their transmission characteristics. Such fibers are used for illuminating otherwise inaccessible regions where space is at a premium. They may, for example, be incorporated into the wall of a vessel to illuminate its inside as illustrated in Fig. 10.14. If there is a requirement for the fiber to be flexible, either to allow for relative movement between the source and the point of application or because the path between these is not a straight line, then either small diameter monofibers or multi-fiber bundles may be used. It is quite straightforward to draw and assemble many thin fibers with diameters of about 50 μm into a bundle or *light guide*. In these devices, the ends of the fibers are bonded together leaving them free to flex over most of their length. The flexible portion is protected by, for example, a plastic sheath, which is bonded to a ferrule at each end of the bundle. Because of the cladding, each individual fiber behaves independently of the other fibers in the bundle and so there is no crosstalk.

The fibers are bonded together at their ends in a close packed, hexagonal, array as illustrated in Fig. 10.15. Only the light entering the fiber cores is transmitted. A simple calculation based on Fig. 10.15 shows that in such an array the core packing fraction

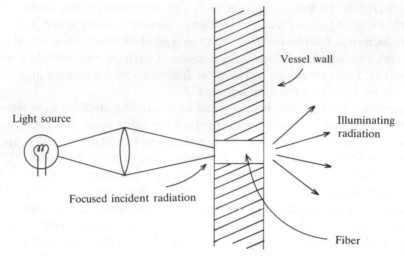

Fig. 10.14 Illumination of the inside of a vessel using a single fiber.

(that is, the area occupied by the cores as a fraction of the total bundle cross-sectional area) is given by (see Problem 10.6)

$$F = \frac{\pi}{2\sqrt{3}} \frac{d^2}{D^2} \simeq \frac{0.9^{d^2}}{D^2} \tag{10.12}$$

where d is the core diameter and D the total fiber diameter. Typically, $F \simeq 0.7$ indicating that the light guide has about 0.7 of the transmission of the individual fibers comprising it.

In addition to simply transferring light between two points, light guides are used for

Fig. 10.15 Close packed hexagonal array of fibers.

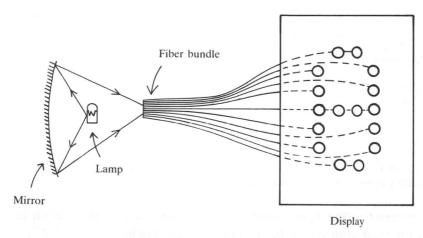

Fig. 10.16 Illumination of the legend on a motorway sign using a fiber bundle beam splitter.

shape converting and beam splitting. In shape converting, the ends of the bundle are bonded into different shapes to match different input and output conditions. A typical application is the collection of light from a spectral source and its guidance to the slit of a spectrometer. Similarly, the output end of a light guide could be shaped to illuminate a single line of script in a document scanner.

In beam splitting guides, the total number of fibers is simply divided among several inputs or outputs (or both). Such beam splitting guides have been used for reading computer punched cards, while another common use is in illuminated signs such as those on motorways. Here a fiber bundle illuminated with a single light source is spread out to the appropriate parts of a matrix display (Fig. 10.16). Different display patterns can be formed by using different fiber bundles, each one being illuminated separately. To ensure that the angular distribution of the emerging light is uniform irrespective of which bundle is being used, a large diameter monofiber is situated at each matrix position to act as a 'mixer' (Fig. 10.17). Finally, a collimating lens reduces the beam divergence of the emerging beam, since a wide angle of view is not required, whereas it is essential to have as high a brightness as possible for the display to be visible in daylight conditions. Beam-splitting guides are widely used as Y-guides in which the light traverses the guide in two directions. As shown in Fig. 10.2(a), light travels along one of the branches to the output end where it is reflected back into the other branch and guided to a detector. The detector signal then provides information about the conditions which exist in the vicinity of the common end. This system has been used for reflectance measurements, counting objects and proximity sensing, as we discussed in section 10.1.1.

10.2.1 Coherent bundles

Coherent bundles are assemblies of optical fibers in which the fibers are positioned accurately within the bundle such that spatial information can be transmitted. This is in

Beam splitter outputs

Large diameter monofiber mixer

Collimating lens

Fig. 10.17 Use of a monofiber to insure an even angular output from a number of beam splitter inputs for a motorway sign.

marked contrast to the light guides discussed earlier where there is no correlation between the fiber positions at the two ends of the bundle.

In coherent bundles it is most important that there is no crosstalk between the constituent fibers, as this would degrade the information being transmitted. Crosstalk is prevented by making the cladding of the fibers sufficiently thick that light cannot leak from one fiber into neighboring fibers. The output of each fiber will then depend solely on the light entering its input end face (within its acceptance angle) and the transmittance of the fiber. Hence, when an image is formed on the end of a coherent bundle the light emerging from each fiber is simply related to the irradiance distribution in the portion of the light illuminating its input face. Now, due to the integrating nature of the individual fibers their outputs are of uniform irradiance so that no detail finer than the fiber diameter will appear in the transferred image.

The information transfer capacity is controlled by the resolution of the bundle, i.e. by the diameter of the individual fibers used. For this reason, the fibers used are of much smaller diameter than those used in light guides; diameters down to 5 μm are common. Such fibers are difficult to handle individually and are often produced as groups where the individual fibers are fused together into a multiple fiber, typically in a 6 × 6 square array. These multiple fibers are then stacked together to form a coherent bundle in their own right, containing up to a million individual fibers in lengths of a few meters with a resolution of about 50 line pairs per millimeter.

Flexible coherent bundles are fabricated by bonding the fibers together only at their extremities. Such bundles are quite widely used as *fiberscopes* for inspection purposes. An object lens is used to form a real image on the input face of the fiberscope. This image is then transferred to an eyepiece. Because the resolution of the fiberscope is limited by the diameter of the individual fibers a certain amount of defocusing of the image formed by the object lens may not be detectable by the observer (see Problem 10.7). Applications of fiberscopes range from the inspection of the inside of pipes and castings to the medical inspection of the human body – in medicine they are often referred to as *endoscopes*.

10.2.1.1 RIGID COHERENT BUNDLES

The individual fibers in a bundle can be fused together by the application of heat and pressure to form a single mechanical unit. Physically, this behaves as a single piece of glass but, owing to the presence of the cladding on each fiber, it retains the optical properties of a coherent fiber bundle. In such bundles, the air gaps are totally removed in the fusing process so that the core packing fraction is simply the square of the ratio of the core diameter to the total diameter. This type of bundle is often made from very small diameter fibers ($\lesssim 5$ μm), and because the cladding has to be thick enough to avoid crosstalk the packing fraction may be as low as 0.5, although packing fractions as high as 0.9 can also be achieved.

There are basically two categories of rigid coherent bundle. These are the image conduit, which has a length much greater than its diameter, and the fused plate, which has a length less than its diameter. Indeed, plates with diameters up to 200 mm are available.

During the manufacturing process, the fibers can be distorted around foreign particles present in the bundle and also any air not expelled may form bubbles, which again distort the fibers. These distortions allow light to escape from the affected fibers, thereby reducing their transmission relative to their neighbors with a consequent reduction in image quality.

Another consequence of the manufacturing process is that, as mentioned above, the bundle becomes effectively a single piece of glass so that stray light becomes much more important. In general, the light which falls on the end face of the cladding of a single fiber is not accepted by the fiber core but may be transmitted by internal reflections at the cladding/air interface. In long flexible fibers, many such reflections occur so that the losses of stray light are high and very little of it reaches the output face. In a rigid bundle, and more especially in fused plates, there are very few such reflections and a large fraction of the stray light may reach the output face. The effect of this is to increase the level of illumination uniformly over the transmitted image, thereby reducing the image contrast.

The most successful way of eliminating this unwanted light is to incorporate a second cladding of absorbing glass around the fiber – this is one form of what is know as *extra mural absorption* (e.m.a.). The absorbing cladding does not affect the light accepted by the core but removes only stray light which may enter the inner cladding. Thus, the light reaching the output face of the bundle is predominantly that which forms the transmitted image.

While image conduits are not widely used, there are many applications for fused plates. Perhaps their major application is as face plates of electro-optic devices where the visual output is formed by the excitation of a phosphor. Fresnel reflection losses at the plate faces can be eliminated by the use of antireflection coatings so that up to 90% of the light emitted by the phosphor may be transmitted. This compares with a figure of less than 25% for lens systems.

A typical use of the fused plate is as a faceplate for photographing traces on a cathode ray tube. The plate must must be vacuum-tight and capable of withstanding the changes of temperature which occur as the plate is coated with the phosphor. The trace excited by the electron beam is transmitted by the fused plate, as illustrated in Fig. 10.18, directly to the photographic film placed in contact with the outer face of the plate. The image intensifier described in Section 7.2.6 also incorporates fused plates. In this case, the plates have a dual role, namely light transmission and image shaping.

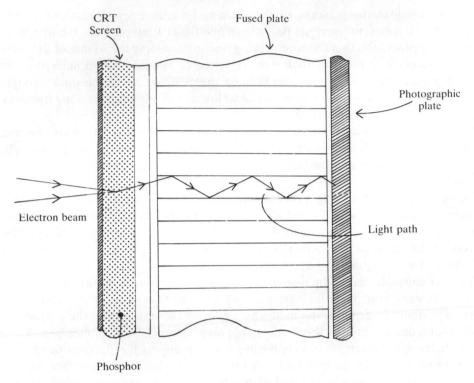

CRT
Screen

Fused plate

Photographic
plate

Electron beam

Light path

Phosphor

Fig. 10.18 Schematic cross-section of a face plate allowing photographs to be taken from a CRT display.

Problems

10.1 Show that the displacement–output characteristic of the fiber displacement sensor shown in Fig. 10.1(c) is proportional to the factor $(\cos^{-1}\delta - \delta \sin (\cos^{-1}\delta))/\pi$ where $\delta = 1 - (d/a)$, a being the radius of the expanded beam and d the displacement measured from the edge of the expanded beam. Assume a constant beam irradiance across the expanded beam and a straight edged shutter.

Sketch this function over the range $d = 0$ to $2a$.

10.2 Calculate the fraction of light transferred from the emitting to the receiving fiber in the two fiber version of the Fotonic sensor shown in Fig. 10.2(b) where two fibers of radius a and separation s are situated a distance z from a reflecting surface. (Hint: as implied in the text, consider the 'coupling' between the image of the emitting fiber and the receiving fiber. There are three regions to be considered: (a) $R < a + s$, (b) $s + 3a \geqslant R \geqslant s + a$, and (c) $R > s + a$, where $R = a + 2z \tan \{\sin^{-1} (\text{NA})\}$. In the second region, where the radiation cone partially overlaps the receiving fiber core, assume that the circular edge of the cone may be replaced by a straight edge, this enables the result of Problem 10.1 to be used.)

Sketch the coupling efficiency as a function of z for typical values of 'a' and 's'.

10.3 It is proposed to use the amount of light transmitted between two fibers whose ends are offset longitudinally (see Fig. 8.27(a) and Problem 8.13) as an optical 'dipstick'.

Investigate how the amount of light transmitted between the fibers varies with the refractive index and the absorption coefficient of the intervening medium.

If the sensor is lowered from air to water, by how much would you expect the coupled radiation to increase?

10.4 Derive the form of the output expected from a Mach – Zehnder interferometer when beam irradiances in the two arms are unequal. What advantage is there in this situation in using both possible outputs from the interferometer?

10.5 Use a simple meridional ray model (i.e. Fig. 8.3) to calculate the phase changes expected in a fiber of length L and radius a for a given mode when there is (a) a length change of dL and (b) a radius change of da. Ignore the phase change on reflection. Hence show that for modes far from cut-off, the same fractional change in L will give rise to a much greater phase change than will the same fractional change in a.

10.6 Show that with a close packed hexagonal bundle of fibers the ratio of core area to the total bundle cross-sectional area is given by $\pi d^2/(2\sqrt{3}D^2)$ where d is the core diameter of an individual fiber and D its total diameter.

10.7 The light from a point object is focused with unit magnification onto the end of a fiber whose core diameter is d. Show that the fiber may be moved backward and forward a distance $fd/2a$ about the point of focus without altering the total amount of light entering the fiber core. When fiber bundles are used for image transmission, this represents the extent to which the lens can be defocused and still give the same resolution in the transmitted image.

References

10.1 J. Gowar, *Optical Communication Systems*, Prentice Hall International, Hemel Hempstead 1984, section 6.6.

10.2 R. S. Longhurst, *Geometrical and Physical Optics* (3rd edn), Longman, 1973, Chapter 21.

10.3 J. C. D. Jones, P. A. Leilabody and D. A. Jackson, 'Monomode fiber-optic sensors: optical processing schemes for recovery of phase and polarization state information', *Int. J. Opt. Sens.*, **1**, 1986, 123.

10.4 (a) T. D. Giallorenzi, J. A. Bucaro, A. Dandridge, G. H. Sigen, J. H. Cole, S. C. Rashleigh and R. G. Priest, 'Optical fiber sensor technology', *I.E.E.E. J. Quant. Electron.*, **18**, 1982, 626.

(b) D. A. Jackson and J. D. C. Jones, 'Fiber optic sensors', *Optica Acta*, **33**, 1986, 1469.

(c) B. Culshaw, *Optical Fiber Sensing and Signal Processing*', Peter Pereginus, London, 1984.

10.5 (a) W. P. Siegmund, 'Fiber optics', In R. Kingslake (ed.) *Applied Optics and Optical Engineering*, Vol. 4, pp 1–29, Academic Press, New York, 1967.

(b) W. B. Allan, *Fibre Optics* (Engineering Design Guides 36), Published for the Design Council, British Standards Institution and the Council for Engineering Institutions by Oxford University Press, 1980.

10.6 A. C. S. Van Heel, 'Optical representation of images with the use of lenses or mirrors', *Ingenieur*, **65**, 1953, 25.

Appendix 1

Answers to numerical problems

1.6 0.8 mm; 47 mm

1.7 1.5×10^{-5} m; for $\lambda = 600$ nm, 35 and 40; for $\lambda = 500$ nm, 42 and 48

1.8 6.25×10^{-8} m; 12.5×10^{-8} m

1.9 6 orders; 409.5 nm; 468 nm; 546 nm; 655.2 nm

1.10 0.77 mm; 15.4 mm

1.11 4.08×10^{-19} J (2.55 eV); 487 nm

1.12 5.80×10^{14} Hz; 2.79×10^{-19} J (1.74 eV)

2.3 5×10^{-11} m

2.4 1.84×10^5 s^{-1}

2.5 6.02×10^{-18} J; 2.41×10^{-17} J; 5.42×10^{-17} J; 4.82×10^{-17} J; 7.27×10^{16} Hz

2.7 1.64×10^{16} m^{-3}; 224 Ω^{-1} m^{-1}; 80 Ω^{-1} m^{-1}

2.8 0.008 eV

2.9 0.127; 0.873

2.11 8.1×10^{11} m^{-3}

2.12 3.5×10^{-3} m^2 s^{-1}; 1.25×10^{-3} m^2 s^{-1}; 4.18×10^{-4} m; 2.5×10^{-4} m

2.15 1×10^{23} m^{-3}

2.16 0.9 V; 4.88×10^{-8} m; 3.68×10^{-7} m

2.17 0.72 μA; 99%

2.18 868 nm; 517.6 nm; 5.52 μm

3.2 Quartz 0.0162 mm; calcite 1.71 μm

3.3 8.4×10^{-5}

3.4 KDP 7.53 kV; KD*P 3.03 kV; ADP 9.21 kV; LiNbO$_3$ 0.74 kV

3.6 1.45 W

4.2 260 MHz

4.4 2.6 nits

4.5 $2\pi AB(0)$; $\pi AB(0)$

4.8 0.5

5.1 41532 K; 69.2 μm

5.2 10.05 m^{-1}; 0.366

5.3 1.39 m^{-1}; 0.098 m^{-1}

5.4 9.30×10^{-29}%; 9.57×10^{-17}%

5.5	5.6×10^8 m^{-3}
5.6	5.19×10^7 Hz; 1.28×10^9 Hz
5.7	90.5% reflectance
5.8	0.64 m^{-1}
5.9	1580278; 3×10^8 Hz; 5
5.11	6.72 MW m^{-3}
5.12	4.22 kW
5.13	0.7
6.1	0.1 m; 0.21 °C
6.2	474.6 m s^{-1}
6.4	4×10^7 Hz
6.5	0.316 mm
6.6	4.05×10^{11} W m^{-2} sr^{-1}; 6.89×10^{13} W m^{-2} sr^{-1}
6.7	0.44 ps; 1.33 ns
6.8	5 m; 122.45 m
6.9	17.59 MeV
7.1	48 μm
7.3	5.5×10^{-11} W
7.5	10^{12} photons s^{-1}
7.9	6.4 mA, 0.346 V
7.11	572 μm
7.12	960 kHz
7.13	10^{-10} W
8.1	52.3 dB km^{-1}
8.2	48.8°
8.4	$\psi = 63.8°$; $\delta = 64.7°$
8.5	0.52%; 1.3×10^{-21}%
8.7	2,230 dB/m
8.8	9.2°; 1.441, 1263, yes to 6.9°
8.9	30 ns km^{-1}
8.10	0.68 μm; 3.6 μm
8.11	150 ps km^{-1}
8.14	0.41 dB/km
8.15	135 MHz-km
8.19	About 4×10^{-3} dB km^{-1}
9.1	0.391% (1 part in 256)
9.2	64 Mb s^{-1}
9.3	6 km (range limited by absorption)
9.4	14.9 μW
9.8	21.5; 10.5 dB
10.3	1.77 times

Appendix 2

Birefringence

Anisotropy in crystals is caused by the arrangement of atoms being different in different directions through the crystal. Consequently, the electric polarization (or dipole moment per unit volume) **P** produced in a crystal by a given electric field $\vec{\mathscr{E}}$ depends on the direction of the field. It follows that the relative permittivity ε_r and the refractive index, which is given by $n = (\varepsilon_r)^{\frac{1}{2}}$ (ref. A2.1, Chapter 11), are also direction dependent.

Because electromagnetic waves are transverse, the refractive index for a given direction of propagation of light through the crystal depends on the value of the electric vector $\vec{\mathscr{E}}$, which is, of course, perpendicular to this direction. In fact, the electric vector oscillates in a plane perpendicular to the wave velocity direction and thus the velocity of the waves through the crystal will be dependent on the plane of polarization of the light. This phenomenon, called *birefringence*, occurs naturally in some materials, and it may be enhanced by the application of external forces such as those due to electric fields. It may also be induced in isotropic materials by the same means.

Natural birefringence

We may appreciate how birefringence arises as follows. Let us consider placing a dielectric medium into a parallel plate capacitor to which a constant voltage is applied. The electric field across the medium polarizes it and dipoles appear throughout the material, lined up in the direction of the field. The resulting polarization can be shown to be related to the electric field and electric displacement, **D**, by the equations (ref. A2.1, Chapter 3)

$$\mathbf{D} = \varepsilon_0 \vec{\mathscr{E}} + \mathbf{P}$$

and

$$\mathbf{D} = \varepsilon_0 \varepsilon_r \vec{\mathscr{E}} = \varepsilon \vec{\mathscr{E}}$$

(A2.1)

The ratio $\mathbf{P}/\varepsilon_0 \vec{\mathscr{E}}$ is called the electric *susceptibility*, χ, and we see from eqs. (A2.1) that

$$\mathbf{P}/\varepsilon_0 \vec{\mathscr{E}} = (\varepsilon_r - 1) = \chi$$

(A2.2)

In isotropic media (and cubic crystals), direction in the medium is not important, the vectors $\vec{\mathscr{E}}$, **D** and **P** are all parallel, and ε_r, n and χ are scalar quantities. The speed of propagation of electromagnetic waves in such media is constant irrespective of the direction of propagation. This is not the case in birefringent materials. In general, in such materials, the polarization occurring in response to an applied electric field depends on both the magnitude and direction of the field and, in addition, the induced polarization may be in a different direction from that of the field. Thus, for example, a field applied in the x-direction will induce not only a polarization P_x parallel to itself, but possibly also polarizations P_y and P_z in the y- and z-directions. Thus, from eq. (A2.2) we may write

$$P_x = \varepsilon_0(\chi_{11}\vec{\mathscr{E}}_x + \chi_{12}\vec{\mathscr{E}}_y + \chi_{13}\vec{\mathscr{E}}_z)$$
$$P_y = \varepsilon_0(\chi_{21}\vec{\mathscr{E}}_x + \chi_{22}\vec{\mathscr{E}}_y + \chi_{23}\vec{\mathscr{E}}_z) \tag{A2.3}$$
$$P_z = \varepsilon_0(\chi_{31}\vec{\mathscr{E}}_x + \chi_{32}\vec{\mathscr{E}}_y + \chi_{33}\vec{\mathscr{E}}_z)$$

where the capital letters represent the complex amplitudes of the corresponding time varying quantities. The 3×3 matrix array χ_{ij} is called the electric susceptibility tensor. Alternatively, we can use the electric permittivity tensor ε_{ij} where $D_i = \varepsilon_{ij}\vec{\mathscr{E}}_j$ to describe the dielectric response of the crystal to an applied field. These tensors are also 3×3 matrix arrays.

The magnitude of the χ_{ij} (and ε_{ij}) coefficients depends on the choice of the x, y and z coordinate axes relative to the crystal structure. It turns out always to be possible to choose x, y and z so that the off-diagonal elements vanish, giving

$$\left. \begin{array}{l} P_x = \varepsilon_0\chi_{11}\mathscr{E}_x \\ P_y = \varepsilon_0\chi_{22}\mathscr{E}_y \\ P_z = \varepsilon_0\chi_{33}\mathscr{E}_z \end{array} \right\} \tag{A2.4}$$

and similarly

$$\left. \begin{array}{l} D_x = \varepsilon_{11}\mathscr{E}_x \\ D_y = \varepsilon_{22}\mathscr{E}_y \\ D_z = \varepsilon_{33}\mathscr{E}_z \end{array} \right\} \tag{A2.5}$$

where from eqs. (A2.1) and (A2.4) we see that $\varepsilon_{11} = \varepsilon_0(1 + \chi_{11})$, etc. These directions are called the *principal axes* of the crystal and the corresponding diagonal terms ε_{11}, ε_{22} and ε_{33} the *principal permittivities*.

An electric field parallel to a principal axis will only produce a parallel electric polarization but in general the three permittivities are different. This variation in permittivity and the corresponding variation in refractive index and hence in wave velocity is the origin of birefringence. In general, anisotropic crystals are thus characterized by the three *principal refractive indices* n_x, n_y and n_z, where the subscripts relate to the *direction of polarization* of the electromagnetic waves and *not* to their direction of propagation, and where $n_x = (\varepsilon_{11}/\varepsilon_0)^{\frac{1}{2}} = (1 + \chi_{11})^{\frac{1}{2}}$, etc.

Thus consider, for example, a wave propagating in the z-direction. If its electric field, i.e. its plane of polarization, is parallel to the x-direction, it will induce only polarization P_x according to eqs. (A2.4) and (A2.5) and will experience the permittivity ε_{11}, i.e. the refractive index n_x. If, on the other hand, the wave has its plane of polarization in the y-direction, it will experience the refractive index n_y. Any unpolarized wave propagating in the z-direction can be resolved into two components with polarizations parallel to the x- and y-directions. These two components will thus travel through the crystal with different velocities, and hence get progressively out of phase, so that the resultant at some point z will, in general, represent elliptically polarized light. The exact nature of the light depends on the phase difference and the relative amplitudes of the two components.

Anisotropic crystals are divided into two classes. The first class, which includes calcite ($CaCO_3$), are characterized by a principal axis (the z-axis by convention) along which the permittivity is ε_{33}, but perpendicular to which the permittivity is independent of direction (i.e. $\varepsilon_{11} = \varepsilon_{22}$). Thus, such crystals only have two principal refractive indices. The z-direction is unique in that the velocity of propagation is independent of the wave polarization; it is termed the *optic axis*. Such materials are said to be uniaxial. The difference between n_x ($= n_y$) and n_z is a measure of the birefringence and is said to be positive or negative depending on whether n_z is greater or smaller than n_x. Thus calcite is a negative uniaxial crystal, while quartz, for example, is a positive uniaxial crystal.

Crystals of lower symmetry possess two optic axes and are said to be *biaxial*. These axes do not coincide with any of the principal axes and consequently n_x, n_y and n_z are all different.

If Maxwell's electromagnetic theory is applied to an isotropic medium, it can be shown that light of any polarization can propagate in any direction. In the case of anisotropic materials, however, this is not the case and it turns out (ref. A2.3) that as a result of birefringence only two states of polarization can propagate for any crystal direction. These planes are perpendicular to each other and hence, except when propagation is parallel to an optic axis, the two components experience different refractive indices and travel at different speeds.

The values of refractive index and the states of polarization for a particular direction of propagation can be determined from a three-dimensional representation of the refractive index known as the *index ellipsoid* (or *optical indicatrix*). The form of this can be obtained as follows: the energy density W in a dielectric is given by

$$W = \tfrac{1}{2}\mathbf{D} \cdot \vec{\mathscr{E}} \tag{A2.6}$$

Thus using eq. (A2.5), we have

$$W = \tfrac{1}{2}\left(\frac{D_x^2}{\varepsilon_{11}} + \frac{D_y^2}{\varepsilon_{22}} + \frac{D_z^2}{\varepsilon_{33}}\right)$$

A constant energy surface then takes the form of an ellipsoid in a three-dimensional space where the dielectric displacement is the coordinate.

Dividing both sides by W and writing $D_x^2/2\varepsilon_0 W$ as x^2 etc., leads to the expression

$$\frac{x^2}{n_x^2} + \frac{y^2}{n_y^2} + \frac{z^2}{n_z^2} = 1 \tag{A2.7}$$

where $n_x^2 = \varepsilon_{11}/\varepsilon_0$ etc.

This equation represents an ellipsoid with semiaxes n_x, n_y and n_z. For a uniaxial crystal, $n_x = n_y \neq n_z$ and the index ellipsoid, which has circular symmetry about the z-axis, is as shown in Fig. A2.1.

It is usual to refer to n_x ($= n_y$) as the *ordinary* refractive index n_o and to n_z as the *extraordinary* refractive index n_e, in which case we may write

$$\frac{x^2}{n_o^2} + \frac{y^2}{n_o^2} + \frac{z^2}{n_e^2} = 1 \tag{A2.7'}$$

Let us consider light propagating in a direction **r**, at an angle θ to the optic (z) axis (Fig. A2.1). Because of the circular symmetry mentioned above, we can choose, without loss of generality, that the y-axis should coincide with the projection of **r** on the

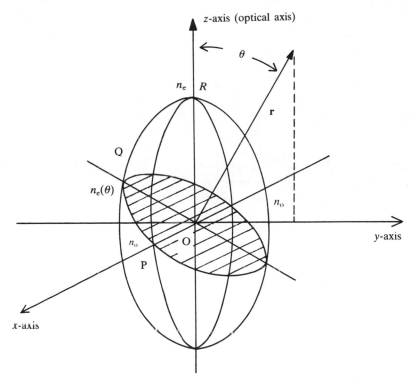

Fig. A2.1 Refractive index ellipsoid or optical indicatrix for a uniaxial crystal. Waves with polarizations parallel to the z-axis experience the refractive index n_e, while those polarized parallel to the x- and y-axes experience the refractive index n_o. For propagation of a wave in the general direction **r**, there are two allowed directions of polarization: parallel to OP, with index n_o, and parallel to OQ, with index $n_e(\theta)$.

xy-plane. The plane normal to **r** intersects the ellipsoid in the shaded ellipse shown. The two allowed directions of polarization are parallel to the axes of this ellipse and thus correspond to OP and OQ; they are thus perpendicular to **r** as well as to each other. The two waves polarized along these directions have refractive indices given by $OP \equiv n_o$ and $OQ \equiv n_e(\theta)$ (the extraordinary wave). In the case of the extraordinary wave, the plane of polarization varies with θ as does the refractive index. We can determine the relationship between $n_e(\theta)$ and n_e, n_o with the aid of Fig. A2.2, which shows the intersection of the index ellipsoid with the yz-plane. From the diagram, we see that

$$n_e^2(\theta) = z^2 + y^2$$

and

$$z = n_e(\theta) \sin \theta$$

Thus, substituting these equations into the equation for the ellipse shown, namely

$$\frac{y^2}{n_o^2} + \frac{z^2}{n_e^2} = 1$$

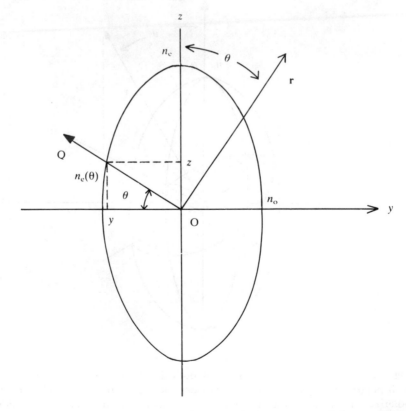

Fig. A2.2 Intersection of the index ellipsoid with the yz-plane. $OQ = n_e(\theta)$ represents the refractive index of the extraordinary wave propagating in the **r** direction.

gives

$$\frac{1}{n_e^2(\theta)} = \frac{\cos^2\theta}{n_o^2} + \frac{\sin^2\theta}{n_e^2}$$

Thus for $\theta = 0°$, i.e. propagation along the optic axis, $n_e(0°) = n_o$ while for $\theta = 90°$, $n_e(90°) = n_e$. The two polarizations which can be propagated correspond to the maximum and minimum refractive indices given by the index ellipsoid (in the case of positive crystals $n_e > n_o$ while in the case of negative crystals $n_e < n_o$). For propagation parallel to the optic axis (i.e. the z-direction), there is no birefringence as the section of the ellipsoid perpendicular to this direction is a circle. For propagation perpendicular to the optic axis, for example the x-direction, the birefringence will be a maximum, the permitted polarizations will be parallel to the y-axis with refractive index n_o and parallel to the z-axis with refractive index n_e.

Induced birefringence

When an electric field is applied to an anisotropic crystal, new principal axes may be introduced (for example, in KDP) while birefringence may be induced in naturally isotropic materials such as GaAs. The electric field changes the shape of the refractive index ellipsoid as a result of changing the permittivity tensor ε_{ij}. After deformation, the axes of the ellipsoid are no longer the original principal axes and the ellipsoid must be represented by an equations of the form

$$\left(\frac{1}{n^2}\right)_1 x^2 + \left(\frac{1}{n^2}\right)_2 y^2 + \left(\frac{1}{n^2}\right)_3 z^2 + 2\left(\frac{1}{n^2}\right)_4 yz + 2\left(\frac{1}{n^2}\right)_5 xz + 2\left(\frac{1}{n^2}\right)_6 xy = 1 \qquad (A2.8)$$

The applied electric field causes a change in the refractive indices of the ordinary and extraordinary refractive indices. This change can be written as

$$\Delta\left(\frac{1}{n^2}\right)_i = \sum_{j=1}^{3} r_{ij}\mathscr{E}_j \qquad (A2.9)$$

where $i = 1, 2, \ldots, 6$, $j = 1, 2, 3$, and r_{ij} is the electro-optic tensor with 6×3 elements. For most materials, however, the symmetry of the crystal reduces the 18 elements of this tensor significantly. We have indicated the values of some of these electro-optic coefficients in Table 3.1 for a representative selection of optical materials. We see that the Pockels effect discussed in Chapter 3 will thus be different in the various crystals. In general, crystal class and symmetry (ref. A2.3) are important and a universal response cannot be predicted. Let us consider as an example crystals similar to KDP. In this class of materials, the only nonzero elements in the electro-optic tensor are r_{41}, r_{52} $(= r_{41})$ and r_{63}. In an arbitrarily directed field, the index ellipsoid becomes

$$\frac{x^2}{n_o^2} + \frac{y^2}{n_o^2} + \frac{z^2}{n_e^2} + 2r_{41}\mathscr{E}_x yz + 2r_{41}\mathscr{E}_y xz + 2r_{63}\mathscr{E}_z xy = 1 \qquad (A2.10)$$

We note that the first three terms are field independent and hence involve n_o and n_e, and that the field induces *cross terms* yz, etc., into the equation for the index ellipsoid so that the major axes of the ellipsoid are no longer parallel to the x-, y- and z-axes as mentioned above. We must now therefore find the magnitudes and directions of the new axes, in the presence of the field, so that we may determine its effect on the propagation of electromagnetic waves. It is usual to restrict the field to the z-direction (the original optical axis) so that $\mathcal{E}_x = \mathcal{E}_y = 0$ and eq. (A2.10) becomes

$$\frac{x^2}{n_o^2} + \frac{y^2}{n_o^2} + \frac{z^2}{n_e^2} + 2r_{63}\mathcal{E}_z xy = 1 \tag{A2.11}$$

The problem now is to find a new coordinate system x', y', z' in which eq. (A2.11) contains no cross term and takes up the form of eq. (A2.7), i.e.

$$\frac{x'^2}{n_{x'}^2} + \frac{y'^2}{n_{y'}^2} + \frac{z'^2}{n_{z'}^2} = 1 \tag{A2.12}$$

The lengths of the major axes of the index ellipsoid are thus $2n_{x'}$, $2n_{y'}$, $2n_{z'}$ and will, in general, depend on the applied field. By inspection of eqs. (A2.11) and (A2.12), the z factor is unchanged if z is parallel to z' and hence $n_{z'} = n_e$, and also eq. (A2.11) is symmetric in x and y so the new axes x', y' bear similar relations to the former axes x and y; they are, in fact, rotated through 45° as shown in Fig. A2.3. The relationships between the two sets of axes are then

$$\left.\begin{array}{l} x = x' \cos 45° + y' \sin 45° \\ y = -y' \sin 45° + y' \cos 45° \end{array}\right\} \tag{A2.13}$$

Substituting eq. (A2.13) into eq. (A2.11) then gives

$$x'^2\left(\frac{1}{n_o^2} - r_{63}\mathcal{E}_z\right) + y'^2\left(\frac{1}{n_o^2} + r_{63}\mathcal{E}_z\right) + \frac{z^2}{n_e^2} = 1 \tag{A2.14}$$

Equation (A2.14) shows that x', y', z' are the new principal axes and that the field has rotated the ellipse and changed the lengths of the major axes. From eqs. (A2.12) and (A2.14), the length of the major axis in the x' direction, $2n_{x'}$, is given by

$$\frac{1}{n_{x'}^2} = \frac{1}{n_o^2} - r_{63}\mathcal{E}_z$$

Hence the change in refractive index may be evaluated by noting that

$$\frac{1}{n_{x'}^2} - \frac{1}{n_o^2} = \frac{(n_o - n_{x'})(n_o + n_{x'})}{n_{x'}^2 n_o^2} = \frac{2(n_o - n_{x'})}{n_o^3}$$

assuming that the change is small so that $n_{x'} \simeq n_o$

Therefore

$$n_{x'} = n_o + \frac{n_o^3}{2} r_{63}\mathcal{E}_z \tag{A2.15}$$

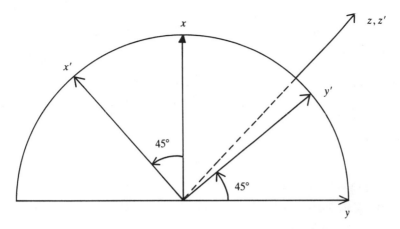

Fig. A2.3 Rotation of the x- and y-axes into the x'- and y'-axes, when crystals such as KDP are subjected to an electric field along the z-axis, in which case z and z' are parallel.

and similarly

$$n_{y'} = n_{\mathrm{o}} - \frac{n_{\mathrm{o}}^3}{2} r_{63} \mathscr{E}_z \tag{A2.16}$$

and, as mentioned above, $n_z = n_{\mathrm{e}}$.

Similar relationships can be derived for other classes of crystal and in general $n = \pm \frac{1}{2} n_{\mathrm{o}}^3 r \mathscr{E}$, but the matrix element r varies according to the crystal class (see Table 3.1).

References

A2.1 P. Lorrain and D. Corson, *Electromagnetic Fields and Waves* (3rd edn.), Freeman, New York, 1988, Chapter 9.

A2.2 (a) M. Born and E. Wolf, *Principles of Optics*, Macmillan, New York, 1964.

(b) A. Yariv, *Quantum Electronics* (2nd edn.), Wiley, New York, 1975.

(c) A. Yariv and P. Yeh, *Optical Waves in Crystals*, Wiley, New York, 1983.

A2.3 (a) See, for example, ref. 3.3(a) or N. W. Ashcroft and N. D. Mermin, *Solid State Physics*, Holt, Rinehart & Winston, New York, 1976, Chapter 4.

Appendix 3

Limitations on LED frequency response due to carrier diffusion and recombination

We consider a relatively low level injection of electrons into a semi-infinite piece of p-type material with unit cross-sectional area. The steady state diffusion equation (eq. (2.41)) for the excess minority carrier density at a point x, $\Delta n(x)$, may be written

$$\frac{\partial}{\partial t} \Delta n(x) = -\frac{\Delta n(x)}{\tau_e} + D_e \frac{\partial^2}{\partial x^2} \Delta n(x) \tag{A3.1}$$

We suppose that the excess population has a time varying component and we write

$$\Delta n(x) = \Delta n^0(x) + \Delta n^1(x) \exp(2\pi i f t) \tag{A3.2}$$

Substituting for $\Delta n(x)$ in eq. (A3.1) and separating out the time-dependent and time-independent parts, we obtain

$$\frac{\partial^2}{\partial x^2} \Delta n^0(x) - \frac{\Delta n^0(x)}{D_e \tau_e} = 0 \tag{A3.3}$$

and

$$\frac{\partial^2}{\partial x^2} \Delta n^1(x) - \frac{\Delta n^1(x)}{D_e \tau_e} (1 + 2\pi i f \tau_e) = 0 \tag{A3.4}$$

Both of these equations have general solutions of the form

$$\Delta n(x) = A \exp(-x/L_e) + B \exp(x/L_e) \tag{A3.5}$$

where A and B are constants and $L_e = \sqrt{D_e \tau_e}$. At $x = \infty$ we must have $\Delta n^0(\infty) = \Delta n^1(\infty) = 0$ and hence $B = 0$. The solutions to eqs. (A3.3) and (A3.4) can therefore be written

$$\Delta n^0(x) = \Delta n^0(0) \exp(-x/L_e) \tag{A3.6}$$

and

$$\Delta n^1(x) = \Delta n^1(0) \exp(-x/L_e^*) \tag{A3.7}$$

where

$$(L_c^*)^2 = \frac{D_c \tau_c}{1 + 2\pi i f \tau_c} \tag{A3.8}$$

We may define a frequency dependent relative quantum efficiency $Q(f)$ by

$$Q(f) = \frac{P(f)}{F(f)} \tag{A3.9}$$

Here $P(f)$ and $F(f)$ are the frequency dependent parts of the photon generation rate within the material and the electron flow rate into the material respectively.

From eq. (A3.2),

$$F(f) = D_c \left| \frac{\partial}{\partial x} \Delta n^1(x) \right|_{x=0} = \frac{D_c \Delta n^1(0)}{L_c^*} \tag{A3.10}$$

The recombination rate in a region of width δx is given by $\Delta n(x) \delta x / \tau_c$ (see eq. (2.38)), and hence the total recombination rate throughout the material is given by

$$\frac{1}{\tau_c} \int_0^\infty \Delta n(x) \, dx$$

If we assume that each recombination gives rise to a photon then

$$P(f) = \frac{1}{\tau_c} \int_0^\infty \Delta n^1(x) \, dx$$

Hence, by substituting for $\Delta n^1(x)$ from eq. (A3.7)

$$P(f) = \frac{1}{\tau_c} \int_0^\infty \Delta n^1(0) \exp\left(-x/L_c^*\right) \, dx = \frac{L_c^* \Delta n^1(0)}{\tau_c} \tag{A3.11}$$

From eqs. (A3.9), (A3.10) and (A3.11) we then have

$$Q(f) = \frac{(L_c^*)^2 \Delta n^1(0)}{\tau_c D_c} = \frac{1}{1 + 2\pi i f \tau_c}$$

The modulus of this expression is $1/(1 + 4\pi^2 f^2 \tau_c^2)^{\frac{1}{2}}$ and hence we see that the light output $R(f)$ resulting when the current is modulated at a frequency f can be written

$$R(f) = \frac{R(0)}{(1 + 4\pi^2 f^2 \tau_c^2)^{\frac{1}{2}}}$$

Appendix 4

Füchtbauer–Ladenburg relation

The Füchtbauer–Ladenburg relation relates the area under an absorption curve to the Einstein coefficients and to the populations of the states responsible for the absorption. We need this relationship to obtain the gain coefficient resulting from the interaction of a highly monochromatic beam with atoms having a relatively broad spectral lineshape in terms of the lineshape function.

The argument leading up the Füchtbauer–Ladenburg relation closely parallels that given in section 5.3, which leads up to eq. (5.13). There, in the interests of simplicity, we maintained the fiction of a perfectly monochromatic beam. Now, however, we must consider a beam of finite spectral width. To do this, we must modify our definition of radiation density. We now take it that the energy of radiation per unit volume with frequencies between ν and $\nu + d\nu$ is $\varrho_\nu d\nu$. We further suppose that there are only dN_1 atoms in level 1 capable of absorbing radiation in this frequency range and only dN_2 atoms in level 2 capable of emitting radiation in this frequency range. The equations for the upward and downward transition rates (corresponding to those at the very start of Chapter 5 and leading up to eq. (5.1)) now read:

$$\frac{d}{dt}(dN_1) = dN_1 \varrho_\nu B_{12}$$

$$\frac{d}{dt}(dN_2) = dN_2 \varrho_\nu B_{21} + N_1 A_{21}$$

Note that we have used ϱ_ν in the right-hand side of both of these equations when by analogy with the previously presented ones it might be expected that we should have used $\varrho_\nu d\nu$. This slight inconsistency arises because of our earlier assumption in Chapter 5 that we were dealing with a perfectly monochromatic beam.

With these modifications, it is easy to see that eq. (5.13) now becomes:

$$\alpha(\nu)d\nu = \left(\frac{g_2}{g_1}dN_1 - dN_2\right)\frac{B_{21}h\nu n}{c} \tag{A4.1}$$

where $\alpha(\nu)$ is the absorption coefficient at frequency ν and we have replaced ν_{21} by ν.

We now integrate this expression over the range of frequencies covered by the absorption line. Thus

$$\int \alpha(v)dv = \frac{B_{21}hn}{c} \int v\left(\frac{g_2}{g_1}dN_1 - dN_2\right)$$ (A4.2)

We suppose that the line width Δv is much less than the line center frequency v_0 and consequently we may remove v from the integral on the right-hand side of eq. (A4.2) and put it equal to v_0. Equation (A4.2) then becomes

$$\int \alpha(v)dv = \frac{B_{21}hv_0n}{c} \int \left(\frac{g_2}{g_1}dN_1 - dN_2\right) = \frac{B_{21}hv_0n}{c}\left(\frac{g_2}{g_1}N_1 - N_2\right)$$ (A4.3)

This is the Füchtbauer–Ladenburg relation.

In section 5.7 we introduced the lineshape function $g(v)$. This gives the shape of the absorption (or emission) curve but it is normalized so that $\int_{-\infty}^{+\infty}g(v)dv = 1$. Thus we may write

$$\alpha(v) = g(v)\int\alpha(v)dv$$

(To see that this must be so, integrate both sides over v.) Hence, substituting for $\int\alpha(v)dv$ from eq. (A4.3) we have

$$\alpha(v) = g(v)\frac{B_{21}hv_0n}{c}\left(\frac{g_2}{g_1}N_1 - N_2\right)$$ (A4.4)

If we have a situation of population inversion, then in terms of the small signal gain coefficient $k(v)$ we may write

$$k(v) = g(v)\frac{B_{21}hv_0n}{c}\left(N_2 - \frac{g_2}{g_1}N_1\right)$$ (A4.5)

This is the result quoted in eq. (5.18).

Appendix 5

Optical bandwidths and pulse broadening

Let us suppose that at the input end of some optical system we have a signal where the optical power is modulated at a frequency f. That is $P_{in} = P_0^i (1 - \cos 2\pi ft)$. If we write the optical power output of the system as $P_{out} = P_0^0(1 - \cos(2\pi ft + \phi))$, then the ratio of P_0^0 to P_0^i gives the system response, which we denote by $H(f)$. We define the *optical bandwidth* of the system as the frequency range Δf_{opt} over which $|H(f)|$ exceeds one half of its maximum value. If we allowed the optical output signal to fall onto a 'perfect' photodetector, then the current output, i_d, would follow the variations in $H(f)$. Electrical power from the detector, being proportional to i_d^2, will follow $|H(f)|^2$. In electrical systems the *electrical bandwidth* is defined as the frequency range over which the *electrical power* exceeds one half of its maximum value. We can therefore define an electrical bandwidth for our optical system, Δf_{el}, as the frequency range over which $|H(f)|$ exceeds $\sqrt{\frac{1}{2}}$ of its maximum value. Similar arguments apply when a voltage output is obtained.

We turn now to pulse broadening. Suppose we launch an optical pulse which, as far as the system response is concerned, is infinitesimally narrow. When the pulse emerges from the system, it will have broadened into a definite shape. This shape may be described by the *normalized impulse response* $h(t)$. The function $h(t)$ is normalized so that:

$$\int_{-\infty}^{\infty} h(t)\, dt = 1$$

The functions $h(t)$ and $H(f)$ are in fact related by a Fourier transform (ref. A5.1), i.e.

$$H(f) = \int_{-\infty}^{\infty} h(t) \exp(-i2\pi ft)\, dt \tag{A5.1}$$

A useful parameter for characterizing the spread of a pulse is the r.m.s. pulse width. This is given by σ where:

$$\sigma^2 = \int_{-\infty}^{\infty} t^2 h(t)\, dt - \int_{-\infty}^{\infty} th(t)\, dt \tag{A5.2}$$

If $h(t)$ is symmetrical about a time t_0, then by measuring time from t_0 we ensure that the second term on the right-hand side of eq. (A5.2) is zero.

σ is a useful quantity because it may be shown (ref. 8.21) that any pulse having an r.m.s. pulse width σ_1 that passes through a linear system whose impulse response has an r.m.s. width σ_2 will emerge with an r.m.s. pulse width σ_3 where:

$$\sigma_3^2 = \sigma_1^2 + \sigma_2^2 \tag{A5.3}$$

By using eq. (A5.1) in conjunction with the definitions for Δf_{opt} and Δf_{el}, it is possible to determine expressions for the two latter quantities provided the function $h(t)$ is known. To give a relatively simple example of this type of calculation, we may take a rectangular normalized impulse response (Fig. A5.1(a)). That is, we have

$$h(t) = \frac{1}{\Delta T}, \qquad -\frac{\Delta T}{2} < t < \frac{\Delta T}{2}$$

and $\quad h(t) = 0, \qquad t < -\dfrac{\Delta T}{2}, t > \dfrac{\Delta T}{2}$

From eq. (A5.1)

$$H(f) = \frac{1}{\Delta T} \int_{-\Delta T/2}^{\Delta T/2} \exp\left(-i2\pi ft\right) dt$$

$$= \frac{i}{2\pi f \Delta T}\left[\exp\left(-\frac{i2\pi f \Delta T}{2}\right) - \exp\left(\frac{i2\pi f \Delta T}{2}\right)\right]$$

$$= \frac{\sin\left(\pi f \Delta T\right)}{\pi f \Delta T}$$

(a)

(b)

(c)

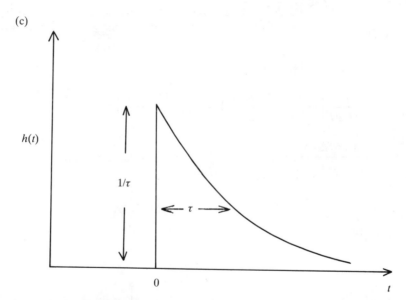

Fig. A5.1 Three 'theoretical' system response functions: (a) rectangular, (b) Gaussian, and (c) exponential decay.

Thus $|H(f)|$ falls to one half of its maximum value (i.e. 1 at $f = 0$) when $\pi f \Delta T = 1.895$, i.e. $\Delta f_{\text{opt}} = 0.603/\Delta T$. Similarly, $\Delta f_{\text{el}} = 1.392/(\pi \Delta T)$ or $\Delta f_{\text{el}} = 0.443/\Delta T$.

From eq. (A5.2) we also have

$$\sigma^2 = \frac{1}{\Delta T} \int_{-\Delta T/2}^{\Delta T/2} t^2 \, dt$$

Thus $\sigma^2 = \Delta T^2/12$ and hence $\sigma = \Delta T/2\sqrt{3}$ we can therefore also write:

$$\Delta f_{\text{opt}} = 0.174/\sigma \quad \text{and} \quad \Delta f_{\text{el}} = 0.128/\sigma$$

When we are dealing with digital signals, the bandwidth is referred to in terms of B, the *bit rate*. We are then concerned with how close a succession of optical pulses can become before they can no longer be resolved (see Fig. A5.2). It may be shown (ref. A5.1) that for a wide range of pulse shapes B should not exceed $1/4\sigma$. Thus we can write for rectangular pulses that

$$\Delta f_{\text{opt}} \simeq 0.7B \quad \text{and} \quad \Delta f_{\text{el}} \simeq 0.5B$$

Of course, all the above relationships between Δf_{opt}, Δf_{el}, σ and B are only valid for rectangular pulses. For other pulse shapes, the relationships will be slightly different. A pulse shape that is often convenient to use is the Gaussian shape, for which

$$h(t) = \frac{1}{\sigma\sqrt{2}} \exp\left(-t^2/2\sigma^2\right)$$

This is illustrated in Fig. A5.1(b). We may proceed as for rectangular pulses, although the integrals involved are more difficult to evaluate, and we merely quote the results:

$$H(f) = \exp\left(-2\pi^2 f^2 \sigma^2\right)$$

and hence

$$\Delta f_{\text{opt}} = 0.187/\sigma$$

$$\Delta f_{\text{el}} = 0.132/\sigma$$

Since we do not often know exactly what pulse shapes we are dealing with, it is convenient to use the following approximations:

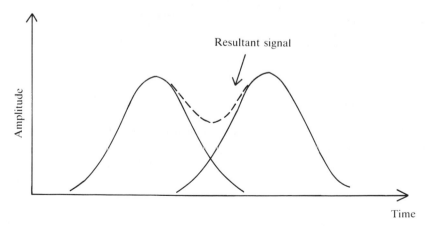

Fig. A5.2 Two overlapping pulses giving rise to a resultant signal that may be difficult to resolve into individual pulses if these are too close together.

$$\Delta f_{opt} \simeq 2\Delta f_{el}$$

$$B \simeq 2\Delta f_{el}$$

$$B \simeq \frac{1}{4\sigma}$$

$$\Delta f_{el} \simeq \frac{1}{8\sigma}$$

Finally, another system response often encountered is that of exponential decay (Fig. A5.1(c)). In this case we may write

$$\left. \begin{array}{ll} h(t) = \dfrac{1}{\tau}\exp\left(-\dfrac{t}{\tau}\right), & t > 0 \\[2mm] h(t) = 0, & t < 0 \end{array} \right\}$$

From eq. (A5.1)

$$\begin{aligned} H(f) &= \frac{1}{\tau}\int_{-\infty}^{\infty} \exp\left(-\frac{t}{\tau}\right)\exp\left(-\mathrm{i}2\pi f t\right)\mathrm{d}t \\[2mm] &= \frac{1}{\tau}\int_{-\infty}^{\infty} \exp\left\{-t\left(\frac{1}{\tau} + \mathrm{i}2\pi f\right)\right\}\mathrm{d}t \\[2mm] &= \frac{1}{1 + \mathrm{i}2\pi f\tau} \end{aligned}$$

whence

$$|H(f)| = \frac{1}{(1 + 4\pi^2 f^2 \tau^2)^{\frac{1}{2}}} \tag{A5.4}$$

and

$$\left. \begin{array}{l} \Delta f_{opt} = 0.276/\tau \\[2mm] \Delta f_{el} = 0.159/\tau \end{array} \right\}$$

References

A5.1   J. Gowar, *Optical Fiber Communication Systems*, Prentice Hall International, Hemel Hempstead, 1984, Chapter 15.

Appendix 6

Physical constants and properties of some common semiconductors at room temperature (300 K)

Rest mass of electron	m	$= 9.110 \times 10^{-31}$ kg $= 0.000549$ u
Charge of electron	e	$= 1.602 \times 10^{-19}$ C
Electron charge/mass ratio	e/m	$= 1.759 \times 10^{11}$ C kg^{-1}
Avogadro's constant	N_A	$= 6.022 \times 10^{23}$ mol^{-1}
Planck's constant	h	$= 6.626 \times 10^{-34}$ J s
	$\hbar = h/2\pi$	$= 1.055 \times 10^{-34}$ J s
Boltzmann's constant	k	$= 1.381 \times 10^{-23}$ J K^{-1}
Speed of light (in vacuum)	c	$= 2.998 \times 10^{8}$ m s^{-1}
Permittivity of a vacuum	ε_0	$= 8.854 \times 10^{-12}$ F m^{-1}
Permeability of a vacuum	μ_0	$= 4\pi \times 10^{-7} = 1.258 \times 10^{-6}$ H m^{-1}
Stefan–Boltzmann constant	σ	$= 5.670 \times 10^{-8}$ W m^{-2} K^{-4}

Property	Si	Ge	GaAs
Atomic (molecular) weight	28.09	72.60	144.6
Energy gap, E_g (eV)	1.12	0.67	1.43
Intrinsic carrier concentration, n_i (m^3)	1.5×10^{16}	2.4×10^{19}	1×10^{13}
Electron mobility, μ_e (m^2 V^{-1} s^{-1})	0.135	0.39	0.85
Hole mobility, μ_h (m^2 V^{-1} s^{-1})	0.048	0.19	0.045
Relative permittivity, ε_r	11.8	16.0	10.9[†]
Electron effective mass[‡], m_e^* ($\times$ m)	0.12	0.26	0.068
Hole effective mass[‡], m_h^* ($\times$ m)	0.23	0.38	0.56

[†] The range of values of relative permittivity of GaAs quoted in the literature is from 10.7 to 13.6. We have adopted the value 10.9.

[‡] Two different 'effective masses' are defined, namely the *density of states* and *conductivity* effective masses. The values given here are representative of those quoted for the conductivity effective mass.

Appendix 7

Laser safety

Most practical lasers emit radiation that is potentially hazardous. The degree of hazard is related to the output characteristics of the laser, the way in which it is used, and the experience of the operator. The types of laser and the ways in which they are used are, as we have seen, many and varied. It would therefore be unreasonable to suggest that all laser applications require the same degree of control. Accordingly, systems of classification of lasers have been introduced, which are given below.

Any classification must be operable and consistent with our knowledge concerning the injuries that laser radiation can cause. The mechanism by which laser radiation causes damage is similar for all biological systems and may involve thermal, thermo–acoustic and photochemical processes. The degree to which any of these mechanisms is responsible for damage depends on the parameters of the laser source, such as wavelength, pulse duration, image size and power, and energy density.

One of the principal characteristics of laser radiation is its beam collimation. This, together with a high energy content, can result in large amounts of energy being transmitted to biological tissues. The primary event causing damage is the absorption of the radiation by the biological system. Absorption occurs at an atomic or molecular level and is thus wavelength specific. Thus it is primarily the laser wavelength which determines which tissue is liable to be damaged. Absorption of radiation leads to a rise in temperature which may then disrupt molecular bonds and hence impair the function of the molecule. The rise in temperature is related to the length of exposure and the irradiance of the beam, and this leads to the notion of a radiant exposure, or dose, measured in joules per square meter. Exposure to Q-switched or mode locked lasers can cause a very rapid rise in temperature of the tissue, so that liquid components may be converted to gas leading to a rupturing of cells. In general, the interrelationships between damage mechanisms and exposure are very complex (ref. A7.1), and we shall not pursue them here.

Probably the most vulnerable part of the body as far as laser radiation is concerned is the eye. This is mainly because the eye lens will focus incident collimated laser radiation to a small point with a radius of the order of the wavelength, and with a corresponding high energy density. The hazards are wavelength dependent so that radiation in the ultraviolet and infrared regions, which is absorbed by the cornea, represents a danger to the cornea, whilst radiation in the visible and near infrared represents a retinal hazard. The increase in irradiance from the cornea to the retina is given approximately by the ratio of the pupil area to that of the image formed on the retina. Typically, the pupil may expand to a diameter of 5 mm with a corresponding retinal image of 10–20 μm diameter. Thus the irradiance increases by a factor of between 2×10^5 and 5×10^5. A laser beam of 50 W m^{-2} incident on the cornea then represents an irradiance of about 25 MW m^{-2} on the retina. About 5 percent of the radiation falling on the retina is absorbed in the pigments within the rods and cones, and this may be sufficient to cause a burn and a loss of vision.

In general terms, skin can tolerate a great deal more exposure than can the eye, although blisters, ulceration and scarring of the skin may occur for high levels of irradiation. Again, the hazards are wavelength dependent, being more pronounced for ultraviolet radiation. Any organization which uses lasers would be advised to draw up a Safety Code of Practice, which should be based on an accepted classification of lasers. The classification accepted in the UK and Europe is described below. There are slight differences in the classification used in the USA. These classifications and related information on laser safety are given in considerable detail in the various publications cited in ref. A7.2.

Laser Classification based on BS 4803

Class 1
Output power is so low as to be inherently safe

Class 2
Such lasers operate in the visible part of the spectrum 400 nm to 700 nm and their output power is limited to 1 mW for continuous wave (CW) operation. Such lasers are not inherently safe, but some eye protection is afforded by the natural aversion response of the eye, including the blink reflex. Hazards can be controlled by relatively simple procedures

Class 3A
These lasers operate in the visible part of the spectrum (400 nm to 700 nm) and their output is limited to 5 mW for CW operation. Some protection is still provided by the aversion responses. Direct intra-beam viewing with optical aids may be hazardous

Class 3B
These lasers operate in any part of the electromagnetic spectrum between wavelengths of 200 nm and 1 mm. Their output power is limited to 500 mW for CW operation. Direct beam viewing could be hazardous and must be avoided. Likewise specular reflections may be hazardous but diffuse reflections will not generally be so. Under no circumstances should the beam be viewed with optical aids. More detailed control measures are necessary

Class 4

These lasers also operate within the wavelength range 200 nm to 1 mm and their output power exceeds 500 mW. Not only is viewing of the direct beam and specular reflections hazardous but, in some cases, viewing of diffuse reflections may also be hazardous to the eye. In addition, there is also a risk of skin burns from the direct beam and from first order specular relections. The beam from such lasers is also capable of igniting flammable material so that care must be taken to minimise the risk of fire. The use of Class 4 lasers requires extreme caution for safety both of the user and of other persons who may be present. If possible the system should be totally enclosed

Safety note

It is stressed that the above table is intended as a guide only and that laser users should consult the publications given in ref. A7.2. Particular attention is drawn to the American *Standards for the Safe Use of Lasers* Number Z136 issued by the National Standards Institute (ref. A7.2b).

Each laser device should carry an appropriate label and notice of warning (ref. A7.2a, Part 2, p. 3).

The safe use of lasers often involves the provision of safety interlocks and warning lights on access doors to rooms where lasers are being used, together with beam stops and enclosures. Materials which act as diffuse reflectors should be used wherever possible. Protective eyewear should also be available for use with Class 3B and Class 4 lasers. Bearing in mind that the protective medium will not afford protection over the whole spectral range, care must be taken to ensure that the eye protection has the correct spectral response to match the laser being used.

Finally, it should be remembered that most lasers are high voltage devices (several kV) often capable of delivering large currents and therefore care must also be taken in this respect.

References

A7.1   A. F. McKinlay and F. Harlon, Biological bases of maximum permissible exposure levels of laser standards. Part 1: Damage mechanisms, *J. Soc. Radiol. Prot.*, **4** (1), 1984.

A7.2   (a) British Standard, *Radiation Safety of Laser Products and Systems*, Parts 1, 2 and 3, BS4803 1983.

(b) American National Standards Institute, 'Standards for the safe use of lasers', *ANSI* **Z136**, 1980.

(c) International Electrotechnical Commission, IEC Standard 'Radiation safety of laser products, equipment classification, requirements and users' guide', Publication 825, 1984.

(d) 'Safety in universities. Part 2:1: Lasers', Committee of Vice-Chancellors and Principals of the Universities of the United Kingdom, 1978.

(e) R. J. Rockwell, 'Analysing laser hazards' *Lasers and Applications*, May 1986, p. 97.

(f) J. Kaufman and R. Tucker, 'Laser absorbers for eye protection', *Lasers and Applications*, October 1985, p. 69.

(g) D. C. Winban, 'Dispelling myths about laser eyewear', *Lasers and Applications*, March 1986, p. 73.

(h) R. J. Rockwell, 'Controlling laser hazards', *Lasers and Applications*, September 1986, p. 93.

(i) M. L. Wolbarsht, and D. H. Sliney 'Laser safety standards move on', *Lasers and Applications*, April 1987, p. 97.

Index